AF832924

Insect Cell Biotechnology

Insect Cell Biotechnology

Edited by

Karl Maramorosch, Ph.D.
Robert L. Starkey Professor of Microbiology
Professor of Entomology
Rutgers — The State University of New Jersey
New Brunswick, New Jersey

Arthur H. McIntosh, Sc.D.
Research Scientist
Biological Control of Insects
Research Laboratory
United States Department of Agriculture
Agricultural Research Services
Columbia, Missouri

CRC Press
Taylor & Francis Group
Boca Raton London New York

CRC Press is an imprint of the
Taylor & Francis Group, an **informa** business

First published 1994 by CRC Press
Taylor & Francis Group
6000 Broken Sound Parkway NW, Suite 300
Boca Raton, FL 33487-2742

Reissued 2018 by CRC Press

© 1994 by CRC Press, Inc.
CRC Press is an imprint of Taylor & Francis Group, an Informa business

No claim to original U.S. Government works

This book contains information obtained from authentic and highly regarded sources. Reasonable efforts have been made to publish reliable data and information, but the author and publisher cannot assume responsibility for the validity of all materials or the consequences of their use. The authors and publishers have attempted to trace the copyright holders of all material reproduced in this publication and apologize to copyright holders if permission to publish in this form has not been obtained. If any copyright material has not been acknowledged please write and let us know so we may rectify in any future reprint.

Except as permitted under U.S. Copyright Law, no part of this book may be reprinted, reproduced, transmitted, or utilized in any form by any electronic, mechanical, or other means, now known or hereafter invented, including photocopying, microfilming, and recording, or in any information storage or retrieval system, without written permission from the publishers.

For permission to photocopy or use material electronically from this work, please access www.copyright.com (http://www.copyright.com/) or contact the Copyright Clearance Center, Inc. (CCC), 222 Rosewood Drive, Danvers, MA 01923, 978-750-8400. CCC is a not-for-profit organization that provides licenses and registration for a variety of users. For organizations that have been granted a photocopy license by the CCC, a separate system of payment has been arranged.

Trademark Notice: Product or corporate names may be trademarks or registered trademarks, and are used only for identification and explanation without intent to infringe.

Library of Congress Cataloging-in-Publication Data

Insect cell biotechnology/edited by Karl Maramorosch, Arthur H. McIntosh.
p. cm.
Includes bibliographical references and index.
ISBN 0-8493-4597-9
1. Insect cell biotechnology. I. Maramorosch, Karl.
II. McIntosh, Arthur H.
TP248.27.156157 1994
632'.951 — dc20 93-33795

A Library of Congress record exists under LC control number: 93033795

Publisher's Note
The publisher has gone to great lengths to ensure the quality of this reprint but points out that some imperfections in the original copies may be apparent.

Disclaimer
The publisher has made every effort to trace copyright holders and welcomes correspondence from those they have been unable to contact.

ISBN 13: 978-1-315-89447-8 (hbk)
ISBN 13: 978-1-351-07357-8 (ebk)

Visit the Taylor & Francis Web site at http://www.taylorandfrancis.com and the
CRC Press Web site at http://www.crcpress.com

PREFACE

Insect Cell Biotechnology provides a clear and up-to-date description of recent major advances in the field. The widespread use of baculoviruses, *in vitro* gene expression systems using insects cells, the construction of transfer vectors containing foreign genes, and the relevant background information have been admirably conceived in the first chapter by Professor S. Maeda. In the following two chapters, the production of baculoviruses in insect cells is explored by Drs. C. L. Goodman and A. H. McIntosh, and baculovirus specificity by Dr. A. H. McIntosh and Mr. J. J. Grasela gives all the relevant information and detailed laboratory methods that will be an invaluable update to researchers in the field. The chapter by Dr. W. J. McCarthy on bacterial toxin studies in cultured insect cells focuses primarily on *Bacillus thuringiensis* toxins, providing the reader with a comprehensive view of the field. The chapter on large-scale production of cultured insect cells by Professor S. Agathos summarizes present knowledge about scale-up operations required in the production of recombinant protein and insect viruses propagated in insect cells. The chapters on growth and nervous system interactions by Drs. L. A. Oland and H. Oberlander and on physiological and developmental capacities of cell lines by Drs. H. Oberlander and S. M. Ferkovich are the first in-depth treatments of these subjects. Transfection in *Drosophila* cells is presented by Professor G. Echalier and his associates in France in a thorough, well structured, and in-depth description of current studies. The chapter by Dr. T. M. Koval on "Intrinsic Stress Resistance of Cultured Insect Cells" places emphasis on stress produced by X-rays, ultraviolet light, chemicals, psoralens and heat, analyzing the mechanisms and discussing the theoretical and practical implications.

The editors expect that this book will make a valuable contribution to the subject of insect cell biotechnology and that it will prove useful to graduate students and researchers at universities, and to industrial laboratories where commercial usage of insect cell culture has found wide applications.

The editors express their sincere gratitude to the contributors for their effort and care with which they have prepared their chapters, and to CRC Press for their part in indexing, proofreading, and other aspects of production of this volume.

Karl Maramorosch
Arthur H. McIntosh

THE EDITORS

Karl Maramorosch, Ph.D., is Robert L. Starkey Professor of Microbiology and Professor of Entomology at Rutgers — The State University of New Jersey, New Brunswick, New Jersey.

Dr. Maramorosch, a native of Vienna, Austria, graduated from the Agricultural University of Warsaw, summa cum laude, and obtained his Ph.D. in 1949 from Columbia University. His scientific career began at Rockefeller University where he worked on plant viruses and insect vectors for 12 years. Later, he became Program Director of Virology and Insect Physiology at the Boyce Thompson Institute, and since 1974, he has been Professor II (Distinguished Professor) at Rutgers University, where he became the Robert L. Starkey Professor in 1983.

Dr. Maramorosch is a Fellow and former Recording Secretary and Vice-President of the New York Academy of Sciences, Fellow of the American Association for the Advancement of Science, the Entomological Society of America, and the American Phytopathological Society, President of the International Association for Medicinal Forest Plants, Honorary Fellow of the Indian Virological Society, and a member of the Tissue Culture Association, Harvey Society, Society for Invertebrate Pathology, American Society for Microbiology, Electron Microscopy Society, and several other professional organizations. Dr. Maramorosch has been the recipient of grants from the National Science Foundation, National Institutes of Health, U.S. Department of Agriculture, Rockefeller Foundation, and private industry. He won the 1980 Wolf Prize in Agriculture and many other awards, including the Ciba Geigy Award in Entomology, Jurzykowski Award in Biology, AIBS Award of Distinction, Cressy Morisson Prize of the N.Y. Academy of Sciences, Waksman Award, and AAAS-Campbell Award. He was nominated by the Entomological Society of America for the National Medal of Science in 1983 and was the Founders Lecturer of the Society for Invertebrate Pathology in 1990. He is an elected member of the Leopoldina, oldest European Academy of Sciences, and a Fellow of the Indian National Academy of Sciences.

Dr. Maramorosch served twice as visiting Fulbright Professor in Yugoslavia and as visiting professor in the People's Republic of China, U.S.S.R., The Netherlands, Germany, Japan, India, and Poland. He was consultant to the Food and Agriculture Organization in the Philippines, India, Nigeria, Thailand, Sri Lanka, and Mauritius, and has held short-term assignments from the Rockefeller Foundation, the Agency for International Development, U.S. Department of Agriculture (FERRO), and Ford Foundation in Mexico, India, Kenya, and the Philippines. Dr. Maramorosch has edited more than 50 volumes on viruses, vectors, plant diseases, and tissue culture, and is the author or co-author of more than 600 research papers. His major research interests include comparative virology, invertebrate cell culture, parasitology, diseases caused by spirochetes, viroids, and spiroplasmas, biotechnology, and international scientific cooperation.

Arthur H. McIntosh, Sc.D., is Laboratory Director and Research Leader of the Biological Control of Insects Research Laboratory with the United States Department of Agriculture, Agricultural Research Services, in Columbia, Missouri.

Dr. McIntosh graduated in 1959 from McMaster University, Hamilton, Ontario, Canada with a B.A. He attended the University of Guelph, in Guelph, Ontario, Canada from 1960–1962 and graduated from the Department of Microbiology with an M.S. Dr. McIntosh worked on his doctorate from 1965–1969 in the Department of Microbiology, Harvard University School of Public Health and graduated in 1969. He then completed postdoctoral studies at Stanford Research Institute from 1969–1971 and accepted a position at Boyce Thompson Institute with Dr. Karl Maramorosch in Insect Virology and Cell Culture from 1971–1974. He was an Assistant Research Professor at Waksman Institute of Microbiology, Rutgers, Piscataway from 1974–1979 before joining the USDA-ARS. He holds a joint appointment in the Department of Entomology, University of Missouri as Adjunct Associate Professor and participates in the teaching of a course in Biological Control. He spent a six month sabbatical (1992–1993) at the Commonwealth Scientific and Industrial Research Organization, in Canberra, Australia, where he initiated cell lines from *Helicoverpa-Heliothis* species for the replication of the pesticidal baculoviruses.

Dr. McIntosh conducts research in the area of Biological Control utilizing insect cell culture to study the biological characteristics of baculoviruses. His interest also includes the study of the specificity of insect viruses. He has published over 60 research papers, reviews, and book chapters.

Dr. McIntosh is a member of the Tissue Culture Association, Society for Invertebrate Pathology, Entomological Society of America, Japanese Society of Applied Entomology and Zoology, and Sigma Xi.

CONTRIBUTORS

Spiros N. Agathos, Ph.D.
Professor
Department of Bioengineering
Catholic University of Louvain
Louvain-la-Neuve, Belgium

Guy Echalier, Ph.D.
Professor
UPMC — CNRS Groupe de Genetique
Cellulaire et Moleculaire
Paris, France

Stephen M. Ferkovich, Ph.D.
Agricultural Research Services
United States Department of Agriculture
Gainesville, Florida

F. Fourcade-Peronnet, Ph.D.
UPMC — CNRS Groupe de Genetique
Cellulaire et Moleculaire
Paris, France

Cynthia Lenz Goodman, Ph.D.
Biological Control
of Insects Research Laboratory
Agricultural Research Services
United States Department of Agriculture
Columbia, Missouri

James J. Grasela, M.S.
Biological Control of Insects Research
Laboratory
Agricultural Research Services
United States Department of Agriculture
Columbia, Missouri

Thomas M. Koval, Ph.D.
Senior Staff Scientist
National Council on Radiation Protection
and Measurements
Bethesda, Maryland

Susumu Maeda, Ph.D.
Associate Professor
Department of Entomology
University of California
Davis, California

William J. McCarthy, Ph.D.
Professor
Pesticide Research Laboratory
Department of Entomology
The Pennsylvania State University
University Park, Pennsylvania

Arthur M. McIntosh, Sc.D.
Laboratory Director and
Research Leader
Biological Control of Insects
Research Laboratory
Agricultural Research Services
United States Department
of Agriculture
Columbia, Missouri

Herbert Oberlander, Ph.D.
Director
Insect Attractants, Behavior and Basic
Biology Research Laboratory
Agricultural Research Services
United States Department of
Agriculture
Gainesville, Florida

Lynne A. Oland, Ph.D.
Arizona Research Laboratories
Division of Neurobiology
University of Arizona
Tucson, Arizona

TABLE OF CONTENTS

Chapter 1
Expression of Foreign Genes in Insect Cells Using Baculovirus Vectors 1
Susumu Maeda

Chapter 2
Production of Baculoviruses for Insect Control Using Cell Culture 33
Cynthia Lenz Goodman and Arthur H. McIntosh

Chapter 3
Specificity of Baculoviruses 57
Arthur H. McIntosh and James J. Grasela

Chapter 4
Application of Insect Cell Culture to the Study of *Bacillus thuringiensis* Toxins 71
William J. McCarthy

Chapter 5
Large Scale Insect Cell Production 89
Spiros N. Agathos

Chapter 6
Growth and Interactions of Cells from the Insect Nervous System *In Vitro* 105
L. A. Oland and H. Oberlander

Chapter 7
Physiological and Developmental Capacities of Insect Cell Lines 127
H. Oberlander and S. Ferkovich

Chapter 8
Differences in Transfection Efficiency between Drosophila Cell Lines 141
G. Echalier and F. Fourcade-Peronnet

Chapter 9
Intrinsic Stress Resistance of Cultured Lepidopteran Cells 157
Thomas M. Koval

Index 187

Chapter 1

EXPRESSION OF FOREIGN GENES IN INSECT CELLS USING BACULOVIRUS VECTORS

Susumu Maeda

TABLE OF CONTENTS

I. Introduction 3

II. Baculovirus Expression Vectors 3
  A. Organization and Expression of the Baculovirus Genome 3
  B. Baculovirus Expression Vector Systems 4
    1. Advantages 4
    2. Construction of Recombinant Baculoviruses 4
    3. Foreign Gene Constructs for Insertion into Baculovirus Transfer Vectors 5
  C. Promoters Used to Drive Foreign Gene Expression 5
    1. Polyhedrin Promoter 5
    2. p10 Promoter 5
    3. Other Baculovirus Late Gene Promoters 6
    4. Baculovirus Early Gene Promoters 6
    5. Chimeric and Modified Promoters 6
    6. Host Insect and Nonbaculovirus Promoters 6
  D. Expression Vectors Generating Polyhedron-Deficient Viruses 7
    1. Traditional Transfer Vectors 7
    2. Polyhedrin Fusion Transfer Vectors 7
    3. Transfer Vectors for Coexpression of Two Genes 8
  E. Improved Strategies for the Isolation and Identification of Recombinant Viruses 8
    1. Isolation by Dilution Methods and DNA Hybridization 8
    2. Linearized Parental Viral DNA 8
    3. β-Galactosidase Marker 9
    4 Polyhedron-Positive Marker 9
    5. Baculovirus-Yeast Shuttle Vectors 9
    6. Other Isolation Strategies 10
  F. Host-Dependent Foreign Gene Expression 10

III. Characteristics of Expressed Foreign Gene Products 11
  A. Proteolytic Processing 11
    1. Signal Sequence Cleavage and Secretion 11
    2. Internal Proteolytic Cleavage 11
  B. Glycosylation 12
  C. Phosphorylation 14
  D. C-Terminus α-Amidation 14
  E. Fatty Acid Acylation 15
  F. Solubility 15

0-8493-4597-9/94/$0.00+$.50
© 1994 by CRC Press Inc.

G. Oligomerization 16
H. Assembly of Viral Particles 16

IV. Expression of Foreign Genes in Larval Insects 18

V. Foreign Gene Expression for Pest Insect Control 19

VI. Concluding Remarks 21

Acknowledgments 22

References 22

Susumu Maeda, Ph.D., is Associate Professor of Entomology at the University of California, Davis. He received his B.S. in Entomology/Agrobiology, in 1975 and Ph.D. at the University of Tokyo, in 1983. Postdoctoral fellow at U.C. Berkeley, 1980–1981; Assistant Professor at Tottori University, Japan, 1978–1988; Visiting Research Scientist, Zoecon Research Institute, Sandoz Crop Protection in Palo Alto, California, 1987–1988; Assistant Professor at U.C. Davis, 1988–1991; Associate Professor at U.C. Davis, 1991–present. American Society of Microbiology, Society of Invertebrate Pathology, American Society of Virology, American Entomological Society, American Society of Biochemistry and Molecular Biology. Molecular biology of insect viruses and their molecular interactions to host insects; application of recombinant insect viruses in medical and agricultural sciences.

I. INTRODUCTION

Baculoviruses (family Baculoviridae) are characterized by circular double-stranded DNA genomes that are encapsulated within rod-shaped enveloped virions. They have been isolated exclusively from insects and crustaceans. Baculoviruses are classified into three genera: nuclear polyhedrosis virus (NPV), granulosis virus, and non-occluded baculovirus. During an early stage of infection, NPVs produce viral particles (budded virions) that are involved in cell-to-cell infection (secondary infection) in both larval tissues and established cell lines. During a late stage of infection, NPVs have a unique feature of producing inclusion bodies called polyhedra in the infected nucleus. Polyhedra protect the numerous progeny viral particles (occluded virions) embedded within them and are required for oral (per os) infection. Upon ingestion, polyhedra are broken down by alkaline proteinases in the insect gut juices and release occluded virions, which subsequently infect midgut tissues by fusion. The major component of polyhedra is a very highly expressed (roughly 30% of total proteins at a late stage of infection) protein called polyhedrin. The polyhedrin gene is nonessential for viral replication. Using the polyhedrin promoter, baculoviruses (NPVs) have been recognized as efficient vectors for the high-level expression of foreign genes in insect cell lines and larvae. *Autographa californica* NPV (AcNPV) followed by *Bombyx mori* NPV (BmNPV) is the most commonly used baculovirus vector for foreign gene expression experiments.

This chapter will emphasize recent (generally post-1989) progress in the use of baculovirus expression vectors, especially in regard to applications in agriculture and medicine, and outline the characteristics of baculovirus expressed proteins. The unique advantages of expression in insect larval systems will also be covered. Many general and in-depth reviews are available in the use[1-11] and methodology[12-16] of baculovirus expression vector systems for those wishing further background.

II. BACULOVIRUS EXPRESSION VECTORS

A. ORGANIZATION AND EXPRESSION OF THE BACULOVIRUS GENOME

The baculovirus genome is composed of 100 to 180 kbp of double-stranded circular DNA. About 50% of the nucleotide sequence of the AcNPV genome has been determined (see Reference 17). Several genes of the *Orgyia pseudotsugata* NPV genome have also been identified and sequenced (see Reference 18), and two laboratories have nearly completed sequencing the entire genomes of AcNPV[19] and BmNPV.[20]

Five discrete (hr1 to hr5) homologous regions (hrs) are found in the NPV genome.[21-23] Hr4 of AcNPV[22] and BmNPV,[23] and hr2 of BmNPV[23] can be subdivided into two smaller hrs. Each hr is 1 to 2 kbp long and possesses 3 to 7 repeats of a relatively conserved, 100 bp-long, AT-rich sequence with a 20 bp-long palindrome structure at its core.[22-23] In the case of AcNPV and BmNPV, a GAATTC sequence is found in the palindrome, which is recognized and cleaved by the restriction endonuclease *Eco*RI. Hrs function to enhance the expression of genes[22] or act as replication origins for viral DNA replication.[24]

Outside of the hrs, the NPV genome is generally covered by unique sequences consisting of coding regions and their 5′ and 3′ noncoding regulatory sequences (promoters, poly(A) signals, etc.). The noncoding regions generally range from a few to several hundred base pairs. Only one intron-possessing gene, IE-0, has been identified in the AcNPV genome.[25] The IE-0 gene product is detected only at an early stage of infection; however, it is unknown whether IE-0 is essential for viral replication. In most cases baculovirus open reading frames that correspond to expressed polypeptides do not overlap. A preferential gene direction and gene clusters with similar functions are not found (see Reference 14). Genes commonly found in viruses including those involved in virus structure (e.g., envelope (gp64) protein), DNA replication (e.g., DNA polymerase), and regulation of gene expression (e.g., IE-1) have been identified in the NPV genome (see Reference 14). Unique or unusual genes identified in the baculovirus genome include: UDP-glucosyltransferase,[26] apoptosis blocking protein,[27] chitinase,[28,29] and ubiquitin.[30]

The expression of baculovirus genes is temporally regulated and characterized by three distinct phases: early, late, and very late. Early genes do not require the expression of other viral genes for basal levels of expression. Late and very late gene expression begin 6 to 8 and 20 to 24 hours postinfection, respectively, and require the expression of early (and/or late) gene products. Early genes are transcribed by the host RNA polymerase II complex, which is α-amanitin sensitive.[31] Late and very late genes are believed to be transcribed by a viral-coded or viral-induced (or modified) host RNA polymerase complex. The expression of most viral early and host genes declines with the onset of late gene expression. Host and viral (both early and late) gene expression is nearly completely terminated during the active expression of very late genes (36 to 72 hours postinfection). In AcNPV the most abundantly expressed very late gene is polyhedrin followed by the p10 gene. All baculovirus late gene promoters identified to date possess the consensus sequence (A/T/G)TAAG (see Reference 32). About half of all known baculovirus early gene promoters possess a CAGT motif 20 to 25 bp downstream of the TATA box (see Reference 32).

B. BACULOVIRUS EXPRESSION VECTOR SYSTEMS

1. Advantages

Baculoviruses (especially NPVs) have been successfully used as highly efficient eucaryotic expression vectors (see reviews of References 14, 15). Baculoviruses have several characteristics that make them ideal vectors for the expression of foreign proteins, including (1) nonessential genes possessing very strong promoters that can efficiently drive foreign gene expression, (2) a double-stranded DNA genome that can be easily modified, (3) a rod-shaped viral capsid that can accommodate foreign DNA inserts of up to 10 kbp, and (4) cell lines that support viral replication. The polyhedrin gene promoter was the first to be used to drive foreign gene expression. As previously mentioned, polyhedrin is extremely highly expressed, yet nonessential for viral replication. Polyhedrin can also serve as a visual selection marker; i.e., recombinant viruses lacking the polyhedrin gene can be identified microscopically, and it is activated at a very late stage of infection, which is advantageous for the expression of genes whose products are toxic to the host insect cells.

2. Construction of Recombinant Baculoviruses

Due to the large size of the baculovirus genome, a foreign gene construct cannot be directly inserted into the viral genome. The construction of recombinant baculoviruses can be divided into three steps. In the first step, the foreign gene of interest is inserted into a baculovirus transfer vector and propagated using standard molecular biological techniques. Baculovirus transfer vectors generally contain an ampicillin resistance gene for selection, bacterial replication origin for propagation in *E. coli,* and a polylinker sequence for insertion of the foreign gene. The polylinker is flanked by the 5′ and 3′ (about 1 to 3 kbp each) regions of the target gene (e.g., the polyhedrin gene) along with an appropriate promoter (e.g., the polyhedrin promoter). In the second step, the foreign gene is transferred to the target site of the viral genome by homologous recombination; i.e., the recombinant transfer vector is cotransfected with wild-type (or other appropriate) viral DNA in a susceptible established insect cell line. Finally, recombinant viruses are isolated (generally by several rounds of plaque assay) from wild-type (nonrecombinant) viruses and propagated in insect cell lines or in insect larvae. If necessary the product can then be purified by traditional biochemical procedures. Once mastered, the entire process takes about 3 to 4 weeks.

3. Foreign Gene Constructs for Insertion into Baculovirus Transfer Vectors

Nearly any type of foreign gene construct can be incorporated into the baculovirus genome; however, the following should be taken into account. The splicing efficacy of foreign genes inserted into the baculovirus genome is extremely poor, especially at a late stage of infection.[33] Therefore, foreign genes without introns (e.g., cDNA) are required for high-level expression by baculovirus vectors. If cDNA is not available, any introns should be removed by an appropriate procedure (e.g., a polymerase chain reaction). One exception, however, is the relatively efficient splicing of cholione gene transcripts expressed by a recombinant BmNPV in the follicle cells of silkworm larvae.[34] It is hypothesized that larval tissues, which support viral replication poorly (e.g., follicle cells, silk gland tissues, and nervous tissues), may be able to splice genes more efficiently than tissues (e.g., fat body) or established cell lines, which strongly support viral replication.

The length of the 5′ non-coding region of the foreign gene is another consideration for high-level expression. In general, if the 5′ non-coding region of the foreign gene is longer than 50 bp (especially if it is GC rich), it will not be highly expressed. A 5′ non-coding region that is shorter than 20 to 30 bp and AT rich, however, should affect expression levels very little. This topic is discussed further in a later section.

A foreign gene whose product is cytotoxic may be expressed inefficiently or may not be inserted into the baculovirus genome. For example, the nonstructural protein, NS-1, of Aleutian mink disease parvovirus induces cytotoxicity in mammalian cells. When NS-1 is expressed in SF cells, viral replication drops by about 90% and dramatic changes in cell morphology are induced, presumably due to cytotoxicity induced by the expressed NS protein.[35]

C. PROMOTERS USED TO DRIVE FOREIGN GENE EXPRESSION

Following the initial development of baculovirus expression vector systems based on AcNPV[36,37] and BmNPV[38,39] using transfer vectors possessing the polyhedrin gene promoter, other transfer vectors have been developed. These transfer vectors use both baculovirus (nonpolyhedrin) and nonbaculovirus (e.g., insect host) promoters to drive foreign gene expression. Several of these vectors have been developed specifically to simplify the isolation of recombinant baculoviruses or to retain the polyhedrin-positive phenotype in the recombinant virus.

1. Polyhedrin Promoter

As mentioned previously, the polyhedrin promoter has several characteristics that make it ideal for driving foreign gene expression. The polyhedrin promoter is also considered to be the most active in the baculovirus genome based on the high percentage of polyhedrin (out of total protein) and polyhedrin mRNA (out of total mRNAs) found in NPV infected cells. As a general rule, however, the expression rate of foreign genes under the polyhedrin promoter is usually only 0.1 to 10% of that of polyhedrin. The rate of foreign gene transcription, however, is similar to the transcriptional rates of polyhedrin, indicating that the low expression rates of foreign genes are rooted at the translational level. This also implies that the coding region of the polyhedrin gene may be important for expression rates comparable to polyhedrin. Also, numerous experiments show that little or no correlation exists between codon usage and expression rates.

2. p10 Promoter

Like polyhedrin, the p10 gene is strongly expressed at late and very late stages of infection. The p10 gene product is a nonessential fibrous protein mainly found in the infected nucleus that helps increase the rate of cell degradation following cell death.[40-41] At a very late stage of infection, the p10 gene is expressed at lower levels than polyhedrin; however, the p10 gene promoter is activated slightly earlier. The p10 product cannot be detected microscopically; therefore, transfer vectors carrying the p10 promoter are used in conjunction with parental viral DNA carrying the β-galactosidase gene at the p10 locus.[42,43] Alternatively, transfer vectors carrying the p10 promoter can also carry a selection marker (e.g., the β-galactosidase[44] or polyhedrin[42] genes). Transfer vectors carrying the polyhedrin gene as a marker must be used in conjunction with polyhedrin-deficient parental virus DNA, as is discussed in a later section.

3. Other Baculovirus Late Gene Promoters

Other baculovirus late gene promoters including those of the basic DNA binding[45] and capsid[46] proteins have been used to drive foreign gene expression. These promoters, however, have activities that are significantly lower than those of polyhedrin and p10 and generally are not very useful for driving foreign gene expression. They may, however, be useful in hybrid promoter constructs.

4. Baculovirus Early Gene Promoters

Foreign genes expressed during an early stage of infection may be subject to more efficient post-translational modifications, since during an early stage of infection the cell machinery critical for post-translational processing is less disturbed compared to a late stage of infection. The promoters of PCNA[47] and IE-0[25] have been used to express β-galactosidase as a selection marker for the identification of recombinant baculoviruses.[48,49] The activities of these promoters, however, are not high. The promoter of the early gene IE-1 has been used to express a tissue plasminogen activator (tPA) gene inserted into the chromosomal DNA of SF-21 cells.[50] Although the tPA gene product is efficiently secreted outside the cell, the expression rate of tPA is extremely low. The tPA gene product is also secreted outside the cell when expressed under a late gene promoter;[51] however, the secretion levels decline significantly as viral infection progresses (i.e., during a late stage of infection).

5. Chimeric and Modified Promoters

A chimeric (hybrid) promoter derived from the baculovirus capsid (late) and polyhedrin (very late) gene promoters has been constructed (see Reference 52). Under this promoter, foreign genes are expressed at very high rates and at times slightly earlier than those under the polyhedrin gene promoter. Chimeric promoters possessing both early and very late gene promoters may be useful for expression during both early and late stages of infection. A modified polyhedrin promoter with a 6 bp substitution upstream of the consensus late gene promoter sequence (ATAAG) has also been constructed by linker scanning mutation experiments. This promoter yields about 50% higher expression rates compared to the unmodified polyhedrin promoter.[53]

6. Host Insect and Nonbaculovirus Promoters

In larval insects, genes are often specifically activated at high levels in response to metamorphosis or other metabolic changes. The promoters of these genes, especially those activated at an early stage of infection without induction by any other viral gene products, have great potential for driving foreign gene expression at specific times in the larval life cycle. Promoters of constitutively expressed host genes and tissue-specific genes may also be useful to drive foreign gene expression. For example, the heat shock promoter of *Drosophila melanogaster* (hsp70) is constitutively activated during viral replication in SF[44] and BmN[54] cells regardless of heat induction and has been used to efficiently drive foreign gene expression. The promoters of actin and chorion are specifically expressed in midgut tissues[55] and follicle cells,[56] respectively, and may be useful for tissue-specific expression. Promoters activated in mammalian cells (e.g., the promoter of long terminal repeat of Rous sarcoma virus),[57] may also be useful for expression of foreign genes in insect cells and larvae.

D. EXPRESSION VECTORS GENERATING POLYHEDRON DEFICIENT VIRUSES

1. Traditional Transfer Vectors

The purification of foreign gene products can be facilitated if they are found at high concentrations. Although baculovirus vectors generally express foreign genes at higher rates than those found in other expression systems, the expression of most foreign genes driven by the polyhedrin promoter is generally only 0.1 to 10% of polyhedrin production. In general, the expression rate of a foreign gene will be similar to the expression rates of similar polypeptides already expressed by a baculovirus vector. The following trends have also been found.

1. If the authentic protein is highly expressed in the original organism, it will be highly expressed by the baculovirus vector.
2. If the protein is highly expressed by yeast or *E. coli*, it will be highly expressed by the baculovirus vector.
3. Viral structural polypeptides are the most highly expressed, followed by secreted proteins.
4. Polypeptides localized in the nucleus and cytoplasm are expressed at relatively low rates.
5. The lowest expression rates are observed for membrane-bounded polypeptides and small peptides, even if they are of baculovirus origin (e.g., conotoxin).[58]
6. Expression of secreted proteins in larvae generally occurs at higher rates, often 100- to 1000-fold, than expression of the same gene *in vitro*.

Deletion experiments have shown that the 5′ non-coding region of the polyhedrin gene is essential for high-level expression in both BmNPV-[59] and AcNPV-based[60] expression vectors. The consensus late gene promoter sequence (A/G/T)TAAG is also essential for late gene transcription.[61] Therefore, most AcNPV (e.g., pAcYM1[62]) and BmNPV (e.g., pBE284[13]) transfer vectors carry nearly all of the 5′ noncoding region of the polyhedrin gene and an ATAAG sequence in order to maximize foreign gene expression.[59,60]

Transfer vectors (e.g., pVL941) that carry the 5′ coding region of the polyhedrin gene as a nontranslated sequence (by inactivation of the original transcriptional start site (e.g., ATG to ATT) upstream of the polylinker site(s)) have also been constructed.[63] It is reported that these transfer vectors (e.g., p36C) can express foreign genes at about 5-fold higher rates than for example pAc373, which lacks 6 bp of the 5′ polyhedrin noncoding region adjacent to the translational start.[64,65] Our extensive analysis of the expression rates of BmNPV transfer vectors, with and without a modified (e.g., ATG to ATA) 5′ polyhedrin coding region, indicate that these sequences do not significantly affect expression rates in established cell lines or in insect larvae (Maeda et al., unpublished data). Although the translational start site is modified in these transfer vectors, a few amino acids of the polyhedrin gene can be translated and fused to the foreign gene product.[66] This can lead to a lack of functional activity or difficulty in purification. A polyhedrin fusion protein will not be produced, however, if an inframe stop codon is inserted upstream of the linker site for insertion of the foreign gene so that a mini-cistron, like that found in the baculovirus gp64 gene,[67] is generated.

2. Polyhedrin Fusion Transfer Vectors

Polyhedrin fusion proteins are highly expressed by baculoviruses.[68,69] The expression rate of many polyhedrin fused polypeptides is close to 50% of polyhedrin expression. The expression level, however, does vary depending upon the characteristics of the expressed polypeptide that is fused to polyhedrin. Secondly, although a few fusion polypeptides have been shown to retain enzymatic activity, enzymatic or biological activity is often lost or greatly reduced. Furthermore, the fused polypeptides are less soluble, and often form precipitates (see Reference 70). Numerous polyhedrin fusion transfer vectors carrying varying lengths of the polyhedrin coding region in different frames are available for the BmNPV expression vector system.[13] Polyhedrin fusion vectors are ideal for the expression of polypeptides (antigens) for use in inducing antibody production.[69,71]

3. Transfer Vectors for Coexpression of Two Genes

A unique advantage of baculovirus expression vector systems is that two or more different genes can be expressed within the same cell. Two or three recombinant viruses, each expressing a different foreign gene, can be coinfected into the same culture at a multiplicity of infection of 5 or more, so that all of the genes are expressed simultaneously. If four or more genes are to be expressed in the same cell, coinfection of recombinant viruses carrying dual promoters for the expression of multiple foreign genes is recommended. For example, by this strategy seven rotavirus structural protein genes have been expressed in the same SF cells (see Reference 72). The use of dual promoter transfer vectors is also strongly recommended for the expression of multiple foreign genes in larval hosts (especially in semipermissive hosts, such as *Manduca sexta* for AcNPV; see Section IV). Several dual promoter vectors with different restriction endonuclease sites for insertion of two genes are available.[73]

E. IMPROVED STRATEGIES FOR THE ISOLATION AND IDENTIFICATION OF RECOMBINANT VIRUSES

1. Isolation by Dilution Methods and DNA Hybridization

A key to the isolation of pure recombinant baculoviruses is the plaque assay. Skill is required to generate clear plaques and to be able to distinguish polyhedron-deficient recombinant plaques from wild-type (polyhedron positive) plaques. Difficulty in generating clear plaques is usually caused by an inappropriate concentration of cells or dilution of virus, or the use of unhealthy cells. If clear plaques cannot be obtained, a dilution method using 96-well plates can be used; however, this method is more labor intensive. The 96-well assays must be done at a relatively high dilution, whereas the initial round of plaque assay is done at relatively low dilutions, allowing higher numbers of viruses to be assayed; i.e., by plaque assay several thousand viruses per 60 mm dish can be assayed, as opposed to only a hundred or so per 96-well plate.

The identification of cells infected by polyhedrin-negative baculoviruses is difficult to the untrained eye. To alleviate this problem, several alternatives have been developed. For example, recombinant viruses can be identified by DNA hybridization using probes with sequences homologous to the inserted foreign DNA fragment.[74] In order to perform this assay, the cells must be propagated in a specially designed, commercially available 96-well plate. The disadvantages of this system are that radioactive materials must be used and that viral stocks are difficult to keep sterile. Specific antibodies (if available) against the inserted gene can also be used to identify recombinant viruses by similar procedures. If the expressed foreign gene product possesses enzymatic activity, enzymatic activity can be assayed for. These procedures (96-well and hybridization assays), however, are not commonly used, since simpler methods for the identification of recombinant viruses have recently become available, as described below.

2. Linearized Parental Viral DNA

The use of linearized parental viral DNA increases the ratio of recombinant viruses to nonrecombinant viruses. In this strategy the circular viral genome is linearized by digestion at a unique restriction endonuclease site, which is generated by site-specific mutation in the target (e.g., polyhedrin) gene. The linearized DNA is then cotransfected with a typical transfer vector so that the linearized DNA is recircularized upon the formation of a recombinant virus. The disadvantage of this system is that the total number of viruses generated following cotransfection is generally low. Furthermore, even if the ratio of recombinant to nonrecombinant viruses is high, recombinant viruses must still be isolated and identified (e.g., by plaque assay). A recent modification of this procedure is the insertion of unique restriction endonuclease sites in the polyhedrin and ORF 1629 genes. ORF 1629 is an essential gene located immediately downstream of the polyhedrin gene. Use of this linearized DNA requires the transfer vector to carry the 3′ end of ORF 1629; however, nearly all of the generated viruses are recombinants.

3. β-Galactosidase Marker

As mentioned above, the identification of polyhedrin negative recombinant viruses is difficult to the untrained eye. This difficulty can be alleviated by using transfer vectors, which express a secondary marker gene product (e.g., β-galactosidase). In general, these transfer vectors carry the β-galactosidase gene construct adjacent to the linker sequence for insertion of the foreign gene. Transfer vectors carrying a marker gene construct and the PCNA (see Reference 14) or p10[44,49] promoter for foreign gene expression are used in a manner similar to typical transfer vectors. Plaques formed by recombinant viruses constructed using these transfer vectors turn blue when 5-bromo-4-chloro-3-indoyl-β-D-galactorylanoside (X-gal) is added to the agarose overlay and are easily identified. In general, recombinant viruses are subjected to routine plaque assay; however, once plaques become visible, a second agarose overlay containing X-gal is used to identify recombinant plaques. This second overlay is necessary since X-gal, which is relatively difficult to dissolve in water, must be suspended in DMSO or *N,N*-dimethyl formamide (DMF), which show toxicity toward insect cells. A disadvantage of this system is that the transfer vector must carry an extra 3 kbp encoding the β-galactosidase gene. This may make isolation of the recombinant transfer vector or recombinant virus more difficult. It has also been reported that deletions occur more frequently in the inserted foreign gene.[49]

4. Polyhedron-Positive Marker

Transfer vectors carrying an intact polyhedrin gene for the construction of polyhedron-positive recombinant viruses are available. These vectors must be cotransfected with polyhedron-negative parental virus DNA. By this strategy, recombinant viruses that produce polyhedra can be easily visualized; however, contamination by polyhedron-negative parent viruses is difficult to detect. This problem can be alleviated by using polyhedron-negative parental virus DNA carrying the β-galactosidase gene (generally by replacement of the polyhedrin gene) and X-gal in the agarose overlay as described above. Another potential problem is that the expression rate of the foreign gene may be reduced in polyhedrin-positive recombinant viruses. Polyhedron-positive viruses, as mentioned previously, are necessary for oral infectivity. Oral infectivity can be a disadvantage if susceptible insect colonies are present in the laboratory or a necessity if the recombinant baculovirus is to be used as an insecticide, as will be described later.

5. Baculovirus-Yeast Shuttle Vectors

Baculovirus DNA can be stably maintained in yeast cells if they possess the essential yeast elements ARS (necessary for autonomous replication) and CEN (for centromere function). A baculovirus-yeast shuttle vector system has been constructed by incorporating these elements plus selectable markers (i.e., the suppressor tRNA gene and URA3) into the polyhedrin coding region of AcNPV by homologous recombination.[75] This baculovirus-yeast parent shuttle vector DNA, YCbv::SUP4-o, which sequentially possesses the polyhedrin gene promoter, suppressor tRNA gene, ARS, and URA3, can replicate in yeast as a large plasmid and in insect cells as a recombinant virus. A special transfer vector (e.g., pAcY1) that carries the foreign gene flanked by polyhedrin coding and yeast ARS sequences is used to replace the suppressor tRNA gene (SUP4-o) in the parent baculovirus-yeast DNA (YCbv::SUP4-o) with the foreign gene in cotransfected yeast cells. Recombinant baculovirus-yeast DNA carrying the foreign gene at the sup4-o gene locus (i.e., after the polyhedrin gene promoter) is selected for by adding suppressor to the yeast growth media. The recombinant baculovirus-yeast DNA is then transfected into SF cells to generate a recombinant baculovirus expressing the foreign gene driven by the polyhedrin promoter. A possible disadvantage of this system is that mutations in the baculovirus genome may occur during the selection process in yeast.

6. Other Isolation Strategies

Genes whose products are conditionally cytotoxic can be used to enhance the number of recombinant viruses. Thymidine-thymidylate kinase of herpes simplex virus type 1 (HSV1-TK) can metabolize nucleoside analogues (e.g., Ganciclovir, 9-(1,3-Dihydroxy-2-propoxymethyl)guanine) into toxic inhibitors of DNA replication. Baculoviruses expressing the HSV1-TK gene in the absence of the Ganciclovir replicate normally. Cotransfection of parent viral DNA carrying the HSV1-TK gene at the target site for foreign gene insertion (e.g., polyhedrin gene) with an appropriate transfer vector (either polyhedron negative or positive) will generate recombinant viruses that lack the HSV1-TK gene and that will replicate normally in the presence of Ganciclovir. It has been reported that such a system results in over 85% recombinant viruses.[48] Two disadvantages of this system are that the presence of Ganciclovir may cause an increase in the mutation rate, and the HSV1-TK gene is relatively easily mutated leading to false positives.

A similar strategy (although not proven experimentally) is the use of endogenous baculovirus genes (e.g., p35 or putative DNA helicase) that can cell line-specifically inhibit viral replication. The p35 gene prevents viral-induced apoptosis in host cells.[27,54] AcNPV lacking the p35 gene replicates poorly in SF-21 cells and normally in TN-368 cells. If a transfer vector possessing a functional p35 gene in addition to a foreign gene is cotransfected with p35-deleted AcNPV DNA, only recombinant viruses (i.e., possessing the p35 gene) can replicate normally in SF-21 cells. The AcNPV DNA helicase gene induces cytotoxicity and premature inhibition of protein synthesis in BmN (*Bombyx mori* derived) cells, resulting in the inhibition of viral replication.[76,77] By this strategy, if genomic DNA of a BmNPV mutant carrying the AcNPV DNA helicase gene at the polyhedrin gene locus is cotransfected with a transfer vector carrying a β-galactosidase marker cassette and foreign gene, only recombinant viruses that lack the AcNPV helicase gene will replicate in BmN cells.

F. HOST-DEPENDENT FOREIGN GENE EXPRESSION

Among baculoviruses AcNPV has a relatively wide host specificity compared to BmNPV. The most commonly used *in vitro* hosts of AcNPV are the cell lines SF-21 and its subclone SF-9 derived from *Spodoptera frugiperda,* followed by TN-368 and High Five, which are derived from *Trichoplusia ni* (see Reference 14). Several other cell lines support the expression of foreign genes by AcNPV, including a coleopteran-derived insect cell line, which has been shown to strongly support AcNPV replication.[78] The most commonly used *in vitro* host of BmNPV is the cell line BmN, derived from *B. mori,* followed by BoMo[79] and other recently established *B. mori*-derived cell lines.[80] The cell line Bm36 supports both AcNPV and BmNPV replication; however, the production of polyhedra is extremely low in this cell line. Although Bm36 is reportedly derived from *B. mori,*[81] its origin is questionable based on chromosomal DNA analysis.[82]

Based on the currently available experimental data, it is difficult to speculate which cell line is the most appropriate for obtaining maximum expression of a particular protein. However, regardless of the cell chosen, the health of the cells is critical for high-level expression. The recommended *in vitro* hosts are SF9, SF21, or High Five for AcNPV, and BmN for BmNPV. Several AcNPV hosts, including SF-9 and SF-21, can be cultured in serum-free medium. Serum-free medium is advantageous for mass production of the foreign gene product or if proteins in the fetal bovine serum supplement interfere in the purification or assay of the foreign gene product. BmN can also be cultured in serum-free medium and in suspension cultures; however, the rate of cell growth is poor.

BmNPV and AcNPV possess greater than 90% DNA sequence homology,[54,83] allowing recombination to occur frequently between these viruses,[84] BmNPV and AcNPV transfer vectors can be cross-used to generate recombinant AcNPVs and BmNPVs, respectively. For example, a recombinant BmNPV has been constructed using an AcNPV transfer vector for expression of immunoglobulin genes in the silkworm, *B. mori.*[85] The efficiency of insertion, however, is poor, and nonspecific insertion and/or lower expression rates often result (Maeda, unpublished). Hybrids of AcNPV and BmNPV capable of replicating in both BmN and SF cells have also been isolated. Using an AcNPV transfer vector, the luciferase gene has been inserted into a hybrid AcNPV-BmNPV virus and used to express luciferase in BmN and SF cells and in silkworm larvae.[86]

III. CHARACTERISTICS OF EXPRESSED FOREIGN GENE PRODUCTS

A. PROTEOLYTIC PROCESSING

1. Signal Sequence Cleavage and Secretion

Secretion is a major pathway for the release of proteins from the cell. Secreted proteins must possess a chain of hydrophobic amino acids (secretion signal) at their N-termini. Ribosomes synthesizing secretion signal amino acid chains are thought to be recognized and bound by signal recognition particles (SRP). The SRP-ribosome complex then attaches to receptor sites in the endoplasmic reticulum (ER) membrane where the SRP is released and the hydrophobic chain is cleaved. At the ER membrane the mature protein is synthesized, released into the ER lumen, and eventually secreted from the cell. In most cases, the signal sequences of foreign genes expressed by baculoviruses are efficiently recognized, cleaved rapidly,[52] and are not considered to affect expression levels. Heterologous signal sequences have also been shown to be precisely cleaved at the expected cleavage site (e.g., see Reference 87). There are, however, two examples of the inefficient cleavage of the signal sequences of baculovirus-expressed foreign proteins: (1) the signal sequence of a *Bacillus anthracis* protective antigen,[88] and (2) the propapain signal sequence.[89] Replacement of the propapain signal sequence with that of melitin (a honeybee protein) increases efficiency of secretion 5-fold.[89]

Although nearly all signal sequences of baculovirus-expressed proteins are cleaved at the proper sites, secretion of the expressed polypeptides following cleavage is generally not as efficient. Temporal inhibition of secretion has also been reported for several baculovirus-expressed proteins. For example, during an early stage of infection, human tissue plasminogen activator (tPA) expressed in SF cells by a recombinant AcNPV is efficiently secreted; however, during a late stage of infection, it is apparently trapped in the Golgi apparatus.[51] In addition to temporal regulation, glycosylation (as will be discussed later) also plays a role in protein secretion. Examples of glycoproteins that are poorly secreted due to inefficient glycosylation include human insulin receptor,[90] propapain,[91] rat fibronectin,[92] human growth hormone receptor (extracellular domain),[93] mouse soluble CD4,[94] and human factor VIII.[95]

2. Internal Proteolytic Cleavage

Many polypeptides are synthesized as large precursor molecules that must be proteolytically processed to become functional proteins. Post-translational proteolytic processing is considered advantageous since (1) it allows the precursor to be specifically activated at the target site, (2) it allows proper folding of polypeptides, (3) it protects small polypeptides from proteinase degradation, and (4) it allows several mature polypeptides to be generated from a single precursor molecule, which can be translated from a single mRNA molecule.

Internal proteolytic cleavage can occur by host proteinases or by autocatalytic proteinase activities. Autocatalytic cleavage of baculovirus-expressed proteins seems to occur efficiently; however, the rate of internal cleavage is generally slower in insect cells. For example, cleavage of the major 116 kDa glycoprotein complex, gI, of bovine herpesvirus-1 occurs significantly less efficiently in SF cells than in mammalian cells; however, it occurs at the same site.[96] Cleavage of gp160 of HIV-1 to generate gp120 occurs in SF cells at the correct site; however, it occurs very poorly[97,98] or is not detectable.[99,100]

The efficiency of host proteinase-mediated cleavage varies depending upon the amino acid sequence that is cleaved. The efficiency of host proteinase-mediated cleavage of the same amino acid sequence can also be different in larval tissues and in cell lines. For example, the hemagglutinin precursor of influenza virus is more efficiently cleaved in larval hosts than in SF cells.[101] Furthermore, if tissue-specific proteinases are involved in proteolytic cleavage, proper processing generally does not occur in insect cell cultures. Insect host proteinases may also recognize and cleave additional sites that are not cleaved in the authentic (e.g., mammalian) cells. For example, the human gastrin-releasing peptide precursor is cleaved differently in insect and human cells.[102] Also, mouse interleukin-3 (IL-3) is highly expressed in silkworm larvae; however, some aberrant proteolytic cleavage is found between alanine 127 and serine 128, which is presumably caused by a host (or viral-induced) proteinase.[103]

B. GLYCOSYLATION

Glycosylation is a post-translational modification of proteins that can impart stability and antigenicity; it may also be involved in intracellular sorting, cell surface attachment, folding, secretion, and solubility. The structure of the oligosaccharide chain attached during glycosylation often depends upon the host (*E. coli,* yeast, mammal, insect, etc.) and tissue type that the protein is expressed in, and often occurs heterogeneously. Two types of glycosylation, *N*-glycosylation and *O*-glycosylation, can occur. In *N*-glycosylation, the oligosaccharide chain is linked to the asparagine residue of the sequence Asn-X-Ser/Thr and processed. *N*-glycosylation occurs predominantly in secreted or membrane-bounded polypeptides in the ER and Golgi apparatus. In *O*-glycosylation, a relatively simple oligosaccharide chain is linked to serine and threonine residues.

N-glycosylation and *O*-glycosylation occur in the baculovirus structural proteins, gp64[104] and gp41,[105] respectively. Although both *N*- and *O*-glycosylations occur in insect cells, in most cases the oligosaccharide structures attached to the expressed polypeptides are not the same as those attached in mammalian cells. The major difference of proteins glycosylated in insect cells is a general lack of complex-oligosaccharide chains. This is presumably due to the lack of appropriate enzymes in the Golgi apparatus involved in the transfer of galactose or sialic-acid residues, which are required for the formation of complex oligosaccharides.[106]

Based on the studies to date, the following generalizations can be made about *N*-glycosylation in insect cells:

1. Glycosylation in insect cell lines and larvae occurs at sites identical to those found in the authentic cells.
2. High mannose oligosaccharides are attached while complex oligosaccharides are not (one exception will be discussed later).
3. The lack of complex oligosaccharides results in proteins of lower molecular weight, which can be detected by SDS-PAGE.
4. Glycosylation patterns often change during the viral replication cycle.
5. Biological activity is not affected by heterogeneity in the attached oligosaccharides (exceptions will be discussed later).

N-glycosylation may also be required for secretion based on the fact that tunicamycin, which blocks the initial attachment of the oligosaccharide chain in the ER, interferes in the secretion process. Tunicamycin treatment also reduces the number of budded virions that are produced, since the budded virion contains the *N*-glycosylated envelope protein gp64.[107]

Despite heterogeneous or inefficient glycosylation, authentic biological activity is generally found even in highly complicated polypeptides expressed in insect cells by baculovirus vectors, as shown by the following examples. Mouse IL-5 exhibits authentic biological activity based on lymphoid cell proliferation assay.[108] The truncated and secreted forms of the interferon gamma receptor show specific binding to interferon gamma.[109] The human corticosteroid-binding globulin binds cortisol with an affinity and specificity equivalent to that of human serum-derived corticosteroid-binding globulin.[110] The cardiac Na^+/Ca^{2+} exchanger is electrophysiologically active based on whole cell patch-clamp experiments.[111] The β-subunit of human choriogonadotropin hormone exhibits binding activity to rat testicular receptors, and induction of steroidogenesis in a mouse Leydig cell bioassay.[112,113] The G protein of rabies virus[114] and major glycoprotein complex, gI, of bovine herpesvirus-1[96] exhibit fusion activity.

Although the majority of baculovirus-expressed proteins are sufficiently glycosylated so that normal biological activity is observed, a decrease or elimination of biological activity due to inefficient glycosylation has also been found in a few proteins. For example, gp120/160 expressed in insect cells does not bind to CD4,[115] and ovine prostaglandin endoperoxide synthase-1 expressed in *SF* cells exhibits lower (56 versus 112 units/mg in cos-1 cells) specific activity of cyclooxygenase.[116]

The oligosaccharide chains of several glycosylated polypeptides expressed by baculoviruses in insect cells have been directly analyzed. The oligosaccharide chains of fowl plaque virus hemagglutinin (HA) expressed in SF cells are mainly composed of (1) a high-mannose type oligosaccharide ($Man_{5-9}GlcNAc_2$), (2) a truncated oligosaccharide core ($Man_3GlcNAc_2$), and (3) a fucose attached core ($Man_3(Fuc)GlnNac_2$).[117] Detailed analysis by mass spectrometry of mouse IL-3 expressed in silkworm larvae indicate that mannose, fucose, and glucosamine are attached to the glycosylation sites.[118] Specifically, asparagine 16 is heterogeneously glycosylated with one of four different oligosaccharides; asparagine 86 is glycosylated with a single type of oligosaccharide or not glycosylated; asparagine 44 and asparagine 51 are not glycosylated; and no *O*-glycosylation is found.[103]

There is one reported example (human plasma plasminogen) of complex oligosaccharides found in a baculovirus-expressed protein.[119] This indicates that modifying enzymes (most likely protein specific) that are required to generate complex oligosaccharides are found in the insect cell. Analysis of the oligosaccharide chains attached to asparagine 289 of human plasma plasminogen reveals that approximately 40% are of the complex type and 60% are of the high-mannose type ($Man_9GlcNAc_2$, $Man_{3-5}GlcNAc_2$).[119] Glycosylation of human plasma plasminogen also changes during the different stages of the viral infection cycle.[120] Between 0 and 20 h p.i., 96% of the oligosaccharides attached to asparagine 289 are of the high mannose type ($Man_{3-9}GlnNAc_2$); however, between 60 and 96 h p.i., 92% are of the complex type (biantenary, triantennary, and tetraantennary oligosaccharides at varying stages of outer arm completion). Between 20 and 60 h p.i., a mixture of complex and high-mannose oligosaccharides are found. These results indicate that (1) the high-mannose oligosaccharide may be processed into a complex one, and (2) insect cells possess the ability to modify oligosaccharides that are already attached to the polypeptide chain by the activation of appropriate glycosyl transferases. Similar results are also found for human plasma plasminogen expressed in *Mamestra brassicae* cells,[121] indicating that lepidopteran cells in general can generate complex oligosaccharides from high mannose oligosaccharides at a late stage of infection. In *M. brassicae* cells (48 h p.i.) the complex oligosaccharides are of the bisialo-biantennary (28%), asialo-biantennary (7%), fucosylated bisialo-biantennary (25%), and fucosylated asialo-biantennary (3%) types, while the high-mannose oligosaccharides are of the $Man_9GlnNAc_2$ (22%), $Man_5GlnNAc_2$ (13%), and $Man_3GlnNAc_2$ (2%) types. Furthermore, α-D-mannosidase, which catalyzes trimming of mannose of $Man_6GlnNAc_2$ to generate $Man_5GlnNAc_2$ and $Man_3GlnNAc_2$, is induced in SF cells by baculovirus infection.[122] This trimming process is essential for converting high-mannose oligosaccharides to complex and hybrid glycans.

O-glycosylation has been shown to occur in a limited number of proteins expressed in insect cells by baculovirus expression vectors. Examples include human choriogonadotropin,[123] adenovirus-2 fiber protein,[124] and the soluble form of human CD8 α.[125]

C. PHOSPHORYLATION

The phosphorylation and dephosphorylation of polypeptides often play major roles in essential cellular processes such as cell differentiation and signal transduction. Although a protein kinase gene has not yet been identified in the baculovirus genome, kinases induced by viral infection and associated with viral particles have been detected.[126,127] In general, however, kinase activity is relatively low in baculovirus-infected SF and Bm cells. It is also important to note that phosphorylation activity changes during the virus replication cycle.[129] Also, an AcNPV gene encoding tyrosine phosphatase, which may affect the phosphorylation of expressed polypeptides, has recently been isolated.[130]

The low kinase activities found in baculovirus-infected insect cells is advantageous for studying phosphorylation that is induced (e.g., by signal transduction). For example, tyrosine phosphorylation of α and β platelet-derived growth factor receptors expressed at an early stage of viral infection is not detectable, allowing ligand-dependent tyrosine phosphorylation to be exhibited.[131] At a late stage of infection, however, the receptors are tyrosine-phosphorylated, due presumably to a high rate of receptor production, resulting in ligand-independent dimer formation. The protein kinase domains of several receptors have also been expressed in insect cells and analyzed. Expression of the cytoplasmic protein kinase domain of the epidermal growth factor (EGF) receptor in SF cells results in an increase in kinase activity in response to sphingosine or ammonium sulfate.[132] Autophosphorylation of the kinase domain also occurs at the correct residue, and the protein exhibits enzyme kinetics similar to those of the authentic activated EGF receptor kinase.[133]

The phosphorylation of foreign gene products does not occur identically in insect and noninsect cells. For example, phosphorylation of the SV40 large T antigen occurs inefficiently in SF cells at threonine 124, indicating that the serine residues surrounding threonine 124 are dephosphorylated.[134] Serine residues of the oncoprotein p53, either complexed or uncomplexed with the SV40 large T antigen, however, are phosphorylated when expressed in SF cells.[135] The p53 expressed in SF cells is also associated with an endogenous casein kinase II; however, this association can be abolished by a serine to alanine substitution at position 389.[136] The middle T antigen of polyoma virus is poorly phosphorylated or not phosphorylated at tyrosine residues when expressed in SF cells,[137] indicating that SF cells lack a sufficient amount of endogenous tyrosine kinase.[138] Phosphorylation, however, occurs efficiently when middle T antigen is coexpressed with $pp60^{c\text{-}src}$ in SF cells.[138]

Other representative examples of poorly and heterogeneously phosphorylated proteins include: L1 of human pappilomavirus-16;[139] bovine protein kinase C-gamma;[140] pp56lck, a member of the src family of tyrosine kinase mainly expressed in T lymphocytes;[141] and mineralocorticoid receptor.[129]

D. C-TERMINUS α-AMIDATION

C-Terminus α-amidation is a post-translational modification that occurs by hydroxylation of the C-terminus glycine residue, resulting in the α-amidation of the adjacent amino acid (N-terminus side). C-terminus amidation imparts stability to small polypeptides and is often required for biological activity. Nearly all established cell lines (including those derived from insects) are considered to lack the enzymes necessary to amidate polypeptides. For example, sarcotoxin IA, an antibacterial peptide of the flesh fly, expressed by a baculovirus-expression vector in BmN cells, is not amidated.[142]

Expression of a frog cDNA encoding a putative C-terminus-amidating enzyme (peptidylglycine α-amidating monooxygenase) in SF cells has revealed that two steps are involved in C-terminus amidation.[143,144] Peptidylglycine α-amidating monooxygenase has been shown to only possess peptidylglycine α-hydroxylating monooxygenase activity; however, successful amidation occurs due to an unknown enzyme found in the SF cell culture fluid that is able to convert α-hydroxyglycine-extended peptides to the amidated forms, thereby completing the amidation process. These results also suggest that C-terminus α-amidation of baculovirus-expressed peptides will occur in insect cells by coinfection of a recombinant virus expressing the foreign gene of interest and another expressing peptidylglycine α-hydroxylating monooxygenase.

In contrast to established cell lines, C-terminus amidation occurs efficiently *in vivo* in insect larvae. The ability of insect larvae to α-amidate foreign proteins is a major advantage of expression in larvae over established cell lines. In silkworm larvae, diuretic hormone of *M. sexta* expressed by a baculovirus vector is considered to be amidated due to the physiological effects it induces.[145] In last instar larvae of *Trichoplusia ni* and diapausing pupae of *Hyalophoa cecropia,* 20 to 30% and 70%, respectively, of the prececropin A found secreted into the hemolymph is α-amidated.[146] Cecropin A can also be expressed in *T. ni* larvae as a fusion protein with protein A of *Staphylococcus aureus* in order to facilitate purification.[147] Biologically active (i.e., properly amidated) cecropin A, which is indistinguishable from the authentic peptide, can then be generated by cleavage of the fusion protein with CNBr.

E. FATTY ACID ACYLATION

Acylation (myristoylation, palmitoylation, and isoplynylation) is the covalent attachment of fatty acid chains to polypeptides. Acylation imparts stability and can allow membrane binding in certain proteins. The baculovirus envelope protein gp64 is acylated.[148] Foreign proteins expressed by baculoviruses have also been found to be correctly acylated in insect cells. Examples include hemagglutinin of the influenza virus,[149] hepatitis B virus surface antigen with the pre-S domain,[150] human aldose reductase,[151] α subunits of Gi3,[152] band 4.2 protein of erythrocyte membrane,[153] and VPO of poliovirus.[154]

The gag precursor (Pr57gag) of the simian immunodeficiency virus (SIV) is highly expressed and myristoylated in SF cells.[155] A point mutation of Pr57gag can inhibit myristoylation, resulting in the improper association Pr57gag with the cell plasma membrane so that virus-like particles are not released from the cell surface. Similar results have been found for gag of feline immunodeficiency virus[156] and HIV-1[157-160] expressed in SF cells. The nonmyristoylated form of gag of HIV-1, however, is reported to be efficiently released from SF cells.[161] The nonmyristoylated form (due to a point mutation) of $pp60^{c\text{-}src}$ is also secreted as a highly expressed (20 mg from 6 liters of SF cell culture) biologically active protein.[162] Human β 2-adrenergic receptor expressed in SF cells is properly palmitoylated by agonist modulation.[163]

Both palmitoylation and isoprenylation have been shown to occur in a 100-kDa maternal CAAX box-containing protein of *Xenopus* expressed in SF cells.[164] Isoprenylation has also been shown to occur properly for hepatitis delta antigen[165] and guanine nucleotide regulatory protein gamma 2[166] expressed in SF cells.

F. SOLUBILITY

A major problem of bacteria-based expression systems is the formation of insoluble protein aggregates that are not functionally active. In most cases polypeptides expressed by recombinant baculoviruses remain soluble under conditions at which the authentic polypeptide is soluble. If insoluble products are formed in insect cells, a functionally active soluble form is also often found. The ratio of insoluble to soluble forms of the expressed proteins generally increases during the viral replication cycle in a manner similar to that found for glycosylation.[115] This change in solubility may be associated with changes in the host cell machinery induced by viral infection. The insoluble forms of some baculovirus-expressed proteins have been solubilized and refolded into their functionally active forms by treatment with 6 M guanidium chloride.[167,168]

It is unknown why some proteins become insoluble when expressed in insect cells; however, the improper formation of disulfide bridges or high expression rates (of relatively hydrophobic proteins) may play a role. In general, polyhedrin-fused polypeptides tend to be more insoluble,[70] possibly due to changes in protein folding. Baculovirus-expressed proteins with insoluble fractions (i.e., solubilities ranging from 5 to 95%) have been reported for the following proteins: the cytoplasmic domain of the epidermal growth factor receptor,[133] envelope proteins of HIV-1,[115] protein phosphatase 1,[167] the α subunit of Gi3,[152] and androgen receptor.[168] The androgen receptor[169] and P1 of the cauliflower mosaic virus[170] are expressed as nearly completely insoluble proteins.

G. OLIGOMERIZATION

Specific subunit associations are often required for functional activity. In general, protein subunits expressed by baculoviruses properly assemble to form functional polypeptides. Proteins requiring two or three subunits for functional activity can be generated by the coinfection of recombinant viruses, each of which expresses one of the subunits. Proteins composed of four or more subunits can be generated by the coinfection of recombinant viruses carrying dual promoters, each of which expresses two of the subunits.

The correct formation of disulfide bridges of oligomerized proteins can be inferred based on the detection of authentic biological activity. Homodimers of the following polypeptides are correctly formed in insect cells: mouse IL-5,[171] human nerve growth factor receptor,[172] human IL-5,[173] human nerve growth factor,[174] vesicular endothelial growth factor,[175] NS1 of Japanese encephalitis virus,[176] soluble form of human CD8 α,[125] and vascular endothelial growth factor.[178] Heterodimers are correctly formed for DNA polymerase (UL30) and UL42 of herpes simplex virus-1,[178,179] and the α and β subunits of human choriogonadotropin.[180]

Trimerization occurs efficiently and correctly for coronavirus glycoprotein S,[181] and gp160 or gp41 of HIV-1[182] expressed in SF cells. Trimerization of influenza virus hemagglutinin, however, occurs relatively poorly compared to trimerization in mammalian cells.[149] Trimerization of heterologous peptides encoded by UL5, UL8, and UL52 of herpes simplex virus occurs correctly, generating functional helicase-primase complexes.[183,184]

Seven different polypeptides involved in DNA replication of herpes simplex virus have been expressed in SF cells.[185] These polypeptides can associate and induce replication of cotransfected DNA fragments containing DNA replication origins. Other highly complex molecules that have been shown to correctly oligomerize include: mouse immunoglobulin heavy and light chains,[186] mouse major histocompatibility complex class I molecule,[187] and human class II histocompatibility protein α and β subunits.[188]

H. ASSEMBLY OF VIRAL PARTICLES

Viral structural proteins expressed by baculoviruses generally assemble correctly to form structures that are morphologically indistinguishable from the authentic viral particle. Viral structural polypeptides are also generally expressed at very high levels in insect cells by baculovirus vectors. High-level expression facilitates their purification by standard purification processes, such as cesium chloride and sucrose gradient centrifugations, and makes baculovirus-expression systems ideal for the production of subunit vaccines.

Numerous studies have used baculovirus-expression vector systems to screen genes speculated by nucleotide sequencing to encode viral structural proteins. For example, the genome of the Norwalk virus, which induces severe diarrhea in humans, encodes three genes. Expression of the 58K protein of Norwark virus in SF cells results in the generation of virus-like particles similar to authentic capsids, which are immunologically reactive to antibodies against Norwark virus, indicating that the 58K protein is a capsid protein.[189] The parvovirus B19 is composed of the structural proteins VP1 and VP2. Coinfection of recombinant baculoviruses expressing VP1 and VP2 generates capsid-like particles with a stoichiometry similar to that found in the authentic capsids.[190,191] Expression of VP2 alone also generates capsid-like particles, which can be crystalized for X-ray crystallography studies.[192] Canine parvovirus VP2 expressed in SF cells forms capsid-like particles that exhibit hemagglutination activity.[193] Adeno-associated virus type-2 possesses three capsid proteins VP1, VP2, and VP3. Single and coexpression experiments reveal that VP2 of this virus is essential for viral assembly.[194] Similar experiments indicate that VP2 of Aleutian mink disease parvovirus is also essential for viral assembly.[35]

Bluetongue virus (BTV) capsids are composed of four major (VP2, VP3, VP5, and VP7) and three minor (VP1, VP4, and VP6) polypeptides. When VP2 and VP5 of BTV are coexpressed in SF cells, core-like particles are formed.[195] Coexpression of VP3 and VP7 by a dual-expression baculovirus vector also forms core-like particles[196] with a stoichiometry (2:15) similar to that found in the authentic cores. Cryoelectron microscopy of the VP3- and VP7-generated core-like particles indicates that they are (1) slightly different from the authentic core particles, and (2) morphologically similar to rotaviruses rather than reoviruses.[197] Coinfection of multiple recombinant AcNPVs expressing VP2, VP3, VP5, and VP7 in SF cells results in the generation of double-shelled, virus-like particles of the same size and appearance as the authentic BTV.[198] Purified, these virus-like particles exhibit high levels of hemagglutination activity and are highly effective in inducing neutralizing antibodies against BTV. Furthermore, coexpression of VP1, VP4, and VP6 (minor structural proteins) with the four major structural proteins generates double-shelled virus-like particles containing small amounts of VP1, VP4, and VP6.[72] VP3 of epizootic hemorrhagic disease virus (serologically related to BTV) expressed in insect cells can also generate chimeric double-shelled, virus-like particles by coexpression with other bluetongue viral structural polypeptides. This indicates that the functional (having the ability to associate with other structural proteins) epitopes of the VP3 protein are conserved.[72] VP2 (corresponding to VP3 of BTV) of group A rotavirus and VP6 of group B rotavirus can also generate single-shelled particles.[199]

VP1 of polyoma virus expressed in SF cells forms virus-like particles in the nucleus; however, VP1 localized in cytoplasm does not.[200] Cytoplasmic VP1, however, can form virus-like particles after treatment with the calcium inophore inomycin, indicating the importance of the calcium ion for assembly.

The core (internal capsid) protein of hepatitis B virus expressed by a recombinant AcNPV in SF cells forms virus-like particles.[201] The precore antigen, however, is not properly processed in SF cells and forms a highly insoluble cytoplasmic aggregate.[202] Hepatitis B surface antigen expressed in BmN cells or silkworm larvae by a recombinant BmNPV also forms virus-like particles.[203] In silkworm larvae the rate of formation of these virus-like particles is about 1000-fold higher than *in vitro.*

Other examples of structural proteins expressed by baculoviruses with particle-like structures that bud from the cell surface into the medium include: the gag precursor proteins of HIV-1,[158,160] simian immunodeficiency virus,[155] bovine immunodeficiency-like virus,[204] and feline immunodeficiency virus.[156,205] Expression of hemagglutinin-neuraminidase of Newcastle disease virus (NDV) by a recombinant AcNPV in SF cells results in the release of a NDV-like envelope structure into the culture medium, indicating that envelope structures possessing hemagglutinin-neuraminidase without a matrix protein can be generated.[206] Structural protein(s) of foot-and-mouth disease virus,[207] *Arabis* mosaic virus,[208] and human papillomavirus-16[139] expressed in insect cells also form capsid-like particles.

IV. EXPRESSION OF FOREIGN GENES IN LARVAL INSECTS

A major advantage of the baculovirus expression vector system is the availability of an *in vivo* system (insect larvae) for the expression of foreign genes. As mentioned previously, foreign gene expression generally occurs at higher levels in insect larvae compared to expression in established cell lines. Also, as mentioned previously, several types of post-translational modifications (e.g., C-terminus amidation) only occur or occur more efficiently in *in vivo* systems as compared to *in vitro* systems. Infection of larvae with polyhedron-deficient recombinant viruses must be done by injection into the hemocoel, whereas polyhedra of polyhedrin-positive recombinant viruses can be administrated orally. Injection, however, is recommended even for polyhedrin-positive recombinant viruses in order to synchronize viral infection.

Silkworm *(B. mori)* larvae were the first to be used as *in vivo* hosts for the expression of foreign genes.[38,39] *B. mori* is the only appropriate host for foreign gene expression by BmNPV. In contrast, AcNPV has a wider host specificity and can efficiently infect *T. ni, S. frugiperda,* and *Heliothis virescens* larvae. The small size and/or cannibalistic tendencies of these AcNPV hosts, however, make them difficult to handle compared to the domesticated larvae of *B. mori. M. sexta* larva is not a natural host of AcNPV (AcNPV causes death, but cannot replicate in *M. sexta*). However, they are useful for foreign gene expression due to their very large size (up to 10 grams).[209] Infection of *M. sexta* larvae is done by repeated injection in order to increase efficacy.[209]

Silkworm larvae offer a number of advantages over other larval systems[6] including (1) large size (5th instar silkworm larvae can weigh more than 5 grams and express proportionately larger amounts of protein), (2) ease of handling (due to their importance in the sericulture industry, silkworms have been domesticated and automated rearing systems are available), and (3) well-characterized genetics (silkworms have been studied and bred for thousands of years and numerous well-characterized strains are available). Furthermore, domesticated silkworm larvae are not able to move more than a few centimeters, they only feed on mulberry leaves or artificial diet, and the adult silkmoth possesses wings, but cannot fly. These characteristics make it nearly impossible for silkworm larvae to survive in the field, which adds an extra element of safety when working with recombinant viruses. Silkworms diapause during the egg stage and can be stored for several months under appropriate conditions (e.g., 5°C). The entire silkmoth life cycle lasts about 7 weeks, of which the five larval stages (hatching to pupation) take about 25 days.

Generally, 5th instar silkworm larvae (about 18 days after hatching) are appropriate for infection with recombinant virus for foreign gene production. Prior to recombinant virus injection, larvae are starved for several hours to overnight and anesthetized by immersion in an ice-water bath. Four to five days after injection, secreted proteins are collected from the hemolymph, and nonsecreted proteins are isolated from fat body tissues. Hemolymph is collected by piercing the larval prolegs with a pin and allowing the hemolymph to drain into a microfuge tube placed in wet ice or dry ice. Generally, 0.5 to 0.8 ml of hemolymph per 5th instar larva (5 gram) can be collected out of a total hemolymph volume of about 1.5 ml. Secreted proteins that are stored in the hemolymph are easily purified by typical biochemical procedures such as chromatography. Fat body tissue must be collected by dissection, which is a somewhat time-consuming process. Expressed polypeptides are extracted from fat body by homogenization in an appropriate buffer. Generally, a single 5th instar larva (5 grams) possesses about 1 gram of fat body tissue. If the expressed protein can easily be purified, infected larvae can also be homogenized whole in order to purify the expressed protein.[203]

One milligram or more of expressed foreign protein can be collected from a single *B. mori* larva. Polypeptides that are localized within the cytosol (i.e., nonsecreted) are generally expressed at 5- to 10-fold higher rates *in vivo* than *in vitro*. This rate is essentially unchanged among larvae susceptible to AcNPV or BmNPV. A single larva of *B. mori* produces about 1 mg of mouse IL-3[210] or hepatitis B surface antigen,[203] about 0.6 mg of immunoglobulin molecules with authentic binding activity,[85] or about 0.1 mg of human interferon α[59] or feline interferon α.[211] The membrane-bounded protein gp41 of HIV-1 has been expressed as a polyhedrin fusion protein in silkworm larvae (see Reference 6). Homogenized fat body cells expressing the gp41 fusion protein have been successfully used without purification in ELISAs to screen AIDS patients for antibodies against HIV-1. The authentic post-translational modification characteristics of larval expression have also been used to analyze oligosaccharide chains attached to mouse IL-3 as described earlier.[103,118,210]

In *T. ni* larvae, hepatitis B surface antigen,[212] human adenosine deaminase,[213] and influenza virus neuraminidase[212] are expressed at 0.15 mg, 0.3 mg, and 2.8 mg per larva, respectively, by recombinant AcNPVs. In *S. frugiperda* larvae, luciferase and choriogonadotropin are expressed simultaneously by coinfection of recombinant AcNPVs.[214] In *M. sexta* larvae human apolipoprotein E (ApoE) is highly expressed by repeated injection of a recombinant AcNPV carrying the apolipoprotein gene.[209] In contrast to ApoE expressed in SF cells, ApoE expressed in *M. sexta* generates lipoprotein particles, which are excellent (at the concentration of 0.2 μg/ml they have a displacement potential of 50%) ligands for low-density lipoprotein receptors. Similar associations have also been observed for human apolipoproteins A and E expressed in silkworm larvae (see Reference 6). These types of associations may, however, cause difficulties in the purification of the foreign gene products. Aberrant proteolytic cleavage (e.g., autoproteolytic cleavage of IL-3[103] as described earlier) can also be a problem when expressing foreign genes in larvae.

V. FOREIGN GENE EXPRESSION FOR PEST INSECT CONTROL

Insect viruses, especially baculoviruses, have been used as agents for pest control since 1930. The success of baculovirus insecticides in limited forest situations has been good, and continuous protection of forests has often been observed.[215] To date, soybean crops in Brazil and palm and coconut trees in the South Pacific have been efficiently controlled by *Anticarsia gammatalis* NPV[216] and *Oryctes rhinoceros* baculovirus,[217] respectively. The use of naturally occurring baculoviruses for the protection of agricultural crops in developed countries, however, has been limited predominantly due to the slow speed of insect killing; i.e., viral-infected pests can continue to feed. Recombinant baculoviruses expressing genes whose products are toxic to the larvae or interfere in the larval feeding process have been considered as alternatives to naturally occurring baculoviruses. Recombinant baculoviruses exhibiting increased speeds of insect killing and chemical pesticide-like symptoms have been constructed. Many genes have also been considered and tested as candidates for insect pest control, including those encoding insect-specific toxins, neuropeptides associated with essential insect processes, and enzymes and enzyme inhibitors involved in insect metabolism.

Recombinant baculoviruses producing polyhedra are probably the most appropriate for pest control since polyhedra provide stability to the virus in the field and are necessary for oral infectivity. However, in order to screen possible pest-controlling genes, polyhedrin-deficient virus constructs are potentially safer. Furthermore, silkworm larvae, although not pest insects, are very similar to common lepidopteran insect pests and are ideal models for the study of gene products with potential insecticidal activity. As mentioned above, the physiology, biochemistry, and genetics of silkworms are very well characterized, and large numbers of synchronized larvae can be easily reared.

The products of several genes have been shown to increase the insecticidal effects of wild-type baculoviruses. A recombinant BmNPV expressing the diuretic hormone gene of *M. sexta* was the first reported case of a recombinant baculovirus exhibiting an increase in insecticidal activity.[145] Silkworm larvae infected with this recombinant BmNPV show a 30% reduction in hemolymph volume, presumably due to the effects of the expressed diuretic hormone and die 20% faster than larvae infected with wild-type BmNPV. However, the body weight of infected larvae does not change (until death), and the release of H_2O outside of the body is not observed. Other neuropeptides, including eclosion hormone[218] and prothoracicotropic hormone,[219] do not seem to be effective for pest control.

Juvenile hormone (JH) esterase plays a key role in the degradation of JH in insects.[220] Recombinant AcNPVs expressing JH esterase under the polyhedrin promoter have been constructed.[221] Injection of purified baculovirus-expressed JH esterase into the body cavity of *M. sexta* induces blackening, indicating that the baculovirus-expressed JH esterase is biologically active. Oral infection of first instar *T. ni* larvae with recombinant AcNPVs expressing JH esterase inhibits growth; however, growth inhibition is not seen in other larval instars. Mutated JH esterases (with amino acid substitutions) have been expressed and characterized using recombinant AcNPVs, and their insecticidal effects are now being tested.[222] A recombinant AcNPV lacking the viral ecdysteroid UDP-glucosyltransferase gene, which inactivates ecdysteroids, and carrying the JH esterase gene has also been constructed.[223] JH esterase is highly expressed by this virus; however, it does not induce observable changes in development, weight gain, time to mortality, etc.

The bacterium *Bacillus thuringienis* (Bt) produces several types of insect-specific toxins associated with a parasporal crystal that it produces during sporulation. Upon ingestion the crystals solubilize in the high pH of the midgut and release protoxin molecules that are cleaved by midgut proteases to form the mature toxin. Toxins from several Bt strains, with specific activities towards lepidopteran, coleopteran, and dipteran insects, have been cloned and characterized. The specific regions involved in insect specificity and toxicity have also been mapped and analyzed. A recombinant AcNPV carrying a Bt toxin gene has been shown to produce functionally active toxin molecules.[224] Compared to wild-type AcNPV, the LD_{50} of this polyhedron-positive recombinant AcNPV is about two-fold higher and the LT_{50} is nearly the same.

The products of insect-specific scorpion and mite toxin genes expressed in insect larvae by baculovirus vectors have shown dramatic insecticidal effects. Scorpion toxin peptides are classified into mammalian- and insect-specific toxins.[225,226] Scorpion toxin molecules are very small and target the sodium channel. Their amino acid sequences have also been determined. There are at least two types of insect-specific scorpion toxins, including (1) an excitatory toxin that causes immediate spastic paralysis, and (2) a depressant toxin that induces progressive flaccid paralysis by blocking action potentials.[225] Excitatory and depressant toxins show up to 35% amino acid sequence homology to the two known types of mammalian-specific scorpion toxins. However, the inability of mammalian- and insect-specific scorpion toxins to cross-react has been proven by numerous electrophysiological and binding experiments (see Reference 225).

The insect-specific scorpion toxin AaIT isolated from *Androctonus australis* Hector has the least amino acid sequence homology, and one of its four disulfide bridges is found at a different position compared to the mammalian-specific toxins (see Reference 225). A recombinant BmNPV expressing the AaIT gene under control of the polyhedrin promoter replicates normally and does not induce any detectable effects in infected BmN cell cultures. Silkworm larvae infected with this recombinant BmNPV, however, show dramatic changes in behavior (40 h p.i.), including dorsal arching, body tremors, and feeding cessation, which are identical to those observed following the injection of authentic purified AaIT.[87] About 1 to 5 μg per ml of AaIT is produced in hemolymph of infected larvae.

Recombinant AcNPVs that produce polyhedra and express AaIT driven by the p10 promoter have been constructed.[227,228] These recombinant AcNPVs are orally infectious and exhibit symptoms similar to those observed in silkworm larvae as described above. The speed of insect killing of these viruses in *H. virescens*[228] and *T. ni*[227] larvae is about 50% faster compared to wild-type virus, and a reduction of crop damage[227] is also observed. These results indicate that recombinant viruses expressing toxin genes that target the sodium channel can induce symptoms in infected larvae similar to those caused by chemical insecticides (e.g., pyrethroids).

The insect-paralytic mite, *Pyemotes tritici,* produces polypeptides that cause involuntary muscle contractions in lepidopteran insects. The true mechanisms and mode of action of these toxins is still unknown; however, they have been shown to be insect specific.[229] One of the *P. tritici* toxin molecules is 25 to 30 kDa in weight and possesses about 8% cysteine residues. cDNA encoding this toxin has been cloned and inserted into AcNPV under a modified polyhedrin promoter. Intrahemocoelic injection of this recombinant AcNPV into *Galleria mellonella* results in symptoms similar to those observed following the injection of purified mite toxin. Oral infection of *T. ni* larvae with polyhedra produced by recombinant AcNPVs expressing the *P. tritici* toxin gene results in a 50% increase in speed of insect killing.[230]

Safety and efficacy are major considerations in the design and use of the recombinant viruses as insecticides. Recombinant baculoviruses such as the AaIT expressing BmNPV have at least two layers of safety. First, mammalian cells are resistant to the incorporation of baculovirus DNA into the nucleus,[57] so that the vector itself is safe. Second, the expressed gene product is insect specific and has no effects on mammals. Field trials (release into the field) of genetically modified recombinant baculoviruses started in England in 1987 and in the United States in 1989. A great deal of information about the survival of recombinant viruses in the field and secondary effects on other insect species has been accumulated from these trials, which further attests to their safely. In the laboratory, the mechanisms of viral replication and host-specificity determination at the molecular levels are also being studied extensively.

The toxicity of AaIT towards lepidopteran insects is about 500-fold lower than the toxicity it induces in dipteran insects.[231] Modifications of AaIT may increase its toxicity towards lepidopteran insects, resulting in a significant increase in speed of insect killing. Other insect-specific toxins isolated from scorpions, wasps, mites, or other organisms with greater toxicity and/or different target sites may also increase the efficacy of baculoviruses for pest control even further. Another potential strategy to increase efficacy is the use of host promoters that can express AaIT or another appropriate foreign gene at an earlier stage of infection.

VI. CONCLUDING REMARKS

Hundreds of foreign genes have been expressed by baculovirus expression vectors, and their utility for the expression of proteins for basic and applied research has been proven time and time again. As discussed in this chapter, baculovirus-expression vector systems offer numerous advantages over other expression systems, including high-level production of biologically active gene product, an easy-to-use *in vivo* system, and the ability to simultaneously express multiple genes in the same cell. Within the last few years, new strategies such as the use of linearized DNA, a β-galactosidase marker, and baculovirus-yeast shuttle vectors have also been introduced, which allow baculovirus-expression systems to be used faster and more easily.

In addition to their use as expression vectors, baculoviruses expressing foreign gene products have been used as tools for the analysis of complex biological phenomena. For example, seven rotavirus structural genes have been successfully expressed in SF cells in order to study rotavirus capsid formation.[72] DNA replication of the herpes simplex virus has also been studied by the expression of seven different proteins involved in DNA replication in insect cells.[185]

Recombinant baculoviruses expressing insect-specific toxins have also been constructed for use as safe viral insecticides. These baculoviruses are able to kill insects much faster than naturally occurring baculoviruses and induce symptoms similar to those induced by chemical insecticides.[87,227,228,230] Further increases in the speed of insect killing by the use of modified toxins or newly identified genes will greatly increase the potential of recombinant baculoviruses to be used in integrated pest management strategies, thereby reducing the use of traditional chemical pesticides.

Many of the drawbacks of baculovirus-expression vector systems that still remain, such as the poor expression of membrane-bounded and low molecular weight polypeptides, formation of insoluble products, and lack of or inefficient post-translational modifications, may be solved by modification of specific genes in the viral genome or by the establishment of new technologies. For example, our laboratory is now studying the mechanisms responsible for envelope formation of occluded viral particles and protein trafficking. This research may be useful in constructing mutated baculovirus vectors that can, for example, express membrane-bounded polypeptides, including receptors at high levels. These improvements will allow baculovirus-expression vectors to be even more useful as eucaryotic vectors in the future.

ACKNOWLEDGMENTS

I would like to acknowledge Shizuo George Kamita for critical reading of the manuscript. Preparation of this manuscript was supported in part by the U.S. Department of Agriculture (91-37302-6186).

REFERENCES

1. **Luckow, V. A. and Summers, M. D.,** Trends in the development of baculovirus expression vectors, *Biotechnology,* 6, 47, 1988.
2. **Summers, M. D.,** Recombinant proteins expressed by baculovirus vectors, in *Concepts in Viral Pathogenesis,* Notkins, A. L., and Oldstone, M. B. A., Eds., Springer-Verlag, Berlin, 1989, 77.
3. **Miller, L. K.,** Baculoviruses as gene expression vectors, *Annu. Rev. Microbiol.,* 42, 177, 1988.
4. **Miller, L. K.,** Insect baculoviruses: powerful gene expression vectors, *Bioessays,* 11, 91, 1989.
5. **Miller, L. K.,** Baculoviruses: high-level expression in insect cells, *Curr. Opinion Genet. Dev.,* 3, 97, 1993.
6. **Maeda, S.,** Expression of foreign genes in insects using baculovirus vectors, *Annu. Rev. Entomol.,* 34, 351, 1989.
7. **Kang, C. Y.,** Baculovirus vectors for expression of foreign genes, *Adv. Virus Research,* 35, 177, 1989.
8. **Fraser, M.,** Expression of eucaryoteic genes in insect cell cultures, *In Vitro Cell. Dev. Biol.,* 225, 1989.
9. **Fraser, M. J.,** The baculovirus-infected insect cell as a eukaryotic gene expression system, *Curr. Top. Microbiol. Immunol.,* 158, 131, 1992.
10. **Vlak, J. M. and Keus, R. J.,** Baculovirus expression vector system for production of viral vaccines, *Adv. Biotechnol. Processes,* 14, 91, 1990.
11. **Bishop, D. H. L.,** Baculovirus expression vectors, *Semin. Virology,* 3, 253, 1992.
12. **Summers, M. D. and Smith, G. E.,** A manual of methods for baculovirus vectors and insect cell culture procedures, *Texas Agricultural Experiment Station Bulletin* No. 1555, 1987.
13. **Maeda, S.,** Gene transfer vectors of a baculovirus, *Bombyx mori* nuclear polyhedrosis virus, and their use for expression of foreign genes in insect cells, in *Invertebrate Cell System Applications,* Vol. 1, Mitsuhashi, J., Ed., CRC Press, Boca Raton, FL, 1989, 167.
14. **O'Reilly, D. R., Miller, L. K., and Luckow, V. A.,** *Baculovirus Expression Vectors: a Laboratory Manual,* W. H. Freeman and Company, New York, NY, 1992.
15. **Possee, R. D. and King, L. A.,** *The Baculovirus Expression System: a Laboratory Guide,* Chapman and Hall, 1992, 229.
16. **Choudary, P., Kamita, S. G., and Maeda, S.,** Expression of foreign genes in insect larvae using baculovirus vectors, in *Methods in Molecular Biology Protocols for Baculovirus Expression,* Humana Press, Clifton, NJ, in press.
17. **Braunagel, S. C., Daniel, K. D., Reilly, L. M., Guarino, L. A., Hong, T., and Summers, M. D.,** Sequence, genomic organization of the EcoRI-A fragment of *Autographa californica* nuclear polyhedrosis virus, and identification of a viral-encoded protein resembling the outer capsid protein VP8 of rotavirus, *Virology,* 191, 1003, 1992.

18. **Rohrmann, G. F.,** Baculovirus structural proteins, *J. Gen. Virol.*, 73, 749, 1992.
19. **Possee, R. D.,** NERC Institute of Virology and Environmental Microbiology, U.K., personal communication, 1992.
20. **Maeda, S.,** in preparation.
21. **Cochran, M. A. and Faulkner, P.,** Location of homologous DNA sequences interspersed at five regions in the baculovirus AcM NPV genome, *J. Virol.*, 45, 961, 1983.
22. **Guarino, L. A., Gonzalez, M. A., and Summers, M. D.,** Complete sequence and enhancer function of the homologous DNA regions of *Autographa californica* nuclear polyhedrosis virus, *J. Virol.*, 60, 224, 1986.
23. **Majima, K., Kobara, R., and Maeda, S.,** Divergence and evolution of homologous regions of *Bombyx mori* nuclear polyhedrosis virus, *J. Virol.*, 67, 7513, 1993.
24. **Pearson, M., Bjornson, R., Pearson, G., and Rohrmann, G.,** The *Autographa californica* baculovirus genome: evidence for multiple replication origins, *Science*, 257, 1382, 1992.
25. **Chisholm, G. E. and Henner, D. J.,** Multiple early transcripts and splicing of the *Autographa californica* nuclear polyhedrosis virus IE-1 gene, *J. Virol.*, 62, 3193, 1988.
26. **O'Reilly, D. R. and Miller, L. K.,** A baculovirus blocks insect molting by producing ecdysteroid UDP-glucosyl transferase, *Science*, 245, 1110, 1989.
27. **Clem, R. J., Fechheimer, M., and Miller, L. K.,** Prevention of apoptosis by a baculovirus gene during infection of insect cells, *Science*, 254, 1388, 1991.
28. **Possee, R. D.,** personal communication, 1991.
29. **Maeda, S.,** in preparation.
30. **Guarino, L. A.,** Identification of a viral gene encoding a ubiquitin-like protein, *Proc. Natl. Acad. Sci. U.S.A.*, 87, 409, 1990.
31. **Huh, N. E. and Weaver, R. F.,** Identifying the RNA polymerases that synthesize specific transcripts of the *Autographa californica* nuclear polyhedrosis virus, *J. Gen. Virol.*, 71, 195, 1990.
32. **Blissard, G. W. and Rohrmann, G. F.,** Baculovirus diversity and molecular biology, *Annu. Rev. Entomol.*, 35, 127, 1990.
33. **Jeang, K. T., Holmgren-Konig, M., and Khoury, G.,** A baculovirus vector can express intron-containing genes, *J. Virol.*, 61, 1761, 1987.
34. **Iatrou, K., Meidinger, R. G., and Goldsmith, M. R.,** Recombinant baculoviruses as vectors for identifying proteins encoded by intron-containing members of complex multigene families, *Proc. Natl. Acad. Sci. U.S.A.*, 86, 9129, 1989.
35. **Christensen, J., Storgaard, T., Bloch, B., Alexandersen, S., and Aasted, B.,** Expression of Aleutian mink disease parvovirus proteins in a baculovirus vector system, *J. Virol.*, 67, 229, 1993.
36. **Smith, G. E., Summers, M. D., and Fraser, M. J.,** Production of human β interferon in insect cells infected with a baculovirus expression vector, *Mol. Cell. Biol.*, 3, 2156, 1983.
37. **Pennock, G. D., Shoemaker, C., and Miller, L. K.,** Strong and regulated expression of Escherichia coli β-galactosidase in insect cells with a baculovirus vector, *Mol. Cell. Biol.*, 4, 399, 1984.
38. **Maeda, S., Kawai, T., Obinata, M., Chika, T., Horiuchi, T., Maekawa, K., Nakasuji, K., Saeki, Y., Sato, Y., Yamada, K., and Furusawa, M.,** Characteristics of human interferon-α produced by a gene transferred by a baculovirus vector in the silkworm, *Bombyx mori*, *Proc. Jpn. Acad.*, 60, Ser. B, 423, 1984.
39. **Maeda, S., Kawai, T., Obinata, M., Fujiwara, H., Horiuchi, T., Saeki, Y., Sato, Y., and Furusawa, M.,** Production of human α-interferon in silkworm using a baculovirus vector, *Nature*, 315, 592, 1985.
40. **Williams, G. V., Rohel, D. Z., Kuzio, J., and Faulkner, P.,** A cytopathological investigation of *Autographa californica* nuclear polyhedrosis virus p10 gene function using insertion/deletion mutants, *J. Gen. Virol.*, 70, 187, 1989.
41. **van Oers, M. M., Malarme, D., Jore, J. M., and Vlak, J. M.,** Expression of the *Autographa californica* nuclear polyhedrosis virus p10 gene: effect of polyhedrin gene expression, *Arch. Virol.*, 123, 1, 1992.
42. **Weyer, U., Knight, S., and Possee, R. D.,** Analysis of very late gene expression by *Autographa californica* nuclear polyhedrosis virus and the further development of multiple expression vectors, *J. Gen. Virol.*, 71, 1525, 1990.
43. **Vlak, J. M., Klinkenberg, F. A., Zaal, K. J., Usmany, M., Klinge-Roode, E. C., Geervliet, J. B., Roosien, J., and van Lent J. W.,** Functional studies on the p10 gene of *Autographa californica* nuclear polyhedrosis virus using a recombinant expressing a p10-β-galactosidase fusion gene, *J. Gen. Virol.*, 69, 765, 1988.
44. **Vlak, J. M., Schouten, A., Usmany, M., Belsham, G. J., Klinge-Roode, E. C., Maule, A. J., Van Lent, J. W., and Zuidema D.,** Expression of cauliflower mosaic virus gene I using a baculovirus vector based upon the p10 gene and a novel selection method, *Virology*, 179, 312, 1990.
45. **Hill-Perkins, M. S. and Possee, R. D.,** A baculovirus expression vector derived from the basic protein promoter of *Autographa californica* nuclear polyhedrosis virus, *J. Gen. Virol.*, 71, 971, 1990.
46. **Thiem, S. M. and Miller, L. K.,** Differential gene expression mediated by late, very late and hybrid baculovirus promoters, *Gene*, 91, 87, 1990.

47. **Crawford, A. M. and Miller, L. K.,** Characterization of an early gene accelerating expression of late genes of the baculovirus *Autographa californica* nuclear polyhedrosis virus, *J. Virol.*, 62, 2773, 1988.
48. **Godeau, F., Saucier, C., and Kourilsky, P.,** Replication inhibition by nucleoside analogues of a recombinant *Autographa californica* multicapsid nuclear polyhedrosis virus harboring the herpes thymidine kinase gene driven by the IE-1(0) promoter: a new way to select recombinant baculoviruses, *Nucleic Acid. Res.*, 20, 6239, 1992.
49. **Vialard, J., Lalumiere, M., Vernet, T., Briedis, D., Alkhatib, G., Henning, D., Levin, D., and Richardson, C.,** Synthesis of the membrane fusion and hemagglutinin proteins of measles virus, using a novel baculovirus vector containing the β-galactosidase gene, *J. Virol.*, 64, 37, 1990.
50. **Jarvis, D. L., Flemming, J. G. W., Kovacs, G. R., Summers, M. D., and Guarino, L. A.,** Use of early baculovirus promoters for continuous expression and efficient processing of foreign gene products in stably transformed lepidopteran cells, *Bio/Technology*, 8, 950, 1990.
51. **Jarvis, D. L. and Summers, M. D.,** Glycosylation and secretion of human tissue plasminogen activator in recombinant baculovirus-infected insect cells, *Mol. Cell. Biol.*, 9, 214, 1989.
52. **Wang, X. Z., Ooi, B. G., and Miller, L. K.,** Baculovirus vectors for multiple gene expression and for occluded virus production, *Gene*, 100, 131, 1991.
53. **Ooi, B. G., Rankin, C., Miller, L. K.,** Downstream sequences augment transcription from the essential initiation site of a baculovirus polyhedrin gene, *J. Mol. Biol.*, 210, 721, 1989.
54. **Kamita, S. G., Majima, K., and Maeda, S.,** Identification and characterization of the p35 gene of *Bombyx mori* nuclear polyhedrosis virus that prevents virus-induced apoptosis, *J. Virol.*, 67, 455, 1993.
55. **Johnson, R., Meidinger, R. G., and Iatrou, K.,** A cellular promoter-based expression cassette for generating recombinant baculoviruses directing rapid expression of passenger genes in infected insects, *Virology*, 190, 815, 1992.
56. **Iatrou, K. and Meidinger, R. G.,** *Bombyx mori* nuclear polyhedrosis virus-based vectors for expressing passenger genes in silkmoth cells under viral or cellular promoter control, *Gene*, 75, 59, 1989.
57. **Carbonell, L. F., Klowden, M. J., and Miller, L. K.,** Baculovirus-mediated expression of bacterial genes in dipteran and mammalian cells, *J. Virol.*, 56, 153, 1985.
58. **Eldridge, R., Li, Y., and Miller, L. K.,** Characterization of a baculovirus gene encoding a small conotoxinlike polypeptide, *J. Virol.*, 66, 6563, 1992.
59. **Horiuchi, T., Marumoto, Y., Saeki, Y., Sato, Y., Furusawa, M., Kondo, A., and Maeda, S.,** High-level expression of the human-α-interferon gene through the use of an improved baculovirus vector in the silkworm, *Bombyx mori*, *Agric. Biol. Chem.*, 51, 1573, 1987.
60. **Possee, R. D. and Howard, S. C.,** Analysis of the polyhedrin gene promoter of the *Autographa californica* nuclear polyhedrosis virus, *Nucleic Acid. Res.*, 15, 10233, 1987.
61. **Rankin, C., Ooi, B. G., and Miller, L. K.,** Eight base pairs encompassing the transcriptional start point are the major determinant for baculovirus polyhedrin gene expression, *Gene*, 70, 39, 1988.
62. **Matsuura, Y., Possee, R. D., Overton, H. A., and Bishop, D. H.,** Baculovirus expression vectors: the requirements for high level expression of proteins, including glycoproteins, *J. Gen. Virol.*, 68, 1233, 1987.
63. **Luckow, V. A. and Summers, M. D.,** High level expression of nonfused foreign genes with *Autographa californica* nuclear polyhedrosis virus expression vectors, *Virology*, 170, 31, 1989.
64. **Page, M. J.,** p36C: an improved baculovirus expression vector for producing high levels of mature recombinant proteins, *Nucleic Acid. Res.*, 17, 454, 1989.
65. **Page, M. J., Hall, A., Rhodes, S., Skinner, R. H., Murphy, V., Sydenham, M., and Lowe, P. N.,** Expression and characterization of the Ha-ras p21 protein produced at high levels in the insect/baculovirus system, *J. Biol. Chem.*, 264, 19147, 1989.
66. **Beames, B., Braunagel, S., Summers, M. D., and Lanford, R. E.,** Polyhedrin initiator codon altered to AUU yields unexpected fusion protein from a baculovirus vector, *Biotechniques*, 11, 378, 1991.
67. **Blissard, G. W. and Rohrmann, G. F.,** Location, sequence, transcriptional mapping, and temporal expression of the gp64 envelope glycoprotein gene of the *Orgyia pseudotsugata* multicapsid nuclear polyhedrosis virus, *Virology*, 170, 537, 1989.
68. **Sekine, H., Fuse, A., Tada, A., Maeda, S., and Simizu, B.,** Expression of human papillomavirus type 6b E2 gene product with DNA-binding activity in insect *(Bombyx mori)* cells using a baculovirus expression vector, *Gene*, 65, 187, 1988.
69. **Tada, A., Fuse, A., Sekine, H., Simizu, B., Kondo, A., and Maeda, S.,** Expression of the E2 open reading frame of papillomaviruses BPV1 and HPV6b in silkworm by a baculovirus vector, *Virus Res.*, 9, 357, 1988.
70. **Nyunoya, H., Akagi, T., Ogura, T., Maeda, S., and Shimotohno, K.,** Evidence for phosphorylation of transactivator p40x of human T-cell leukemia virus type I produced in insect cells with a baculovirus expression vector, *Virology*, 167, 538, 1988.
71. **Sekine, H., Fuse, A., Inaba, N., Takamizawa, H., and Simizu, B.,** Detection of the human papillomavirus 6b E2 gene product in genital condyloma and laryngeal papilloma tissues, *Virology*, 170, 92, 1989.

72. **Le Blois, H., Fayard, B., Urakawa, T., and Roy, P.,** Synthesis and characterization of chimeric particles between epizootic hemorrhagic disease virus and bluetongue virus: functional domains are conserved on the VP3 protein, *J. Virol.*, 65, 4821, 1991.
73. **Weyer, U. and Possee, R. D.,** A baculovirus dual expression vector derived from the *Autographa californica* nuclear polyhedrosis virus polyhedrin and p10 promoters: co-expression of two influenza virus genes in insect cells, *J. Gen. Virol.*, 72, 2967, 1991.
74. **Fung, M. C., Chiu, K. Y., Weber, T., Chang, T. W., and Chang, N. T.,** Detection and purification of a recombinant human B lymphotropic virus (HHV-6) in the baculovirus expression system by limiting dilution and DNA dot-blot hybridization, *J. Virol. Methods*, 19, 33, 1988.
75. **Patel, G., Nasmyth, K., and Jones, N.,** A new method for the isolation of recombinant baculovirus, *Nucleic Acid. Res.*, 20, 97, 1992.
76. **Kamita, S. G. and Maeda, S.,** Inhibition of *Bombyx mori* nuclear polyhedrosis virus (NPV) replication by the DNA helicase gene of *Autographa californica* NPV, *J. Virol.*, 67, 6239, 1993.
77. **Maeda, S., Kondo, A., and Kamita, S. G.,** Host range expansion of *Autographa californica* nuclear polyhedrosis virus (NPV) following recombinant of a 0.6-kilobase-pair DNA fragment originating from *Bombyx mori* NPV, *J. Virol.*, 67, 6234, 1993.
78. **McIntosh, A. H., Barcenas, N., and Cate, J. R.,** Replication of *Autographa californica* baculovirus (AcMNPV) in a coleopteran cell line, *In Vitro Cell. Dev. Biol.*, 28A, 557, 1992.
79. **Inoue, H. and Mitsuhashi, J.,** A *Bombyx mori* cell line susceptible to a nuclear polyhedrosis virus, *J. Seric. Sci. Jpn.*, 53, 108, 1984.
80. **Ninaki, O., Nagayasu, K., Kameoka, H., Fujiwara, H., Miyajima, N., Takeda, N., and Maekawa, H.,** Establishment of cell lines from *Bombyx mori* and *B. mandarina*, and attempt of cell fusion, in *Invertebrate and Fish Tissue Culture*, Kuroda, Y., Kurstak, E., and Maramorosch, K., Eds., 1988, 243.
81. **Quiot, J. M.,** Establishment of a cell line (S.P.C. Bm36) from the ovaries of *Bombyx mori* L. (Lepidotpera), *Sericologia*, 22, 25, 1982.
82. **Ninaki, O. and Maekawa, H.,** personal communication, 1988.
83. **Maeda, S., Mukohara, Y., and Kondo, A.,** Characteristically distinct isolates of the nuclear polyhedrosis virus from *Spodoptera litura*, *J. Gen. Virol.*, 71, 2631, 1990.
84. **Kondo, A. and Maeda, S.,** Host range expansion by recombination of the baculoviruses *Bombyx mori* nuclear polyhedrosis virus and *Autographa californica* nuclear polyhedrosis virus, *J. Virol.*, 65, 3625, 1991.
85. **Reis, U., Bllum, B., von Specht, B.-U., Domdey, H., and Collins, J.,** Antibody production in silkworm cells and silkworm larvae infected with a dural recombinant *Bombyx mori* nuclear polyhedrosis virus, *Bio/Technology*, 10, 910, 1992.
86. **Mori, H., Nakazawa, H., Shirai, N., Shibata, N., Sumida, M., and Matsubara, F.,** Foreign gene expression by a baculovirus vector with an expanded host range, *J. Gen. Virol.*, 73, 1877, 1992.
87. **Maeda, S., Volrath, S. L., Hanzlik, T. N., Harper, S. A., Majima, K., Maddox, D. W., Hammock, B. D., and Fowler, E.,** Insecticidal effects of an insect-specific neurotoxin expressed by a recombinant baculovirus, *Virology*, 184, 777, 1991.
88. **Iacono-Connors, L. C., Schmaljohn, C. S., and Dalrymple, J. M.,** Expression of the *Bacillus anthracis* protective antigen gene by baculovirus and vaccinia virus recombinants, *Infect. Immun.*, 58, 366, 1990.
89. **Tessier, D. C., Thomas, D. Y., Khouri, H. E., Laliberte, F., and Vernet, T.,** Enhanced secretion from insect cells of a foreign protein fused to the honeybee melittin signal peptide, *Gene*, 98, 177, 1991.
90. **Sissom, J. and Ellis, L.,** Secretion of the extracellular domain of the human insulin receptor from insect cells by use of a baculovirus vector, *Biochemical J.*, 261, 119, 1989.
91. **Vernet, T., Tessier, D. C., Richardson, C., Laliberte, F., Khouri, H. E., Bell, A. W., Storer, A. C., and Thomas, D. Y.,** Secretion of functional papain precursor from insect cells. Requirement for N-glycosylation of the pro-region, *J. Biol. Chem.*, 265, 16661, 1990.
92. **Sottile, J., Selegue, J., and Mosher, D. F.,** Recombinant 70-kDa protein from the amino-terminal region of rat fibronectin inhibits binding of fibronectin to cells and bacteria, *Protein Express. Purif.*, 1, 104, 1990.
93. **Ota, Y., Asakura, A., Matsuura, Y., Kondo, H., Hitoshio, A., Iwane, A., Tanaka, T., Kikuchi, M., and Ikehara, M.,** High-level secretion of the extracellular domain of the human growth hormone receptor using a baculovirus system, *Gene*, 106, 159, 1991.
94. **Kupsch, J., Saizawa, K. M., and Eichmann, K.,** Expression of murine soluble CD4 protein in baculovirus infected insect cells, *Immunobiology*, 186, 254, 1992.
95. **Webb, E., Tkalcevic, J., Edwards, S., Hocking, D., and Nisbet, I.,** Expression of biologically active human factor VIII using a baculovirus vector, *Biochem. Biophys. Res. Commun.*, 190, 536, 1993.
96. **van Drunen Littel-van den Hurk, S., Parker, M. D., Fitzpatrick, D. R., van den Hurk J. V., Campos, M., Babiuk, L. A., and Zamb, T.,** Structural, functional, and immunological characterization of bovine herpesvirus-1 glycoprotein gI expressed by recombinant baculovirus, *Virology*, 190, 378, 1992.

97. **Cochran, M. A., Ericson, B. L., Knell, J. D., and Smith, G. E.,** Use of baculovirus recombinants as a general method for the production of subunit vaccines, in *Vaccines 87, Modern Approaches to Vaccines Including Prevention of AIDS,* Chanock, R. M., Lerner, R. A., Brown, F., and Ginsberg, H., Eds., Cold Spring Harbor Laboratory, Cold Spring Harbor, New York, 1987, 384.
98. **Wells, D. E. and Compans, R. W.,** Expression and characterization of a functional human immunodeficiency virus envelope glycoprotein in insect cells, *Virology,* 176, 575, 1990.
99. **Rusche, J. R., Lynn, D. L., Robert-Guroff, M., Langlois, A. J., Lyerly, H. K., Carson, H., Krohn, K., Ranki, A., Gallo, R. C., and Bolognesi, D. P.,** Humoral immune response to the entire human immunodeficiency virus envelope glycoprotein made in insect cells, *Proc. Natl. Acad. Sci. U.S.A.,* 84, 6924, 1987.
100. **Hu, S. L., Kosowski, S. G., Schaaf, K. F.,** Expression of envelope glycoproteins of human immunodeficiency virus by an insect virus vector, *J. Virol.,* 61, 3617, 1987.
101. **Kuroda, K., Groner, A., Frese, K., Drenckhahn, D., Hauser, C., Rott, R., Doerfler, W., and Klenk, H. D.,** Synthesis of biologically active influenza virus hemagglutinin in insect larvae, *J. Virol.,* 63, 1677, 1989.
102. **Lebacq-Verheyden, A. M., Kasprzyk, P. G., Raum, M. G., van Wyke Coelingh, K., Lebacq, J. A., and Battey, J. F.,** Posttranslational processing of endogenous and of baculovirus-expressed human gastrin-releasing peptide precursor, *Mol. Cell. Biol.,* 8, 3129, 1988.
103. **Knepper, T. P., Arbogast, B., Schreurs, J., and Deinzer, M. L.,** Determination of the glycosylation patterns, disulfide linkages, and protein heterogeneities of baculovirus-expressed mouse interleukin-3 by mass spectrometry, *Biochemistry,* 31, 11651, 1992.
104. **Whitford, M., Stewart, S., Kuzio, J., and Faulkner, P.,** Identification and sequence analysis of a gene encoding gp67, an abundant envelope glycoprotein of the baculovirus *Autographa californica* nuclear polyhedrosis virus, *J. Virol.,* 63, 1393, 1989.
105. **Whitford, M. and Faulkner, P.,** A structural polypeptide of the baculovirus *Autographa californica* nuclear polyhedrosis virus contains *O*-linked *N*-acetylglucosamine, *J. Virol.,* 66, 3324, 1992.
106. **Hsieh, P. and Robbins, P. W.,** Regulation of asparagine-linked oligosaccharide processing: oligosaccharide processing in *Aedes albopictus* mosquito cells, *J. Biol. Chem.,* 259, 2375, 1984.
107. **Charlton, C. A. and Volkman, L. E.,** Effect of tunicamycin on the structural proteins and infectivity of budded *Autographa californica* nuclear polyhedrosis virus, *Virology,* 154, 214, 1986.
108. **Kunimoto, D. Y., Allison, K. C., Watson, C., Fuerst, T., Armstrong, G. D., Paul, W., and Strober, W.,** High-level production of murine interleukin-5 (IL-5) utilizing recombinant baculovirus expression. Purification of the rIL-5 and its use in assessing the biologic role of IL-5 glycosylation, *Cytokine,* 3, 224, 1991.
109. **Fountoulakis, M., Schlaeger, E. J., Gentz, R., Juranville, J. F., Manneberg, M., Ozmen, L., and Garotta, G.,** Purification and biochemical characterization of a soluble mouse interferon-gamma receptor produced in insect cells, *Eur. J. Biochem.,* 198, 441, 1991.
110. **Ghose-Dastidar, J., Ross, J. B., and Green, R.,** Expression of biologically active human corticosteroid binding globulin by insect cells: acquisition of function requires glycosylation and transport, *Proc. Natl. Acad. Sci. U.S.A.,* 88, 6408, 1991.
111. **Li, Z., Smith, C. D., Smolley, J. R., Bridge, J. H., Frank, J. S., and Philipson, K. D.,** Expression of the cardiac Na(+)-Ca2+ exchanger in insect cells using a baculovirus vector, *J. Biol. Chem.,* 267, 7828, 1992.
112. **Chen, W. Y. and Bahl, O. P.,** Recombinant carbohydrate variant of human choriogonadotropin β-subunit (hCG β) descarboxyl terminus (115–145). Expression and characterization of carboxyl-terminal deletion mutant of hCG β in the baculovirus system, *J. Biol. Chem.,* 266, 6246, 1991.
113. **Nakhai, B., Sridhar, P., Pal, R., Talwar, G. P., and Hasnain, S. E.,** Over-expression and characterization of recombinant β subunit of the human chorionic gonadotropin hormone synthesized in insect cells infected with a genetically engineered baculovirus, *Indian J. Biochem. Biophys.,* 29, 315, 1992.
114. **Tuchiya, K., Matsuura, Y., Kawai, A., Ishihama, A., and Ueda, S.,** Characterization of rabies virus glycoprotein expressed by recombinant baculovirus, *Virus Res.,* 25, 1, 1992.
115. **Murphy, C. I., Lennick, M., Lehar, S. M., Beltz, G. A., and Young, E.,** Temporal expression of HIV-1 envelope proteins in baculovirus-infected insect cells: implications for glycosylation and CD4 binding, *Genet. Anal.,* 7, 160, 1990.
116. **Shimokawa, T. and Smith, W. L.,** Expression of prostaglandin endoperoxide synthase-1 in a baculovirus system, *Biochem. Biophys. Res. Commun.,* 183, 975, 1992.
117. **Kuroda, K., Geyer, H., Geyer, R., Doerfler, W., and Klenk, H. D.,** The oligosaccharides of influenza virus hemagglutinin expressed in insect cells by a baculovirus vector, *Virology,* 174, 418, 1990.
118. **Svoboda, M., Przybylski, M., Schreurs, J., Miyajima, A., Hogeland, K., and Deinzer, M.,** Mass spectrometric determination of glycosylation sites and oligosaccharide composition of insect-expressed mouse interleukin-3, *J. Chromatography,* 562, 403, 1991.
119. **Davidson, D. J., Fraser, M. J., and Castellino, F. J.,** Oligosaccharide processing in the expression of human plasminogen cDNA by lepidopteran insect *(Spodoptera frugiperda)* cells, *Biochemistry,* 29, 5584, 1990.

120. **Davidson, D. J. and Castellino, F. J.**, Asparagine-linked oligosaccharide processing in lepidopteran insect cells. Temporal dependence of the nature of the oligosaccharides assembled on asparagine-289 of recombinant human plasminogen produced in baculovirus vector infected *Spodoptera frugiperda* (IPLB-SF-21AE) cells, *Biochemistry*, 30, 6165, 1991.
121. **Davidson, D. J. and Castellino, F. J.**, Structures of the asparagine-289-linked oligosaccharides assembled on recombinant human plasminogen expressed in a *Mamestra brassicae* cell line (IZD-MBO503), *Biochemistry*, 30, 6689, 1991.
122. **Davidson, D. J., Bretthauer, R. K., and Castellino, F. J.**, α-Mannosidase-catalyzed trimming of high-mannose glycans in noninfected and baculovirus-infected *Spodoptera frugiperda* cells (IPLB-SF-21AE). A possible contributing regulatory mechanism for assembly of complex-type oligosaccharides in infected cells, *Biochemistry*, 30, 9811, 1991.
123. **Chen, W. Y., Shen, Q. X., and Bahl, O. P.**, Carbohydrate variant of the recombinant β-subunit of human choriogonadotropin expressed in baculovirus expression system, *J. Biol. Chem.*, 266, 4081, 1991.
124. **Novelli, A. and Boulanger, P. A.**, Deletion analysis of functional domains in baculovirus-expressed adenovirus type 2 fiber, *Virology*, 185, 365, 1991.
125. **Alcover, A., Herve, F., Boursier, J. P., Spagnoli, G., Olive, D., Mariuzza, R. A., and Acuto, O.**, A soluble form of the human CD8 α chain expressed in the baculovirus system: biochemical characterization and binding to MHC class I, *Mol. Immun.*, 30, 55, 1993.
126. **Miller, L. K., Adang, M. J., and Browne, D.**, Protein kinase activity associated with extracellular and occluded forms of the baculovirus *Autographa californica* nuclear polyhedrosis virus, *J. Virol.*, 46, 275, 1983.
127. **Consigli, R. A., Russell, D. L., and Wilson, M. E.**, The biochemistry and molecular biology of the granulosis virus that infects *Plodia interpunctella*, *Curr. Top. Microbiol. Immunol.*, 131, 69, 1986.
128. **Wilson, M. E. and Consigli, R. A.**, Characterization of a protein kinase activity associated with purified capsids of the granulosis virus infecting *Plodia interpunctella*, *Virology*, 143, 516, 1985.
129. **Alnemri, E. S., Maksymowych, A. B., Robertson, N. M., and Litwack, G.**, Overexpression and characterization of the human mineralocorticoid receptor, *J. Biol. Chem.*, 266, 18072, 1991.
130. **Sheng, Z. and Charbonneau, H.**, The baculovirus *Autographa californica* encodes a protein tyrosine phosphatase, *J. Biol. Chem.*, 268, 4728, 1993.
131. **Jensen, R. A., Beeler, J. F., Heidaran, M. A., and LaRochelle, W. J.**, Characterization of baculovirus-expressed human α and β platelet-derived growth factor receptors, *Biochemistry*, 31, 10887, 1992.
132. **Wedegaertner, P. B. and Gill, G. N.**, Activation of the purified protein tyrosine kinase domain of the epidermal growth factor receptor, *J. Biol. Chem.*, 264, 11346, 1989.
133. **Hsu, C. Y., Mohammadi, M., Nathan, M., Honegger, A., Ullrich, A., Schlessinger, J., and Hurwitz, D. R.**, Generation of recombinant cytoplasmic domain of epidermal growth factor receptor with intrinsic protein tyrosine kinase activity, *Cell Growth and Differentiation*, 1, 191, 1990.
134. **Hoss, A., Moarefi, I., Scheidtmann, K. H., Cisek, L. J., Corden, J. L., Dornreiter, I., Arthur, A. K., and Fanning, E.**, Altered phosphorylation pattern of simian virus 40 T antigen expressed in insect cells by using a baculovirus vector, *J. Virol.*, 64, 4799, 1990.
135. **Kraiss, S., Barnekow, A., and Montenarh, M.**, Protein kinase activity associated with immunopurified p53 protein, *Oncogene*, 5, 845, 1990.
136. **Herrmann, C. P., Kraiss, S., and Montenarh, M.**, Association of casein kinase II with immunopurified p53, *Oncogene*, 6, 877, 1991.
137. **Forstova, J., Krauzewicz, N., and Griffin, B. E.**, Expression of biologically active middle T antigen of polyoma virus from recombinant baculoviruses, *Nucleic Acid. Res.*, 17, 1427, 1989.
138. **Piwnica-Worms, H., Williams, N. G., Cheng, S. H., and Roberts, T. M.**, Regulation of pp60c-src and its interaction with polyomavirus middle T antigen in insect cells, *J. Virol.*, 64, 61, 1990.
139. **Xi, S. Z. and Banks, L. M.**, Baculovirus expression of the human papillomavirus type 16 capsid proteins: detection of L1-L2 protein complexes, *J. Gen. Virol.*, 72, 2981, 1991.
140. **Patel, G. and Stabel, S.**, Expression of a functional protein kinase C-gamma using a baculovirus vector: purification and characterisation of a single protein kinase C iso-enzyme, *Cellular Signalling*, 1, 227, 1989.
141. **Carrera, A. C., Li, P., and Roberts, T. M.**, Characterization of an active, non-myristylated, cytoplasmic form of the lymphoid protein tyrosine kinase pp56lck, *Int. Immunol.*, 3, 673, 1991.
142. **Yamada, K., Nakajima, Y., and Natori, S.**, Production of recombinant sarcotoxin IA in *Bombyx mori* cells, *Biochem. J.*, 272, 633, 1990.
143. **Suzuki, K., Shimoi, H., Iwasaki, Y., Kawahara, T., Matsuura, Y., and Nishikawa, Y.**, Elucidation of amidating reaction mechanism by frog amidating enzyme, peptidylglycine α-hydroxylating monooxygenase, expressed in insect cell culture, *Embo J.*, 9, 4259, 1990.
144. **Shimoi, H., Kawahara, T., Suzuki, K., Iwasaki, Y., Jeng, A. Y., and Nishikawa, Y.**, Characterization of a *Xenopus laevis* skin peptidylglycine α-hydroxylating monooxygenase expressed in insect-cell culture, *Eur. J. Biochem.*, 209, 189, 1992.

145. **Maeda, S.,** Increased insecticidal effect by a recombinant baculovirus carrying a synthetic diuretic hormone gene, *Biochem. Biophys. Res. Commun.*, 165, 1177, 1989.
146. **Hellers, M., Gunne, H., and Steiner, H.,** Expression of post-translational processing of preprocecropin A using a baculovirus vector, *Eur. J. Biochem.*, 199, 435, 1991.
147. **Andersons, D., Engstrom, A., Josephson, S., Hansson, L., and Steiner, H.,** Biologically active and amidated cecropin produced in a baculovirus expression system from a fusion construct containing the antibody-binding part of protein A, *Biochem. J.*, 280, 219, 1991.
148. **Roberts, T. E. and Faulkner, P.,** Fatty acid acylation of the 67K envelope glycoprotein of a baculovirus: *Autographa californica* nuclear polyhedrosis virus, *Virology*, 172, 377, 1989.
149. **Kuroda, K., Veit, M., and Klenk, H. D.,** Retarded processing of influenza virus hemagglutinin in insect cells, *Virology*, 180, 159, 1991.
150. **Lanford, R. E., Luckow, V., Kennedy, R. C., Dreesman, G. R., Notvall, L., and Summers, M. D.,** Expression and characterization of hepatitis B virus surface antigen polypeptides in insect cells with a baculovirus expression system, *J. Virol.*, 63, 1549, 1989.
151. **Nishimura, C., Yamaoka, T., Mizutani, M., Yamashita, K., Akera, T., and Tanimoto, T.,** Purification and characterization of the recombinant human aldose reductase expressed in baculovirus system, *Biochim. Biophys. Acta*, 1078, 171, 1991.
152. **Graf, R., Mattera, R., Codina, J., Estes, M. K., and Birnbaumer, L.,** A truncated recombinant α subunit of Gi3 with a reduced affinity for β γ dimers and altered guanosine 5′-3-O-(thio)triphosphate binding, *J. Biol. Chem.*, 267, 24307, 1992.
153. **Risinger, M. A., Dotimas, E. M., and Cohen, C. M.,** Human erythrocyte protein 4.2, a high copy number membrane protein, is *N*-myristylated, *J. Biol. Chem.*, 267, 5680, 1992.
154. **Brautigam, S., Snezhkov, E., and Bishop, D. H.,** Formation of poliovirus-like particles by recombinant baculoviruses expressing the individual VP0, VP3, and VP1 proteins by comparison to particles derived from the expressed poliovirus polyprotein, *Virology*, 192, 512, 1993.
155. **Delchambre, M., Gheysen, D., Thines, D., Thiriart, C., Jacobs, E., Verdin, E., Horth, M., Burny, A., and Bex, F.,** The GAG precursor of simian immunodeficiency virus assembles into virus-like particles, *Embo J.*, 8, 2653, 1989.
156. **Morikawa, S., Booth, T. F., and Bishop, D. H.,** Analyses of the requirements for the synthesis of virus-like particles by feline immunodeficiency virus gag using baculovirus vectors, *Virology*, 183, 288, 1991.
157. **Overton, H. A., Fujii, Y., Price, I. R., and Jones, I. M.,** The protease and gag gene products of the human immunodeficiency virus: authentic cleavage and post-translational modification in an insect cell expression system, *Virology*, 170, 107, 1989.
158. **Gheysen, D., Jacobs, E., de Foresta F., Thiriart, C., Francotte, M., Thines, D., De Wilde M.,** Assembly and release of HIV-1 precursor Pr55gag virus-like particles from recombinant baculovirus-infected insect cells, *Cell*, 59, 103, 1989.
159. **Royer, M., Cerutti, M., Gay, B., Hong, S. S., Devauchelle, G., and Boulanger, P.,** Functional domains of HIV-1 gag-polyprotein expressed in baculovirus-infected cells, *Virology*, 184, 417, 1991.
160. **Wagner, R., Fliessbach, H., Wanner, G., Motz, M., Niedrig, M., Deby, G., von Brunn, A., and Wolf, H.,** Studies on processing, particle formation, and immunogenicity of the HIV-1 gag gene product: a possible component of a HIV vaccine, *Arch. Virol.*, 127, 117, 1992.
161. **Royer, M., Hong, S. S., Gay, B., Cerutti, M., and Boulanger, P.,** Expression and extracellular release of human immunodeficiency virus type 1 Gag precursors by recombinant baculovirus-infected cells, *J. Virol.*, 66, 3230, 1992.
162. **Lydon, N. B., Gay, B., Mett, H., Murray, B., Lieβnz, J., Gutzwiller, A., Piwnica-Worms, H., Roberts, T. M., and McGlynn, E.,** Purification and biochemical characterization of non-myristoylated recombinant pp60c-src kinase, *Biochem. J.*, 287, 985, 1992.
163. **Mouillac, B., Caron, M., Bonin, H., Dennis, M., and Bouvier, M.,** Agonist-modulated palmitoylation of β 2-adrenergic receptor in Sf9 cells, *J. Biol. Chem.*, 267, 21733, 1992.
164. **Kloc, M., Reddy, B., Crawford, S., and Etkin, L. D.,** A novel 110-kDa maternal CAAX box-containing protein from *Xenopus* is palmitoylated and isoprenylated when expressed in baculovirus, *J. Biol. Chem.*, 266, 8206, 1991.
165. **Hwang, S. B., Lee, C. Z., and Lai, M. M.,** Hepatitis delta antigen expressed by recombinant baculoviruses: comparison of biochemical properties and post-translational modifications between the large and small forms, *Virology*, 190, 413, 1992.
166. **Wildman, D. E., Tamir, H., Leberer, E., Northup, J. K., and Dennis, M.,** Prenyl modification of guanine nucleotide regulatory protein gamma 2 subunits is not required for interaction with the transducin α subunit or rhodopsin, *Proc. Natl. Acad. Sci. U.S.A.*, 90, 794, 1993.
167. **Berndt, N. and Cohen, P. T.,** Renaturation of protein phosphatase 1 expressed at high levels in insect cells using a baculovirus vector, *Eur. J. Biochem.*, 190, 291, 1990.

168. **Xie, Y. B., Sui, Y. P., Shan, L. X., Palvimo, J. J., Phillips, D. M., and Janne, O. A.,** Expression of androgen receptor in insect cells. Purification of the receptor and renaturation of its steroid- and DNA-binding functions, *J. Biol. Chem.*, 267, 4939, 1992.
169. **Janne, O. A., Palvimo, J. J., Kallio, P., Mehto, M., Xie, Y. B., and Sui, Y. P.,** Production of recombinant androgen receptor in a heterologous expression system, *Clin. Chem.*, 39, 346, 1993.
170. **Maule, A. J., Usmany, M., Wilson, I. G., Boudazin, G., and Vlak, J. M.,** Biophysical and biochemical properties of baculovirus-expressed CaMV P1 protein, *Virus Genes*, 6, 5, 1992.
171. **Tavernier, J., Devos, R., van der Heyden, J., Hauquier, G., Bauden, R., Fache, I., Kawashima, E., Vandekerckhove, J., Contreras, R., and Fiers, W.,** Expression of human and murine interleukin-5 in eukaryotic systems, *DNA*, 8, 491, 1989.
172. **Vissavajjhala, P. and Ross, A. H.,** Purification and characterization of the recombinant extracellular domain of human nerve growth factor receptor expressed in a baculovirus system, *J. Biol. Chem.*, 265, 4746, 1990.
173. **Ingley, E., Cutler, R. L., Fung, M. C., Sanderson, C. J., and Young, I. G.,** Production and purification of recombinant human interleukin-5 from yeast and baculovirus expression systems, *Eur. J. Biochem.*, 196, 623, 1991.
174. **Barnett, J., Chow, J., Nguyen, B., Eggers, D., Osen, E., Jarnagin, K., Saldou, N., Straub, K., Gu, L., and Erdos, L.,** Physicochemical characterization of recombinant human nerve growth factor produced in insect cells with a baculovirus vector, *J. Neurochem.*, 57, 1052, 1991.
175. **Cohen, T., Gitay-Goren, H., Neufeld, G., and Levi, B. Z.,** High levels of biologically active vascular endothelial growth factor (VEGF) are produced by the baculovirus expression system, *Growth Factors*, 7, 131, 1992.
176. **Flamand, M., Deubel, V., and Girard, M.,** Expression and secretion of Japanese encephalitis virus nonstructural protein NS1 by insect cells using a recombinant baculovirus, *Virology*, 191, 826, 1992.
177. **Fiebich, B. L., Jager, B., Schollmann, C., Weindel, K., Wilting, J., Kochs, G., Marme, D., Hug, H., and Weich, H. A.,** Synthesis and assembly of functionally active human vascular endothelial growth factor homodimers in insect cells, *Eur. J. Biochem.*, 211, 19, 1993.
178. **Gottlieb, J., Marcy, A. I., Coen, D. M., and Challberg, M. D.,** The herpes simplex virus type 1 UL42 gene product: a subunit of DNA polymerase that functions to increase processivity, *J. Virol.*, 64, 5976, 1990.
179. **Stow, N. D.,** Sequences at the C-terminus of the herpes simplex virus type 1 UL30 protein are dispensable for DNA polymerase activity but not for viral origin-dependent DNA replication, *Nucleic Acid. Res.*, 21, 87, 1993.
180. **Chen, W. and Bahl, O. P.,** Recombinant carbohydrate and selenomethionyl variants of human choriogonadotropin, *J. Biol. Chem.*, 266, 8192, 1991.
181. **Godet, M., Rasschaert, D., and Laude, H.,** Processing and antigenicity of entire and anchor-free spike glycoprotein S of coronavirus TGEV expressed by recombinant baculovirus, *Virology*, 185, 732, 1991.
182. **Schawaller, M., Smith, G. E., Skehel, J. J., and Wiley, D. C.,** Studies with crosslinking reagents on the oligomeric structure of the env glycoprotein of HIV, *Virology*, 172, 367, 1989.
183. **Dodson, M. S., Crute, J. J., Bruckner, R. C., and Lehman, I. R.,** Overexpression and assembly of the herpes simplex virus type 1 helicase-primase in insect cells, *J. Biol. Chem.*, 264, 20835, 1989.
184. **Calder, J. M. and Stow, N. D.,** Herpes simplex virus helicase-primase: the UL8 protein is not required for DNA-dependent ATPase and DNA helicase activities, *Nucleic Acid. Res.*, 18, 3573, 1990.
185. **Stow, N. D.,** Herpes simplex virus type 1 origin-dependent DNA replication in insect cells using recombinant baculoviruses, *J. Gen. Virol.*, 73, 313, 1992.
186. **Hasemann, C. A. and Capra, J. D.,** High-level production of a functional immunoglobulin heterodimer in a baculovirus expression system, *Proc. Natl. Acad. Sci. U.S.A.*, 87, 3942, 1990.
187. **Godeau, F., Luescher, I. F., Ojcius, D. M., Saucier, C., Mottez, E., Cabanie, L., and Kourilsky, P.,** Purification and ligand binding of a soluble class I major histocompatibility complex molecule consisting of the first three domains of H-2Kd fused to β 2-microglobulin expressed in the baculovirus-insect cell system, *J. Biol. Chem.*, 267, 24223, 1992.
188. **Stern, L. J. and Wiley, D. C.,** The human class II MHC protein HLA-DR1 assembles as empty α β heterodimers in the absence of antigenic peptide, *Cell*, 68, 465, 1992.
189. **Jiang, X., Wang, M., Graham, D. Y., and Estes, M. K.,** Expression, self-assembly, and antigenicity of the Norwalk virus capsid protein, *J. Virol.*, 66, 6527, 1992.
190. **Brown, C. S., van Lent, J. W., Vlak, J. M., and Spaan, W. J.,** Assembly of empty capsids by using baculovirus recombinants expressing human parvovirus B19 structural proteins, *J. Virol.*, 65, 2702, 1991.
191. **Kajigaya, S., Fujii, H., Field, A., Anderson, S., Rosenfeld, S., Anderson, L. J., Shimada, T., and Young, N. S.,** Self-assembled B19 parvovirus capsids, produced in a baculovirus system, are antigenically and immunogenically similar to native virions, *Proc. Natl. Acad. Sci. U.S.A.*, 88, 4646, 1991.
192. **Martinez, C., Dalsgaard, K., Lopez de Turiso, J. A., Cortes, E., Vela, C., and Casal, J. I.,** Production of porcine parvovirus empty capsids with high immunogenic activity, *Vaccine*, 10, 684, 1992.

193. **Saliki, J. T., Mizak, B., Flore, H. P., Gettig, R. R., Burand, J. P., Carmichael, L. E., Wood, H. A., and Parrish, C. R.,** Canine parvovirus empty capsids produced by expression in a baculovirus vector: use in analysis of viral properties and immunization of dogs, *J. Gen. Virol.*, 73, 369, 1992.
194. **Ruffing, M., Zentgraf, H., and Kleinschmidt, J. A.,** Assembly of viruslike particles by recombinant structural proteins of adeno-associated virus type 2 in insect cells, *J. Virol.*, 66, 6922, 1992.
195. **Liu, H. M., Booth, T. F., and Roy, P.,** Interactions between bluetongue virus core and capsid proteins translated *in vitro, J. Gen. Virol.*, 73, 2577, 1992.
196. **French, T. J. and Roy, P.,** Synthesis of bluetongue virus (BTV) corelike particles by a recombinant baculovirus expressing the two major structural core proteins of BTV, *J. Virol.*, 64, 1530, 1990.
197. **Hewat, E. A., Booth, T. F., Wade, R. H., and Roy, P.,** 3-D reconstruction of bluetongue virus tubules using cryoelectron microscopy, *J. Struct. Biol.*, 108, 35, 1992.
198. **French, T. J., Marshall, J. J., and Roy, P.,** Assembly of double-shelled, viruslike particles of bluetongue virus by the simultaneous expression of four structural proteins, *J. Virol.*, 64, 5695, 1990.
199. **Tosser, G., Labbe, M., Bremont, M., and Cohen, J.,** Expression of the major capsid protein VP6 of group C rotavirus and synthesis of chimeric single-shelled particles by using recombinant baculoviruses, *J. Virol.*, 66, 5825, 1992.
200. **Montross, L., Watkins, S., Moreland, R. B., Mamon, H., Caspar, D. L., and Garcea, R. L.,** Nuclear assembly of polyomavirus capsids in insect cells expressing the major capsid protein VP1, *J. Virol.*, 65, 4991, 1991.
201. **Hilditch, C. M., Rogers, L. J., and Bishop, D. H.,** Physicochemical analysis of the hepatitis B virus core antigen produced by a baculovirus expression vector, *J. Gen. Virol.*, 71, 2755, 1990.
202. **Lanford, R. E. and Notvall, L.,** Expression of hepatitis B virus core and precore antigens in insect cells and characterization of a core-associated kinase activity, *Virology*, 176, 222, 1990.
203. **Higashihashi, N., Arai, Y., Enjo, T., Horiuchi, T., Saeki, Y., Sakano, K., Sato, Y., Takeda, K., Takashina, S., and Takahashi, T.,** High-level expression and characterization of hepatitis B virus surface antigen in silkworm using a baculovirus vector, *J. Virol. Methods*, 35, 159, 1991.
204. **Rasmussen, L., Battles, J. K., Ennis, W. H., Nagashima, K., and Gonda, M. A.,** Characterization of virus-like particles produced by a recombinant baculovirus containing the gag gene of the bovine immunodeficiency-like virus, *Virology*, 178, 435, 1990.
205. **Thomsen, D. R., Meyer, A. L., and Post, L. E.,** Expression of feline leukaemia virus gp85 and gag proteins and assembly into virus-like particles using the baculovirus expression vector system, *J. Gen. Virol.*, 73, 1819, 1992.
206. **Nagy, E., Huber, P., Krell, P. J., and Derbyshire, J. B.,** Synthesis of Newcastle disease virus (NDV)-like envelopes in insect cells infected with a recombinant baculovirus expressing the haemagglutinin-neuraminidase of NDV, *J. Gen. Virol.*, 72, 753, 1991.
207. **Roosien, J., Belsham, G. J., Ryan, M. D., King, A. M., and Vlak, J. M.,** Synthesis of foot-and-mouth disease virus capsid proteins in insect cells using baculovirus expression vectors, *J. Gen. Virol.*, 71, 1703, 1990.
208. **Bertioli, D. J., Harris, R. D., Edwards, M. L., Cooper, J. I., and Hawes, W. S.,** Transgenic plants and insect cells expressing the coat protein of *Arabis* mosaic virus produce empty virus-like particles, *J. Gen. Virol.*, 72, 1801, 1991.
209. **Gretch, D. G., Sturley, S. L., Friesen, P. D., Beckage, N. E., and Attie, A. D.,** Baculovirus-mediated expression of human apolipoprotein E in *Manduca sexta* larvae generates particles that bind to the low density lipoprotein receptor, *Proc. Natl. Acad. Sci. U.S.A.*, 88, 8530, 1991.
210. **Miyajima, A., Schreurs, J., Otsu, K., Kondo, A., Arai, K., and Maeda, S.,** Use of the silkworm, *Bombyx mori*, and an insect baculovirus vector for high-level expression and secretion of biologically active mouse interleukin-3, *Gene*, 58, 273, 1987.
211. **Sakurai, T., Ueda, Y., Sato, M., and Yanai, A.,** Feline interferon production in silkworm by recombinant baculovirus. *J. Vet. Med. Sci.*, 54, 563, 1992.
212. **Price, P. M., Reichelderfer, C. F., Johansson, B. E., Kilbourne, E. D., and Acs, G.,** Complementation of recombinant baculoviruses by coinfection with wild-type virus facilitates production in insect larvae of antigenic proteins of hepatitis B virus and influenza virus, *Proc. Natl. Acad. Sci. U.S.A.*, 86, 1453, 1989.
213. **Medin, J. A., Hunt, L., Gathy, K., Evans, R. K., and Coleman, M. S.,** Efficient, low-cost protein factories: expression of human adenosine deaminase in baculovirus-infected insect larvae, *Proc. Natl. Acad. Sci. U.S.A.*, 87, 2760, 1990.
214. **Jha, P. K., Pal, R., Nakhai, B., Sridhar, P., Hasnain, S. E.,** Simultaneous synthesis of enzymatically active luciferase and biologically active β subunit of human chorionic gonadotropin in caterpillars infected with a recombinant baculovirus, *FEBS Lett.*, 310, 1992.
215. **Granados, R. R. and Federici, B. A.,** *The Biology of Baculoviruses*, Vol. 1 and 2, *Biological Properties and Molecular Biology*, CRC Press, Boca Raton, FL, 1986.

216. **Moscardi, F.**, Use of viruses for pest control in Brazil: The case of the nuclear polyhedrosis virus of the soybean caterpillar, *Anticarsia gemmatalis, Mem. Inst. Osw. Cruz, Rie de Janeiro,* 84, 51, 1989.
217. **Bedford, G. O.**, Biological control of the rhinoceros *(Oryctes rhinoceros)* in the South Pacific by baculovirus, *Agric. Exosystems Environ.*, 15, 141, 1986.
218. **Eldridge, R., Horodyski, F. M., Morton, D. B., O'Reilly, D. R., Truman, J. W., Riddiford, L. M., and Miller, L. K.**, Expression of an eclosion hormone gene in insect cells using baculovirus vectors, *Insect Biochem.*, 21, 341, 1991.
219. **Ishizaki, H., Suzuki, A., and Maeda, S.**, unpublished data.
220. **Hammock, B. D.**, *Comparative Insect Physiology, Biochemistry and Pharmacology,* Vol. 7, Pergamon Press, New York, NY, 1985, 431.
221. **Hammock, B. D., Bonning, B. C., Possee, R. D., Hanzlik, T. N., and Maeda, S.**, Expression and effects of the juvenile hormone esterase in a baculovirus system, *Nature,* 344, 458, 1990.
222. **Ward, V. K., Bonning, B. C., Huang, T., Shiotsuki, T., Griffeth, V. N., and Hammock, B. D.**, Analysis of the catalytic mechanism of juvenile hormone esterase by site-directed mutagenesis, *Int. J. Biochem.*, 24, 1933, 1992.
223. **Eldridge, R., O'Reilly, D. R., Hammock, B. D., and Miller, L. K.**, Insecticidal properties of genetically engineered baculoviruses expressing an insect juvenile hormone esterase gene, *Appl. Environ. Microbiol.*, 58, 1583, 1992.
224. **Merryweather, A. T., Weyer, U., Harris, M. P., Hirst, M., Booth, T., and Possee, R. D.**, Construction of genetically engineered baculovirus insecticides containing the *Bacillus thuringiensis* subsp. kurstaki HD-73 delta endotoxin, *J. Gen. Virol.*, 71, 1535, 1990.
225. **Zlotkin, E.**, Venom neurotoxins — models for selective insecticides, *Phytoparasitica,* 19, 177, 1991.
226. **Maeda, S. and Hammock B. D.**, Recombinant baculoviruses expressing foreign genes for improved insect control, in *Newer Pest Control Agents and Technology with Reduced Environmental Impact,* Duke, S. O., Menn, J. J., and Plimmer, J. R., Eds., ASC Symposium Series, Vol. 524, American Chemical Society, Washington, D.C., 1993, 281.
227. **Stewart, L. M., Hirst, M., Lopez Ferber, M., Merryweather, A. T., Cayley, P. J., and Possee, R. D.**, Construction of an improved baculovirus insecticide containing an insect-specific toxin gene, *Nature,* 352, 85, 1991.
228. **McCutchen, B. F., Choudary, P. V., Crenshaw, R., Maddox, D., Kamita, S. G., Palekar, N., Volrath, S. L., Fowler, E., Hammock, B. D., and Maeda, S.**, Development of a recombinant baculovirus expressing an insect-selective neurotoxin: potential for pest control, *Bio/Technology,* 9, 848, 1991.
229. **Tomalski, M. D., Kutney, R., Bruce, W. A., Brown, M. R., Blum, M. S., and Travis, J.**, Purification and characterization of insect toxins derived from the mite, *Pyemotes tritici, Toxicon,* 27, 1151, 1989.
230. **Tomalski, M. D. and Miller, L. K.**, Insect paralysis by baculovirus-mediated expression of a mite neurotoxin gene, *Nature,* 352, 82, 1991.
231. **Herrmann, R., Fishman, L., and Zlotkin, E.**, The tolerance of lepidopterous larvae to an insect selective neurotoxin, *Insect Biochem.*, 20, 625, 1990.

Chapter 2

PRODUCTION OF BACULOVIRUSES FOR INSECT CONTROL USING CELL CULTURE*

Cynthia Lenz Goodman and Arthur H. McIntosh

TABLE OF CONTENTS

I. Introduction 35
A. Viral Pesticides 35
B. Baculovirus Classification 35
C. Baculovirus Replication Cycle 35

II. Effects of Biological Factors on *In Vitro* Production 36
A. Comparison of *In Vitro* Production Systems 36
1. Virus Production in Different Cell Lines 36
a. Cell Lines from the Same Tissue 36
b. Cell Lines from Different Tissues 40
c. Cell Lines from Different Species 40
2. Production of Different Baculoviruses 40
a. Viral Strains 40
b. Infection Types 41
B. Comparison of Production Procedures 42
1. Multiplicity of Infection 42
2. Passage Effect 43
3. Cell Parameters 43
4. Harvesting and Storage Conditions 44

III. Effect of Chemical and Physical Factors on *In Vitro* Production 44
A. Temperature 44
B. Media 44
1. General Attributes 44
2. Specific Components 45
a. Crude Biologicals 45
b. Purified Biochemicals 46
c. Commercial Media Preparations 46
C. Production Vessels 47
1. Cell Growth Mode 47
2. Aeration 47
3. Configuration 48

IV. Genetic Engineering of Viral Pesticides 48
A. Engineering Goals and Examples 48
1. Enhance Virulence 49
2. Expand Host Range 49
B. Engineering Considerations 49
1. Recombinant Virus and Protein Effects 49

* Mention of a propietary product in this chapter does not constitute a recommendation for use by the USDA.

2. Promoter Influence 49
3. Processing Concerns 50

V. Conclusions 50

References 50

Cynthia Lenz Goodman, Ph.D., is a Research Biologist with the United States Department of Agriculture, Agricultural Research Services, at the Biological Control of Insects Research Laboratory (USDA-ARS-BCIRL), Columbia, Missouri. She received her B.S. in Biological Sciences from the University of Connecticut (1981), her M.S. in Toxicology from the University of Kentucky (1983), and her Ph.D. in Entomology (Insect Physiology) from the University of Missouri (1986). Research Associate, USDA-ARS-BCIRL, 1987. Research Biologist, USDA-ARS-BCIRL, 1988–1990. Research Associate, University of Missouri, 1990–1992. Research Biologist, USDA-ARS-BCIRL, 1992–present. Entomological Society of America; Tissue Culture Association; Sigma Xi. Baculovirus production in insect cells; toxicity and mechanism of action of insect growth inhibitors from plants; biochemical and physiological properties of insect proteins.

I. INTRODUCTION

An important application of insect cell culture is the production of viral insecticides. These biorational pesticides have been produced primarily in whole insects, but ongoing *in vitro* studies have shown the feasibility of generating viruses in insect cell lines. The optimization of insect cell culture systems for the production of insect viruses, primarily those from the Baculoviridae family, has been the focus of many studies. This chapter will overview the biological, chemical, and physical factors involved in enhancing the ability of *in vitro* systems to produce baculoviruses for insect pest control purposes. Before addressing specific cell culture studies, we will briefly examine the use of baculoviruses as biological control agents, the advantages of producing these pathogens in cell culture, and the replication cycle of baculoviruses in cell culture.

A. VIRAL PESTICIDES

Baculoviruses have been isolated from a wide variety of pest insects from several orders, including Lepidoptera (with the most isolates) and Coleoptera (see adjoining chapter by McIntosh and Grasela).[1-5] The agricultural and forestry sectors have been interested in viruses as an alternative for chemical insecticides because of their safety and specificity.[6] Individual baculoviruses have a limited host range, usually only infecting the target insects and a few closely related insect species in particular ecosystems. Therefore, they have a minimal potential for damaging the environment. Yet, few of these viruses have been commercially developed for the control of pest insects of crop systems or forests.[6-8] One reason for the limited commercialization of baculoviruses is that they are usually produced in living insects, which is a complex, expensive, and labor-intensive process.[6] The feasibility of automated scale-up of insect cell culture systems has been shown, which makes the *in vitro* production of viral pesticides possible.[6-24] Generating insect viruses in cell culture systems has the advantages over *in vivo* systems of being more highly controlled and reproducible, which therefore produces cleaner preparations. Increasing the purity of viral preparations generates higher activity/concentration ratios and minimizes the presence of contaminating biotic/abiotic agents capable of altering activity.[25] Since studies have shown that the baculoviruses produced *in vitro* are comparable in virulence, morphology, and molecular structure to those produced in whole insects,[24,26,27] the optimization of cell culture systems will be an important accomplishment for the pest control industry.

B. BACULOVIRUS CLASSIFICATION

Baculoviruses have circular, covalently closed double-stranded DNA and are members of the Baculoviridae family.[2-5] This family is divided into two subfamilies, *Eubaculovirinae* and *Nudibaculovirinae*.[5] The *Eubaculovirinae* consist of the occluded baculoviruses, which include the nuclear polyhedrosis virus (NPV) and granulosis virus (GV) genera. The NPV replicate within the nuclei of invertebrate cells and occlude virions (virus particles) within occlusion bodies (OB), also known as polyhedra. The virions, which are enveloped nucleocapsids, can be packaged into the OB either as singly embedded viruses (SNPV) or multiply embedded viruses (MNPV),[2] with these two groups being considered subgenera. The GV replicate partially within the nucleus and cytoplasm and are individually occluded and singly enveloped.[3] The *Nudibaculovirinae* consist of the nonoccluded baculoviruses, which are singly enveloped and do not produce OB.[4] We will emphasize the NPVs in this chapter since they have been the most intensely studied baculoviruses in cell culture.

C. BACULOVIRUS REPLICATION CYCLE

The replication cycle of the NPV is biphasic.[2,8] In cell culture, cells are initially infected by the nonoccluded form of the virus. These virus particles enter the cells primarily by receptor-mediated endocytosis[28] and harness the cell's protein synthesis machinery for replication of their own proteins. After a genetically programmed time period, which varies among cell-virus systems, infected cells produce viral progeny by budding through the cell membrane. These forms of the virus are called budded viruses or extracellular viruses (ECV), and are the form involved in the secondary infection of tissues in insects. The second phase of replication involves the occlusion of virions into OB. The gene encoding the protein matrix of the OB, referred to as polyhedrin, is one of the primary genes replaced by foreign gene sequences for the production of recombinant proteins, since the polyhedrin promoter is noted for its high level of expression.[8] Cells release OB upon lysis, and the virus particles contained within them are only infectious to cells when they are liberated, which occurs naturally in larval midguts or artificially by the use of alkali.[29,30] OB are the most environmentally stable form of the virus and are used in field applications. OB are also primarily involved in the horizontal transmission of the virus among the target pests within the field.

II. EFFECTS OF BIOLOGICAL FACTORS ON *IN VITRO* PRODUCTION

Numerous biological factors affect the level of *in vitro* production and virulence of baculovirus ECV and OB. These include the cell culture-baculovirus system selected, the virus inoculation protocol, and the harvesting and storage procedures.

A. COMPARISON OF *IN VITRO* PRODUCTION SYSTEMS

The two components of an *in vitro* baculovirus production system are the insect cell line and the viral inoculum. Selection of the appropriate combination of cell line and inoculum is necessary for the optimization of the production of viruses in cell culture.

1. Virus Production in Different Cell Lines

Cell lines exhibit varying degrees of permissiveness towards specific viral strains or species. Using the same viral strain, direct comparisons of viral susceptibility have been made of cell lines from the same tissue of an insect species (including clonal lines of the same parent), those of different tissues of the same species, and those of tissues from different insect species. Table 1 summarizes many of the studies that directly compare virus replication in different cell lines. The viral host species given in this table is the insect species from which the virus was first isolated and from which its designation was derived (for host range see Chapter 3). The parameters compared include the production of OB (determined either quantitatively or qualitatively using light or electron microscopy), titers of ECV (using tissue culture infective dose assays, $TCID_{50}$, or plaque assays), and the virulence of OB or ECV to larvae (by determining either lethal doses, time to death, or percent mortality). Many reasons may explain the degree of permissiveness of cell lines to specific viruses. These include the viruses' ability to bind to and enter the cell, the ability of their DNA to enter the nucleus and initiate viral replication, and the ability of the cell to properly package and release the progeny.[2]

a. Cell Lines from the Same Tissue

Viral replication in clonal and parental cell lines has been studied using the *Helicoverpa zea* SNPV (HzSNPV) and clonal lines from a variety of parental *H. zea* cell lines.[31-33] Over 40-fold differences were noted in the levels of OB and ECV generated, with some differences in OB virulence to larvae being observed. Likewise, Volkman and Summers[34] compared the production of OB and ECV between the parental cell line TN-368 and its clones and found 2- to 4-fold differences. Heterogeneous embryonic cell lines of *Trichoplusia ni* were also reported to differentially replicate the *T. ni* GV (TnGV) or NPV (TnSNPV).[43] Both Rice et al.[32]

TABLE 1
Studies Comparing Viral Production in Different Insect Cell Lines

Category for comparison	Species origin of cell line	Tissue origin of cell line(s)	Cell line designation	Virus (Host species)	Reference
Clonal cell lines	*Helicoverpa zea*	Ovary/fat body	IPLB-Hz-1075 (parent) UND-A to X (clones)	HzSNPV; *H. zea*	31
		Ovary	BCIRL-Hz-AM1 (parent) AM1-B3 (clone)	HzSNPV; *H. zea*	32
		Ovary	BCIRL-Hz-AM1/AM3 (parents) AM1-A4/A7/A11/B3, AM3-F9 (clones)	HzSNPV; *H. zea*	33
	Trichoplusia ni	Ovary	Tn-368 (parent) 3/5/8/10 (clones)	AcMNPV; *Autographa californica*	34
		Ovary	Tn-368 (parent) 1–3 (clones/subclones)	TnMNPV; *T. ni*	35
Cell lines of same tissues	*Adoxophyes orana fasciata*	Neonate	FTRS-AoL1/AoL2	AoNPV; *A. o. fasciatia* PxNPV; *Plutella xylostella*	36
	Helicoverpa zea	Ovary/fat body	IPLB-Hz-100/1240/ 1074/1075/1079	HzSNPV; *H. zea*	37
	Helicoverpa zea	Ovary	BCIRL-Hz-AM1/AM3 UND-K	HzSNPV; *H. zea*	32
		Ovary	BCIRL-Hz-AM1/ AM2/AM3	HzSNPV; *H. zea*	38
	Estigmene acrea	Hemocyte	Ea-1174-A/H	AcMNPV; *A. californica*	39
	Lymantria dispar	Ovary	IPLB-Ld-64/65/66/67 (and sublines)	LdMNPV; *L. dispar*	40
	Orgyia leucostigma	Neonates	IPRL-01-4/9/12/13	O1SNPV; *O. leucostigma*	41
	Pieris rapae	Embryo	BTI-Pr8A1/8A2/9A/ 10B/4AB	AcMNPV; *A. californica* TnMNPV; *T. ni*	42
	Trichoplusia ni	Embryo	BTI-Tn4B1/TN5G33	TnGV, TnSNPV; *T. ni*	43
Cell lines of different tissues	*Cydia pomonella*	Larva/embryo	IZD-Cp2022/0508/ 1508 (81 Total)	CmNPV, CpGV; *C. pomonella*	44
	Lymantria dispar	Ovary/embryo/ fat body	IPLB-Ld-652Y/EIt/FB	LdMNPV; *L. dispar*	45

TABLE 1 (continued)
Studies Comparing Viral Production in Different Insect Cell Lines

Category for comparison	Species origin of cell line	Tissue origin of cell line(s)	Cell line designation	Virus (Host species)	Reference
Cell lines of different tissues	*Lymantria dispar*	Embryo/fat body	IPLB-LdEP IPLB-LdEG IPLB-LdEI IPLB-LdEIt IPLB-LdFB	LdMNPV AcMNPV OpNPV; *Orgyia pseudotsugata* HzSNPV	46
	Spodoptera frugiperda	Ovary/wing disc	IPLB-Sf21AE, IAL-SfD1	AcMNPV; *A. californica* SfMNPV; *S. frugiperda*	47
	Trichoplusia ni	Ovary/wing disc	Tn-368, IAL-TnD1	AcMNPV; *A. californica* SfMNPV; *S. frugiperda*	47
Cell lines of different species	*Adoxophyes orana fasciatia*	Neonate	FTRS-AoL1/AoL2	AoNPV; *A. o. fasciatia*	36
	Adoxophyes sp. *Archippus breviplicanus* *Homana magnanima* *Hoshinoa longicellana* *Pandemis heparana*		FTRS-AfL FTRS-AbL81 FTRS-HmL45 FTRS-HlL1/HlL2 FTRS-PhL	PxNPV; *Plutella xylostella*	
	Aedes albopictus *Heteronychus arator* *Spodoptera frugiperda*	Ovary/larva/ embryo	ATC-15 DSIR-Ha-1179 IPLB-Sf21	OrNOV; *Oryctes rhinocerus*	48
Cell lines of different species	*Antheraea eucalypti* *Bombyx mori* *Mamestra brassicae* *Samia cynthia ricini* *Spilosoma inparilis*	Ovary/embryo/ fat body	(not available) SES-BoMo-15A SES-MaBR-4 NISES-SaCy-12 FRI-SpIm-1229	AyMNPV; *Antheraea yamamai*	49
	Anthonomus grandis *Trichoplusia ni*	Embryo/ovary	AgE TN-CL1	AcMNPV; *A. californica*	50
Cell lines of different species	*Anticarsia gemmatalis* *Heliothis subflexa* *Heliothis virescens* *Helicoverpa armigera* *Helicoverpa zea* *Plutella xylostella*	Ovary	BCIRL-Ag-AM1 BCIRL-Hs-AM1 BCIRL-Hv-AM1 BCIRL-Ha-AM1 BCIRL-Hz-AM1 BCIRL-Px2-HNU3	SfaMNPV; *Anagrapha (Syngrapha) falcifera*	51

Cell lines of different species	*Spodoptera frugiperda*		IPLB-Sf21		
	Trichoplusia ni		Tn-CL1		
	Anticarsia gemmatalis	Ovary	UFL-Ag-286,	AgMNPV; *A. gemmatalis*	52
	Spodoptera frugiperda		IPLB-Sf-21AE	AcMNPV; *A. californica*	
				SfMNPV; *S. frugiperda*	
	Estigmene acrea	Hemocyte/ovary	BTI-EaA	AcMNPV;	53
	Lymantria dispar		IPLB-Ld64BA	*A. californica*	
	Mamestra brassicae		Mb0503		
	Spodoptera frugiperda		IPLB-Sf1254		
	Trichoplusia ni		Tn-368		
	Heliothis virescens	Ovary	BCIRL-Hv-AM1	AcMNPV;	54
	Helicoverpa armigera		BCIRL-Ha-AM1	*A. californica*	
	Helicoverpa zea		BCIRL-Hz-AM1	GmMNPV;	
	Spodoptera frugiperda		IPLB-Sf21	*Galleria mellonella*	
	Trichoplusia ni		Tn-CL1	HzSNPV; *H. zea*	
				SfMNPV; *S. frugiperda*	
				TnMNPV; *T. ni*	
	Heliothis virescens	Ovary	BCIRL-Hv-AM1	HzSNPV; *H. zea*	38
	Helicoverpa zea		BCIRL-Hz-AM1		
	Lymantria dispar	Embryo/fat body[d]	IPLB-LdEP	LdMNPV	46
	Heliothis virescens	testes	IPLB-LdEG	AcMNPV	
			IPLB-LdEI	OpNPV;	
			IPLB-LdEIt	*O. pseudotsugata*	
			IPLB-LdFB	HzSNPV	
			IPLB-HvT1		
	Plodia interpunctella	Wing disc/ovary	IAL-PiD2	AcMNPV;	47
	Spodoptera frugiperda		IAL-SfD1, IPLB-SF21AE	*A. californica*	
	Trichoplusia ni		IAL-TnD1, TN-368	SfMNPV; *S. frugiperda*	
	Spodoptera littoralis	Ovary	UIV-S1-573/HPB-SL-26	S1MNP;	55
	Spodoptera frugiperda		IPLB-Sf21	S. littoralis	
	Spodoptera littoralis	Ovary	HPB-S1-26	AcMNPV;	56
	Spodoptera frugiperda		IPLB-Sf21	*A. californica*	
	Spodoptera littoralis	Ovary	CLS-79	XcNPV;	57
	Spodoptera frugiperda		IPLB-Sf21	*Xestia c-ngrum*	
	Spodoptera frugiperda	Wing disc	IAL-SfD1	AgMNPV;	58
	Trichoplusia ni		IAL-TnD1	*A. gemmatalis*	
				AcMNPV;	
				A. californica	
				TnMNPV; *T. ni*	

and McIntosh and Ignoffo[38] found that ovarian cell lines from *H. zea* pupae differed in some cases over 100-fold in the levels of HzSNPV OB and ECV produced. The resulting OB also varied in virulence to *H. zea* larvae. In addition, Sohi et al.[41] reported that the levels of OB of the *Orgyia leucostigma* NPV (OlSNPV) varied over 10-fold when produced in several neonate cell lines of *O. leucostigma.*

b. Cell Lines from Different Tissues

Cell lines from the same species but different tissues vary in their ability to produce baculoviruses as well. For example, *Spodoptera frugiperda* or *T. ni* cell lines from imaginal wing discs and ovarian tissues exhibited differing levels of ECV production for either the *Autographa californica* baculovirus (AcMNPV) or the *S. frugiperda* baculovirus (SfMNPV), with up to 1000-fold variation occurring.[47] Lynn et al.[45] also observed that cell lines from ovaries, fat bodies, and embryos of *Lymantria dispar* varied significantly in their ability to produce both ECV and OB of the homologous MNPV (LdMNPV). In addition, Miltenburger et al.[44] reported the same phenomenon when comparing the replication of *Cydia pomonella* GV (CpGV) or NPV (CpNPV) in cell lines derived from larval and embryonic tissues of *C. pomonella.*

c. Cell Lines from Different Species

Most studies comparing the efficiency of production of a baculovirus strain between cell lines have involved cell lines of different species. Tsuda et al.[57] compared the ability of *S. frugiperda* and *Spodoptera littoralis* cell lines to replicate the *Xestia c-nigrum* baculovirus (XcNPV) and found the *S. frugiperda* line to be the better producer. Using cell lines from the same species, Kislev[55] found that the *S. littoralis* lines generated the highest levels of the *S. littoralis* baculovirus (SlMNPV) ECV and OB. Alternatively, Roberts[56] reported that *S. frugiperda* and *S. littoralis* cell lines produced similar titers of AcMNPV ECV. McIntosh and co-workers[50,51,54] compared the production of numerous NPVs in homologous and heterologous cell lines and observed a wide range of variation (in some cases over 100,000-fold differences) in the production of OB and ECV, and in the ability of the OB to infect larvae. A study of cell lines from *Estigmene acrea, L. dispar, Mamestra brassicae, S. frugiperda,* and *T. ni* also resulted in the production of differing levels of AcMNPV OB/cell and ECV/cell.[53] Likewise, OB of *Antheraea yamamai* MNPV were generated at varying concentrations in cell lines from *Samia cynthia ricini, Spilosoma inparilis, M. brassicae,* and *Antheraea eucalypti.*[49]

2. Production of Different Baculoviruses

The quantity and quality of baculovirus production in cell culture is also affected by the strain of the virus and the type of infection process it undergoes. The same cell line can vary in its ability to replicate viruses from different sources, such as geographic isolates, clonal strains, species, and genera/subfamilies. In addition, different types of viral infections (e.g., productive and persistent) also generate differing levels of virus product within the same cell line.

a. Viral Strains

Variable production of viral progeny from geographic isolates of the same virus using the same cell lines is exemplified by the production of LdMNPV strains in *L. dispar* cell lines.[45] Ten- to 100-fold differences in OB and ECV of the variants were reported for some cell lines. Kislev[55] reported that SlMNPV geographic isolates replicate at varying levels in *S. frugiperda* and *S. littoralis* cell lines. In addition, two isolates from commercial LdMNPV preparations have been purified and replicated in the homologous cell line IPLB-Ld-652Y.[59] Large differences in ECV/ml, OB/cell, and potency of OB in larvae were reported.

TABLE 2
Replication of Plaque Purified Clones from Wild-Type *Helicoverpa zea* Nuclear Polyhedrosis Virus (HzSNPV) in a Cloned Cell Line

Virus clone	PFU/ml ($\times 10^5$)	% Cells with OB	OB/ml ($\times 10^6$)	OB/cell
wt	37.5 ± 11.8 ab	86.9 ± 2.9 def	26.7 ± 0.9 def	81.7 ± 3.3 bcde
21A	38.8 ± 10.1 ab	74.4 ± 1.2 a	3.7 ± 0.1 a	20.5 ± 0.3 a
21B	52.6 ± 17.1 ab	88.8 ± 0.1 ef	24.6 ± 1.5 cdef	73.8 ± 4.5 acde
21C	57.8 ± 2.4 b	81.8 ± 0.5 bc	8.3 ± 0.6 ab	30.5 ± 3.3 ac
23A	27.2 ± 13.5 a	86.8 ± 0.2 def	11.0 ± 5.9 abd	47.0 ± 26.9 acde
31A	51.3 ± 13.9 ab	85.0 ± 0.8 cd	17.8 ± 1.3 abdef	43.2 ± 2.7 acd
31B	33.5 ± 1.0 ab	90.1 ± 0.6 fg	16.6 ± 5.0 abde	41.4 ± 14.6 acd
31C	32.9 ± 12.3 ab	80.6 ± 1.2 b	13.1 ± 1.1 abd	59.0 ± 12.9 acde
31D	41.2 ± 4.2 ab	94.8 ± 0.1 h	15.0 ± 6.7 abde	63.9 ± 32.7 acde
32A	29.5 ± 5.2 ab	93.0 ± 1.9 gh	8.8 ± 3.9 abc	43.7 ± 15.3 acd
32B	40.8 ± 1.9 ab	80.5 ± 0.7 b	6.6 ± 0.3 ab	25.8 ± 4.1 ab
32C	45.0 ± 3.3 ab	83.3 ± 0.9 bc	11.7 ± 4.0 abd	56.8 ± 17.9 acde
33A	37.7 ± 18.0 ab	89.5 ± 1.4 f	21.6 ± 0.3 bdef	50.7 ± 1.1 acde
33B	51.8 ± 6.2 ab	86.7 ± 0.6 de	23.1 ± 4.9 df	74.2 ± 17.0 cde
33C	34.3 ± 8.3 ab	88.3 ± 0.7 ef	33.5 ± 4.6 f	99.1 ± 2.5 de
33D	48.8 ± 1.8 ab	87.3 ± 0.3 def	29.7 ± 7.1 ef	101.5 ± 23.8 e

Note: An *H. zea* clonal cell line of pupal ovarian origin, BCIRL-HZ-AM1-CLA7, was used in this study. 1×10^6 cells were inoculated with 0.5 MOI in TC-199-MK (10% FBS) and harvested on Day 6. Methods used are described in Lenz et al.[33] Means (±SE) with the same letters in a column indicate no significant differences according to Fisher's LSD (n = 2 to 6).

Plaque-purified clones of the same viral strain also replicate in varying degrees in the same cell line. For example, an *H. zea* cell line produced varying amounts of ECV and OB from a series of *H. zea* SNPV clones — none of which exceeded those of the parental strain (Table 2). Cusack and McCarthy[60] also reported that the number of OB/cell of plaque-purified viral clones of an LdMNPV strain did not exceed that of the wild-type infection.

Cell lines also exhibit varying abilities to replicate baculoviruses of different species, genera, and/or subfamilies, none of which may be related to their *in vivo* specificities (see chapter 3). Examples of this phenomenon include the production of AcMNPV and SfMNPV by *S. frugiperda* or *T. ni* cell lines. Different levels of ECV were generated in these cell lines, with the OB virulence varying for each virus species.[47] Sieburth and Maruniak[52] found a difference in the total number of infected cells when *A. gemmatalis* and *S. frugiperda* cell lines were exposed to *Anticarsia gemmatalis* MNPV and AcMNPV. Five different species of baculoviruses were also shown to produce greatly varying titers of ECV in lepidopteran cell lines by McIntosh et al.[54] In addition, different levels of OB were produced in *Cydia pomonella* and *T. ni* cell lines for baculoviruses of the NPV and GV genera.[43,44]

b. Infection Types

When choosing a cell culture-baculovirus production system, one must be aware of the types of infection processes. All virus-cell interactions discussed so far have involved productive or lytic infections in which the virus completes its course of infection in the cells, resulting in the generation of ECV and OB and the lysis of infected cells. Alternatively, persistent viral infections also occur in insect cells. This process involves the continuous production of viral progeny from cell passage to passage, without the introduction of new inocula. Persistent baculovirus infections have been reported for lepidopteran cell lines such as *S. frugiperda, T. ni, Helicoverpa armigera,* and *H. zea* (with the latter being from the *Nudibaculovirinae* subfamily).[2,4,61-64] Furthermore, the semipermissive replication of baculoviruses in insect cells has also been observed.[65,66] In this type of replication, limited viral macromolecular synthesis occurs and cytopathic effects or OB are not observed.

B. COMPARISON OF PRODUCTION PROCEDURES

In addition to the selection of cell line and virus inoculum, a number of other considerations must be made when designing an *in vitro* production system. These include various aspects of the inoculation procedure, such as the amount of virus to be added to a given number of cells (the multiplicity of infection, or MOI), the number of times the virus inoculum is to be passaged in cell culture, and several characteristics of the individual cells and the cell population. In addition, the effects of harvesting and storage conditions on virus yield, either OB or ECV, must also be considered. Furthermore, the virus yield is partly determined by the interaction of all of these factors; i.e., the modification of one factor can alter the overall effect of another factor. In this section, we will emphasize the effects of the individual factors on virus production, with some interactions being noted.

1. Multiplicity of Infection

Cell lines differ in their responses to variations in the amount of virus inoculum. An MOI is commonly based on the number of plaque-forming units (PFU) (or infectious virus particles) per cell, as determined by plaque assays.[33,53,57,67,68] Roberts[56] reported that the yield of AcMNPV ECV from *S. frugiperda* cells (IPLB-Sf-21) was higher when an MOI of 0.01 to 0.1 PFU/cell was used than when an MOI of 7 PFU/cell was employed. By contrast, the AcMNPV ECV yield in *S. littoralis* cells did not increase with a lower MOI, but was 10- to 100- fold lower than the yield of *S. frugiperda* cells at the same MOI. In a study by Knudson and Tinsley,[67] *S. frugiperda* cells produced elevated levels of ECV as the MOI was raised from 0.2 to 4.5 to 25. On the other hand, Lenz et al.[33] infected an *H. zea* clonal cell line with HzSNPV at MOIs of 1.0 and 10 and found no differences in the levels of ECV and OB produced.

Recent studies on the *in vitro* mass production of recombinant proteins from AcMNPV using the *S. frugiperda* cell line Sf-9 (a clone of IPLB-Sf-21AE) have also examined the effects of MOI on protein yield. Although in most of these studies no OB were produced since the polyhedrin gene was replaced by genes coding for foreign proteins, the level of recombinant proteins produced might suggest how the level of OB from a wild-type virus infection would be affected. Furthermore, since the following studies used different production schemes for their analyses (e.g., differing in media, culture vessels, and/or recombinant AcMNPV strains), their final yields cannot be directly compared. Scott et al.[69] reported that cell viability and recombinant protein yields were enhanced by a progression of the MOI from 0.01 to 0.1 to 1.0. In another study, MOIs ranging from 0.1 to 20 resulted in varying effects on cell viability: MOIs of 0.1 to 1.0 initially exhibited an increase in cell numbers, whereas MOIs of 5 and 20 produced a decrease in cell numbers from day 1 postinfection.[70] These investigators found that the highest level of recombinant protein was produced at an MOI of 5. Licari and Bailey[71] reported that as the MOI was progressively raised from 0.01 to 100, higher yields of recombinant protein were produced and faster declines in cell viability were observed. These researchers also found a logarithmic relationship between MOI and product yield and suggested that this phenomenon was due to system saturation, with the highest level of recombinant protein being produced at an MOI of 100 for cells infected during the late exponential phase. Overall, Licari and Bailey[71] concluded that the most cost-effective virus production strategy to use is a combination of a low MOI (e.g., 0.1) and the infection of cells early in their exponential phase of growth, thus depending on secondary infections to increase the final yield. This same strategy was used when Licari and Bailey[72] designed a computer model for optimizing the production of recombinant proteins in cell culture systems. In addition, Wu et al.[73] concluded from a kinetic characterization of cell death that an increase in MOI (using MOIs of 0.5 to 10) leads to a more rapid decrease in cell viability, with maximal recombinant protein production being achieved at MOIs of 1 to 5. Lastly, in a study with a temperature-sensitive AcMNPV strain (which replicates in cells at 27°C) an increase in MOI (0.02 to 2.0) resulted in a decrease in ECV and recombinant protein titer.[74] This inverse relationship between MOI and virus/protein yield is thought to have been due to the maintenance of the cells at 30°C for 3 to 4 days postinfection.

2. Passage Effect

When baculoviruses (ECV) are serially passaged in cell lines numerous times, a variety of mutations or defective viral populations may come to dominate the cell culture, resulting in effects such as a decrease in the percent of cells producing OB, a decrease in the number of OB/cell, a decrease in the number of virions/OB, and an increase in the number of less virulent, morphologically abnormal OB.[2,35,43,53,75-78] These aberrant baculoviruses may either themselves produce low levels of OB/cell (referred to as FP for "few polyhedra"),[79-81] or they may interfere with the replication of the high-producing viral strains (referred to as MP for "many polyhedra") and require the presence of the latter for their replication (referred to as "defective interfering particles" or DIF).[60,76,82,83] Genomic mutations, such as insertions or deletions, have been shown to be the cause of many of these baculovirus aberrations.[60,79-81,83] The time of onset, type, and degree of the passage effect is dependent upon the cell-virus system in question, but often is initially seen by the 10th *in vitro* passage. Media formulations may also affect the rate at which these virus mutations accumulate. Tompkins et al.[84] reported that supplementing cell culture media with lipids stabilized or enhanced the *in vivo* virulence of OB for three NPVs after five passages in *T. ni* cells. The use of computer models of bioreactor systems has indicated that the configuration of the reaction vessels can either accelerate or delay the passage effect for virus *in vitro* mass-production systems.[16,21] These studies indicated that the larger the number of bioreactors in a continuous production system, the more rapid the onset of the passage effect.

3. Cell Parameters

The final yield of virus progeny can also be affected by the population growth phase, the density of the cells, and/or the cell cycle phase during the time of virus inoculation. The population growth phase at inoculation determines whether or not a secondary infection can occur in the cells not infected by the original inoculum, and thus cause an increase in virus yield. Licari and Bailey[72] found that infection during the early exponential phase minimized the effect of MOI on recombinant protein titer, which they indicated was due to the viral progeny's having time to infect the uninfected cells while in the exponential phase. On the other hand, the product titers for late-exponential infections were strongly dependent on MOI, with a high MOI resulting in an elevated protein titer. Neutra et al.[70] also found that infecting cells earlier in their population growth cycle generated higher recombinant protein yields. Sherman[77] reported that infection during the log phase of growth produced the highest yield of OB, whereas Stockdale and Gardiner[85] found that the cell growth phase did not affect the yield of OB, but that cell density was the deciding factor.

Concerning cell density, Hink and Strauss[86] noted that higher cell densities resulted in a higher percentage of infected cells and ECV/cell, whereas no relationship between OB/cell and initial cell density was observed. By contrast, Wood et al.[87] reported that virus inoculation of high-density cell cultures produced 98% fewer OB and ECV as compared to low-density cultures. This phenomenon was reversible and required cell-to-cell contact. Caron et al.[88] found that increasing the cell density caused a decrease in recombinant protein yield, which was reversed by the replacement of spent media with fresh media. Stockdale and Gardiner[85] also observed that increasing cell density decreased yield of OB, indicating that density affects the nutritional condition of the media, which in turn affects the ability of the cells to replicate virus. Sherman reported in his review[77] that the cell density during infection directly affects OB/ml and OB/cell. Wu et al.[73] found that overconfluent cell populations tended to decrease recombinant protein yield. In addition, Scott et al.[69] reported that there was a relationship between cell density, MOI, and recombinant protein production.

The effect of cell cycle phases on virus proliferation has also been investigated. Lynn and Hink[89] observed that virus inoculation of cells in middle to late S phase produced higher percentages of infected cells than the inoculation of cells in the G_2 phase. Alternatively, the amount of virus generated per individual cell did not differ between cells in the varying phases.

4. Harvesting and Storage Conditions

To maximize virus yield, the optimal harvesting time can be determined by performing a virus growth curve.[15,27,52,57,86] Virus titers reach a time-dependent plateau, which may decline after a few days due to several factors such as degradation (e.g., by catabolic enzymes from lysed cells) or adsorption. Since the stability of ECV is affected by pH, with a loss of infectivity being observed above pH 7.5 and below pH 6.0,[90,91] changes in pH may also compromise the overall virus yield prior to harvesting. In addition to harvesting conditions, storage conditions may also affect virus yield. Ignoffo and McIntosh[92] reported that the virulence of ECV and OB of HzSNPV to larvae is not affected by storage at 4°C for over 1.5 years. They compared this result with several studies and found that several baculoviruses are less likely to remain as virulent after long periods of storage.

III. EFFECT OF CHEMICAL AND PHYSICAL FACTORS ON *IN VITRO* PRODUCTION

In addition to biological factors, numerous chemical and physical factors also affect virus replication and yield in cell culture. These include the temperature, the media composition, and the production vessel(s) employed during viral replication.

A. TEMPERATURE

The temperature of cells during the viral replication process affects their ability to generate virus. Knudson and Tinsley[67] compared yields of AcMNPV ECV in *S. frugiperda* cells at temperatures ranging from 17°C to 37°C and found that the optimal temperature was 27°C. Knittel and Fairbrother[90] compared the yields of AcMNPV (ECV and OB) in *T. ni* cells at 28°C and 37°C and reported a lower yield at the higher temperature. In a review by Cho et al.,[13] the optimal temperature for viral growth was also reported to be between 27°C and 30°C. On the other hand, Hink and Strauss[86] reported the optimal temperature for cell growth (28 to 30°C) was not necessarily the same as that for the production of OB, since higher numbers of OB were generated at 25°C and 32°C than at 28°C and 30°C.

B. MEDIA

The medium composition used for cells directly affects their ability to generate viral progeny. Numerous studies have described the relationship between media composition and cell growth,[93] with the more recent studies examining the direct relationship between media composition and virus yield.

1. General Attributes

Both pH and osmolarity of the media affect cell viability and replication, but few, if any, studies have directly examined their effects on virus production. The optimal growth of insect cells has been reported by various investigators to occur between pH 6.0 and 6.7,[13,94] but this may or may not be true for virus replication in these cells. In fact, ECV and OB of several baculoviruses have been reproducibly produced at pH 7.0 using the medium TC-199-MK.[32,33,38,50,51,54] Studies have determined that optimal cell growth is produced in media having osmolarities between 290 and 360 mOsm/kg.[6,95] Stockdale and Gardiner[85] compared the effects of various media on yield of OB and indicated the osmotic pressure of each medium. Although not specifically addressed, the osmotic pressure may have been a factor contributing to the differences in yields of OB in this study. The freshness of the media is also an important factor in viral replication, which may be due more to a decrease in available nutrients in the used medium rather than to an increase in toxic metabolites.[85] Lindsay and Betenbaugh[96] found that the replacement of spent medium with new medium increased the yield of recombinant proteins 20-fold. Likewise, Caron et al.[88] reported that the decrease in protein production due to an increase in cell density was ameliorated by the renewal of the medium. Medium exchange has also been reported to increase the overall production of AcMNPV OB by 28%.[15]

2. Specific Components

Several authors have reviewed the importance of media composition in cell growth and viral replication from both a historical and a practical perspective.[6,10,13,14,18,97] Recent studies have highlighted advances in this area and will be emphasized in this section. The main goal of media optimization studies is to reduce the cost of viral pesticide production by developing formulations that enhance virus production yet utilize components that are easily obtained, reproducible, microbe-free, and compatible with mass production in large bioreactors (e.g., are easily sterilized and do not readily foam).[6,13,14,18,97,98] Unlike media formulations for the production of recombinant proteins for medical, veterinary, and pharmaceutical applications, media for viral pesticide production need not be protein-free or chemically defined.

One of the primary ways of developing less expensive media for virus production is the replacement of the mammalian sera, primarily fetal bovine serum (FBS), present in many media. When used, FBS accounts for 90 to 95% of the media cost[6,13,99] and is a major source for virus and mycoplasma contamination. The availability and quality of sera can vary between lots. In addition, FBS can generate foam when used in large bioreactors and can only be sterilized through filtration or gamma-radiation, both of which are time-consuming and costly. FBS also can inhibit the growth of some cell lines and, at high concentrations, can interfere with viral replication. Broussard and Summers[100] reported that cells produce more AcMNPV polyhedrin in medium containing 0.5% FBS than 10% FBS (FBS is routinely added to many media at a 10% concentration). Likewise, Zhang et al.[101] observed increased levels of *Bombyx mori* MNPV OB when the amount of FBS in their medium was decreased from 10% to 5%. Therefore, the replacement of FBS in media is a major thrust in the development of insect cell culture media for virus production.

a. Crude Biologicals

The use of crude biologicals for the *in vitro* replication of viruses and/or the replacement of FBS has been one area of study, but these media supplements may have some of the same drawbacks as FBS. Potential biologicals include tryptose phosphate broth, bovine embryonic fluid, and hydrolysates of lactalbumin, casein, soybeans, liver, and meat.[93,97,98,102] Vaughn et al.[103] compared virus production in media containing 10% FBS to media supplemented with 10% CPSR-1 or CPSR-3, the commercially produced derivatives of bovine plasma which contain "appropriate whole fluids of bovine origin."[104] Findings indicated that fewer ECV were produced by cells grown in the media containing either of the plasma derivatives. A decrease in OB/cell was also observed for the virus produced in media supplemented with CPSR-1, although a decrease in the LC_{50} (but not LC_{70} or LC_{90}) values for the OB generated in the CPSR-3 containing media was observed. In another study, supplementing media with insect hemolymph improved the level of production of OB from BmMNPV in *B. mori* cells.[105] Egg yolk emulsions can also replace FBS while maintaining cell viability and/or baculovirus permissiveness.[14,106,107] In addition, a commercially available aqueous lipid mixture originating from bovine serum, ExCyte VLE, elevated the level of recombinant protein production in insect cells as compared to a serum-supplemented medium.[108]

b. Purified Biochemicals

Biochemicals used in insect cell culture media for the optimization of virus production have included lipids, vitamins, amino acids, carbohydrates, trace metals, and other cell metabolites. According to Goodwin,[97,109,110] tissue-specific functions in cell lines, such as baculovirus permissiveness, may be strongly affected by the availability of lipids important for the structure and function of cell membranes. Goodwin's work with liposomes[109,111] indicated that the addition of specific lipids to serum-free media enhances different aspects of the baculovirus replication cycle. For example, cholesterol and methyl oleate enabled the *L. dispar* cell line to produce *L. dispar* MNPV virions and OB, although these OB were devoid of virions.[111] Addition of α-tocopherol acetate and an increase in cholesterol, along with the addition of other nonlipid components, further increased the levels of OB/cell.[111] Occlusion of virion bundles was initiated after supplementing the medium with phosphatidyl choline and increasing the cholesterol concentration.[97] A lipid emulsion has also been developed that contains cod liver oil fatty acid methyl esters, cholesterol, α-tocopherol acetate, ethanol, and Tween 80. This emulsion has enabled *S. frugiperda* cells to replicate AcMNPV in the medium IPL-41 without the addition of FBS.[11,112] IPL-41 supplemented with the above lipid emulsion has also been used to generate recombinant AcMNPV in large-scale bioreactors using insect cells.[11,19,88] As previously mentioned, the addition of liposomes to media can maintain the virulence of baculoviruses produced over numerous passages in cell culture.[84]

Supplementing cell culture media with an array of nonlipid biochemicals important in cellular metabolism also enhances baculovirus production and reduces cost. Reuveny et al.[113] lowered the cost of media by substituting glucose for a mixture of carbohydrates common in insect cell culture media. This procedure maintained the level of viral replication, as evidenced by the amount of recombinant protein produced. Supplementing media with the primary nonlipid choline derivative present in Lepidoptera, acetyl-β-methylcholine, in combination with fresh glutamine, enhanced cell viability and/or encouraged the production of OB of HzSNPV.[114] Folic acid added to media, along with lipids and glutamine, also increased the levels of OB/cell of LdMNPV in a homologous cell line.[111] Furthermore, the addition of $AlCl_3$ and $ZnSO_4$ to media increased the amounts of AcMNPV ECV and OB.[10,115] α-Glycerol-phosphate and glycerol additions have also aided in the occlusion process of single virions into LdMNPV OB, with occasional multicapsid inclusions being observed.[111]

c. Commercial Media Preparations

Several serum-free media have recently been developed that support the replication of baculoviruses in insect cells. ExCell 400™ (JRH Biosciences, Lenexa, KS), which contains a proprietary formulation of lipids, supported replication of AcMNPV in *S. frugiperda, T. ni,* and *P. xylostella* cell lines.[6,96,116-118] In these studies, comparisons were made with serum-containing media, with varying results being seen as to the level of virus and/or recombinant protein yield generated. The BmMNPV has also been successfully replicated in an ovarian *B. mori* cell line using ExCell 400™. Similar titers of ECV were generated as with the serum-containing medium but higher levels of OB were produced in the serum-free medium.[101] In addition, McIntosh et al.[119] found that a parental *H. zea* cell line produced similar levels of ECV and lower amounts OB of HzSNPV in ExCell 400™ as compared to a serum-containing medium. In this study, the clonal *H. zea* line produced lower titers of ECV and similar levels of OB in ExCell 400™ as compared to a serum-containing medium. A powdered form of this medium (ExCell 401™) has also been developed for the replication of baculoviruses in cell culture. Sf-900™ (GIBCO/BRL Life Technologies Inc., Grand Island, NY) is another commercial serum-free medium available for *in vitro* virus replication. This medium has supported AcMNPV replication at various levels in *S. frugiperda, T. ni,* and *P. xylostella* cell lines.[118] ECV of HzSNPV were generated at equivalent or lower levels by parental and clonal cell lines of *H. zea,* respectively, in Sf-900™ as compared to a serum-containing medium.[119] In this study, production of OB was lower or higher for the parental and clonal cell line, respectively, in Sf-900™ as compared to a serum-containing medium. On the other hand, Zhang et al.[101] reported that Sf-900™ did not support the *B. mori* cell line Bm5, rendering this medium's effect on BmMNPV replication untestable. In addition, Weiss et al.[24] have reported the development of "BioPesticide SFM," which supports the replication of AcMNPV and/or OpMNPV in several insect cell lines.

C. PRODUCTION VESSELS

A major goal in the research involving the replication of baculoviruses in insect cell culture is to scale-up the overall process for commercial mass production of biopesticides. To accomplish this, the optimal vessel for production must be selected based on the following considerations: manner of cell growth, aeration needs and effects, and bioreactor configuration. Reviews[6,10,13,14,18,20] have already provided detailed analyses of these topics, especially with respect to cell viability and growth. This section will briefly summarize bioreactor considerations for virus production schemes from the most recent studies.

1. Cell Growth Mode

Insect cell lines grow either in suspension or attached to vessel surfaces. A particular cell line may grow in either manner or loosely attached, and some can be induced to grow in either manner by altering the formulation of the media. To generate baculoviruses, either method of growth is feasible, although some data suggest that adhesive interactions may affect virus proliferation.[120] Many initial baculovirus infection studies were performed in small T-flasks, and the attachment of cells was often promoted. Scaled-up versions of static culture have included roller bottles, spiral-core film bottles, glass bead packed-bed reactors, microcarriers, fiber-disk carriers, and hollow-fiber reactors,[121] as reviewed elsewhere.[10,13,18] AcMNPV has been produced in many of these vessels, including roller bottles,[9,10,122] spiral-core film bottles,[10] glass bead packed-bed reactors,[120] fiber-disk carriers,[123] and microcapsules.[124] The main disadvantage to attached or immobilized culture systems is that cell density is limited by the surface area. This directly affects virus yield, although marked improvements have been made in the latest designs. On the other hand, static cultures avoid problems associated with shear stress seen in suspension vessels.

The use of suspension cultures for production scale has been of interest because of the proven successes of fermentors in the mass production of plant and mammalian cells, yeast, and bacteria. Numerous culture vessels are available for suspension growth and include smaller specialized flasks and larger agitated tank or airlift bioreactors, as detailed previously.[6,10,13,14,18,20] Recent studies have described the production of AcMNPV in spinner flasks,[11,69,96,116] shake flasks,[17,70] stirred tank bioreactors (standard agitators,[12,15] or supplied with marine impellers[16,19,21,69,88,125]), and airlift fermentors.[11,12,17] Although bioreactor technology has been well advanced, problems specific to insect cells are still being worked out, such as how to aerate the medium while producing minimal shear stress.

2. Aeration

Providing insect cells with an adequate supply of oxygen for maintenance and baculovirus replication can be performed by a variety of methods,[6,10,13,17,18,20] none of which are without their drawbacks. Because cells increase their need for dissolved oxygen during virus replication,[10,11,18,69,126] the method of aeration is an important element of virus production systems. Methods employed for culture aeration include the direct introduction of air into the bioreactor through air sparging and surface aeration (often in combination with stirring), or the indirect introduction through the use of semipermeable tubing or membrane bioreactors.[6,12,20] Insect cells are sensitive to damage from hydrodynamic or shear stress caused by agitation or air bubble formation. The possible mechanisms of this process are currently being studied.[20,127] Sensitivity of insect cells to shear stress is dependent on cell line, population growth phase, viral infection (which increases sensitivity), and media formulation.[20] Media components such as pluronic polyols (e.g., Pluronic F-68), methyl cellulose, and dextran are known to protect insect cells from shear stress.[6,11,17,18,20] Since damaged cells cannot efficiently generate baculoviruses, the effects of aeration techniques on cell viability and virus yield need to be considered.

Recent studies involving the scale-up of *in vitro* baculovirus production have indicated that AcMNPV can be successfully generated in suspension bioreactors employing surface aeration[16,19,88,125] or spargers,[11,12,17,69,128] and in spinner flasks employing silicone closures[116] or air filters.[96] In addition, Shuler et al.[120] employed an external aeration loop in glass bead packed-bed bioreactors to minimize exposure of cells to shear stress and to maximize virus yield.

3. Configuration

The *in vitro* production of baculovirus pesticides involves two stages: (1) cell production, and (2) virus replication. This biphasic process can be accomplished by either the use of one bioreactor for both phases[11,12,19,69,88] or the use of separate bioreactors for each stage.[15,16,21,82,125,128,129] The employment of separate bioreactors can involve either non-continuous[15] or continuous formats.[16,21,125,129] Klöppinger et al.[15] employed a repeated batch mode in which they optimized both stages of production separately. Continuous-production schemes have been designed using either two or three linked bioreactors,[16,21,125,129] with the three-bioreactor system employing two bioreactors for the virus generation phase. The studies of the continuous configurations have indicated that the two-reactor system has the longest run time and generates the highest virus yield. The differing results obtained from these continuous bioreactor schemes have been attributed to an earlier onset of the passage effect in the three-reactor configuration.[16,21]

IV. GENETIC ENGINEERING OF VIRAL PESTICIDES

Optimizing the production of viral pesticides involves not only enhancing their yield but also their efficacy. We have already discussed the selection of viral strains based on yield and virulence. However, studies employing baculovirus expression vectors to express a variety of foreign genes[130-133] have indicated the feasibility of using genetic engineering to enhance viral biocide performance. Several considerations must be made when using genetic engineering for the generation of improved baculovirus pesticides. One concern is for the safety of these recombinant baculoviruses to the environment, but field releases of genetically engineered baculoviruses have indicated that these agents should prove to be an ecologically safe means of controlling insects.[134-137] The primary safety concerns involving environmental and ecological issues have already been reviewed.[134,138,139] In this section, we will outline the main goals for the genetic engineering of viral pesticides and the primary engineering concerns.

A. ENGINEERING GOALS AND EXAMPLES

Recombinant baculoviruses can be designed to modify at least two properties of the wild-type baculovirus: their speed to kill can be increased and their host range can be expanded. Both of these modifications would improve the efficacy of viral pesticides and would enhance their marketability.

1. Enhance Virulence

Since baculoviruses require several days to kill insects, during which time they are still actively feeding and causing crop damage, the development of faster-acting viruses enhances their overall effectiveness as insect control agents in agricultural settings.[8] A variety of genes have been inserted into the baculovirus genome that have the potential for disrupting physiological processes such as digestion or molting. Examples of these genes include: *Bacillus thuringiensis* delta endotoxin genes,[137,140-142] insect hormone genes (diuretic hormone,[137,143,144] eclosion hormone[145]), insect enzyme genes (juvenile hormone esterase[137,146,147]), insect venom genes (from hornets[148]), and invertebrate neurotoxin genes (from predatory mites,[137,149] and scorpions[137,147,150-152]). Many of these recombinant baculoviruses or their gene products cause an earlier cessation of feeding or the premature death of larvae.[137,140-144,147,149-152] Deletion mutants of baculoviruses may also enhance their efficacy,[153] such as the deletion of *egt* gene from AcMNPV, which shortens the feeding time of infected insects and causes an earlier mortality.[154]

2. Expand Host Range

Another goal in recombinant baculovirus development is the generation of baculoviruses with expanded host ranges. Many baculoviruses infect only insects within one family or genus (see Chapter 3), which limits their marketability. Kondo and Maeda[155] were able to expand the host range of the BmNPV in cell culture and larvae by coinfecting cells with AcMNPV. Using DNA cotransfection, Mori et al.[156] established a hybrid recombinant from AcMNPV and BmNPV that was able to infect cells and larvae refractory to their parental strains.

B. ENGINEERING CONSIDERATIONS

When engineering baculoviruses for insect control, numerous considerations must be made, including the effects of the following components on insect cells and whole insects: the recombinant proteins and/or viruses, the foreign gene promoters, and posttranslational processing events.

1. Recombinant Virus and Protein Effects

Developing more effective viral biological control agents with *in vitro* systems involves different considerations than those made when optimizing cell culture systems devoted solely to the generation of recombinant proteins for pharmaceutical, medical, and veterinary applications. In the latter case, emphasis is placed on protein yield and processing. For insecticidal applications, the recombinant proteins are often insect toxins. Therefore, in addition to yield and processing concerns, one must be aware of the effects of the gene products on the insect cells. If the recombinant proteins are cytotoxic, they could affect the production of the virus in cell culture systems by interfering with cell viability and/or directly with viral replication and, thus, reduce the overall yield of the viral pesticide.[160] Various cell lines could have differing levels of sensitivity to the recombinant proteins, which would need to be determined.

Another important criteria to investigate is the effect of the recombinant virus on living insects. Bioassays need to be performed to determine the toxicity (e.g., LC_{50} and LT_{50}) of the virus, and not merely that of the recombinant protein. The physiological concentrations of the recombinant protein in the hemolymph and/or the target tissues should also be determined if there is any question as to the efficacy of the recombinant virus. Using this approach, the reason(s) for the ineffectiveness to whole larvae of a recombinant AcMNPV containing a juvenile hormone esterase gene was made, which gave insights into potential avenues for improvement.[147,157,158]

2. Promoter Influence

Selection of the proper promoter for the expression of insecticidal or physiological inhibitory proteins for insect control is an important consideration. For baculovirus expression vectors of nonagricultural applications, the promoter that produces the highest recombinant protein yield overall is the promoter of choice. Alternatively, in field applications the level and timing of expression of the recombinant proteins will be determined by the promoter. Thus, the promoter will affect the degree and timing of feeding cessation and insect mortality of infected insects and, in this way, its overall efficacy as a biological control agent. Indeed, several promoters have been developed which vary in their level of expression and/or effect on timing of insect mortality.[141,146,147,149]

The host range of the viral pesticide may also be determined by the promoter. Morris and Miller[159] showed that promoters from various temporal transcription classes (i.e., early, late, and very late transcription) differently affected the ability of permissive and nonpermissive cell lines to express foreign gene products. They concluded that the use of early promoters and late promoters of viral origin, as well as of insect promoters, could enable recombinant viruses to infect a wider range of insects, possibly including nontarget insects, with the extent of the effect being determined by other outlined parameters.

3. Processing Concerns

In nonagricultural applications, proper posttranslational processing of recombinant proteins is a primary concern, with the desire to mimic as closely as possible the authentic protein.[130-132] Incorrectly processed recombinant proteins for medical or pharmaceutical applications may have altered functions and undesirable side effects, rendering them unmarketable. For viral pesticides, the overall concern is the effectiveness of the virus on insect growth and feeding, which may or may not mean the generation of an accurately processed recombinant protein. As long as the resulting recombinant baculovirus is more toxic to insects than the wild-type virus (and innocuous to nontarget organisms), the engineered virus may be considered a marketable improvement.

V. CONCLUSIONS

The production of baculovirus pesticides in cell culture is a viable means of generating an effective insect control agent. The optimization of *in vitro* production systems for viral biocide generation involves:

1. The selection of appropriate insect cell lines and virus strains (including recombinant baculoviruses)
2. The determination of the optimal viral inoculum amount (MOI)
3. The understanding of the virus passage history
4. The employment of the proper cell population growth phase, cell cycle phase, and/or density
5. The maintenance of optimal harvesting and storage conditions
6. The selection of the appropriate temperature and media during cell infection
7. The employment of appropriate production vessels and configurations

As research into the understanding of basic virus-cell interactions is furthered, this too will lead to the improvement of viral pesticides and their *in vitro* production schemes.

REFERENCES

1. **Martignoni, M. E. and Iwai, P. J.,** *A Catalog of Viral Diseases of Insects Mites, and Ticks,* Washington, D.C., 1986.
2. **Adams, J. R. and McClintock, J. T.,** Baculoviridae. Nuclear polyhedrosis virus. Part 1. Nuclear polyhedrosis viruses of insects, in *Atlas of Invertebrate Viruses,* Adams, J. R. and Bonami, J. R., Eds., CRC Press, Boca Raton, FL, 1991, 87.

3. **Tanada, Y. and Hess, R. T.,** Baculoviridae. Granulosis viruses, in *Atlas of Invertebrate Viruses,* Adams, J. R. and Bonami, J. R., Eds., CRC Press, Boca Raton, FL, 1991, 227.
4. **Huger, A. M. and Krieg, A.,** Baculoviridae. Nonoccluded baculoviruses, in *Atlas of Invertebrate Viruses,* Adams, J. R. and Bonami, J. R., Eds., CRC Press, Boca Raton, FL, 1991, 257.
5. **Francki, R. I. B., Fauquet, C. M., Knudson, D. L., and Brown, F.,** Classification and nomenclature of viruses. *Arch. Virol. Suppl.,* 2, 117, 1991.
6. **Agathos, S. N.,** Mass production of viral insecticides, in *Biotechnology for Biological Control of Pests and Vectors,* Maramorosch, K., Ed., CRC Press, Boca Raton, FL, 217, 1991.
7. **Huber, J.,** Use of baculoviruses in pest management programs, in *The Biology of Baculoviruses,* Vol. II, Granados, R. R. and Federici, B. A., Eds., CRC Press, Boca Raton, FL, 1986, 181.
8. **Wood, H. A. and Granados, R. R.,** Genetically engineered baculoviruses as agents for pest control, *Annu. Rev. Microbiol.,* 45, 69, 1991.
9. **Vaughn, J. L.,** The production of nuclear polyhedrosis viruses in large-volume cell cultures, *J. Invert. Pathol.,* 28, 233, 1976.
10. **Weiss, S. A. and Vaughn, J. L.,** Cell culture methods for large-scale propagation of baculoviruses, in *The Biology of Baculoviruses, Vol. II,* Granados, R. R. and Federici, B. A., Eds., CRC Press, Boca Raton, FL, 1986, 63.
11. **Maiorella, B., Inlow, D., Shauger, A., and Harano, D.,** Large-scale insect cell-culture for recombinant protein production, *Bio/Technology,* 6, 1406, 1988.
12. **Murhammer, D. W. and Goochee, C. F.,** Scaleup of insect cell cultures: effects of pluronic F-68, *Bio/Technology,* 6, 1411, 1988.
13. **Cho, T., Shuler, M. L., and Granados, R. R.,** Current developments in new media and cell culture systems for the large-scale production of insect cells, in *Advances In Cell Culture, Vol. 7,* Academic Press, New York, 1989, 261.
14. **Wu, J., King, G., Daugulis, A. J., Faulkner, P., Bone, D. H., and Goosen, M. F. A.,** Engineering aspects of insect cell suspension culture: a review, *Appl. Microbiol. Biotechnol.,* 32, 249, 1989.
15. **Klöppinger, M., Fertig, G., Fraune, E., and Miltenburger, H. G.,** Multistage production of *Autographa californica* nuclear polyhedrosis virus in cell cultures, *Cytotechnology,* 4, 271, 1990.
16. **van Lier, F. L. J., van den End, E. J., de Gooijer, C. D., Vlak, J. M., and Tramper, J.,** Continuous production of baculovirus in a cascade of insect-cell reactors, *Appl. Microbiol. Biotechnol.,* 33, 43, 1990.
17. **Wu, J., King, G., Daugulis, A. J., Faulkner, P., Bone, D. H., and Goosen, M. F. A.,** Adaptation of insect cells to suspension culture, *J. Ferment. Bioeng.,* 70, 90, 1990.
18. **Agathos, S. N.,** Production scale insect cell culture, *Biotechnol. Adv.,* 9, 51, 1991.
19. **Kamen, A. A., Tom, R. L., Caron, A. W., Chavarie, C., Massie, B., and Archambault, J.,** Culture of insect cells in a helical ribbon impeller bioreactor, *Biotechnol. Bioeng.,* 38, 619, 1991.
20. **Murhammer, D. W.,** The use of insect cell cultures for recombinant protein synthesis engineering aspects, *Appl. Biochem. Biotech.,* 31, 283, 1991.
21. **de Gooijer, C. D., Koken, R. H. M., van Lier, F. L. J., Kool, M., Vlak, J. M., and Tramper, J.,** A structured dynamic model for the baculovirus infection process in insect-cell reactor configurations, *Biotechnol. Bioeng.,* 40, 537, 1992.
22. **Vaughn, J. L.,** Low cost cell culture media, *In Vitro Cell. Dev. Biol.,* 28A, 43A, 1992.
23. **Delisle, D. W., Walls, E. L., Celeri, C., Knoch, C., Singer, V., and Tang, K.,** From shake flask to industrial fermentor: challenges of large scale cell culture, *In Vitro Cell. Dev. Biol.,* 28A, 43A, 1992.
24. **Weiss, S. A., Godwin, G. P., Whitford, W. G., Dougherty, E. M., and Rohrman, C. F.,** Viral pesticides: production in serum-free insect cell culture, *In Vitro Cell. Dev. Biol.,* 28A, 87A, 1992.
25. **McIntosh, A. H. and Ignoffo, C. M.,** Effect of larval extract and proteinase K on the *in vitro* infectivity of *Heliothis zea* nonoccluded and alkali-liberated virions, *Appl. Ent. Zool.,* 23, 175, 1988.
26. **Ignoffo, C. M. and Hostetter, D. L.,** Efficacy of insect viruses propagated *in vivo* and *in vitro, J. Invert. Pathol.,* 24, 184, 1974.
27. **Granados, R. R., Lawler, K. A., and Burand, J. P.,** Replication of *Heliothis zea* baculovirus in an insect cell line, *Intervirology,* 16, 71, 1981.
28. **Wickham, T. J., Shuler, M. L., Hammer, D. A., Granados, R. R., and Wood, H. A.,** Equilibrium and kinetic analysis of *Autographa californica* nuclear polyhedrosis virus attachment to different insect cell lines, *J. Gen. Virol.,* 73, 3185, 1992.
29. **Ignoffo, C. M. and McIntosh, A. H.,** Comparison of *in vivo* infectivity of tissue-cultured nonoccluded virus and alkali-liberated occluded virus of *Baculovirus heliothis, J. Invert. Pathol.,* 48, 289, 1986.
30. **Keddie, A. and Volkman, L. E.,** Infectivity difference between the two phenotypes of *Autographa californica* nuclear polyhedrosis virus: importance of the 64K envelope glycoprotein, *J. Gen. Virol.,* 66, 1195, 1985.
31. **Corsaro, B. G. and Fraser, M. J.,** Characterization of clonal populations of the *Heliothis zea* cell line IPLB-HZ 1075, *In Vitro Cell. Dev. Biol.,* 23, 855, 1987.

32. **Rice, W. C., McIntosh, A. H., and Ignoffo, C. M.,** Yield and activity of the *Heliothis zea* single nuclear polyhedrosis virus propagated in cloned and uncloned lines of *Heliothis* cells, *In Vitro Cell. Dev. Biol.*, 25, 201, 1989.
33. **Lenz, C. J., McIntosh, A. H., Mazzacano, C., and Monderloh, U.,** Replication of *Heliothis zea* nuclear polyhedrosis virus in cloned cell lines, *J. Invert. Pathol.*, 57, 227, 1991.
34. **Volkman, L. E. and Summers, M. D.,** Comparative studies with clones derived from a cabbage looper ovarian cell line, TN-368, in *Invertebrate Tissue Culture, Applications in Medicine, Biology and Agriculture*, Kurstak, E. and Maramorosch, K., Eds., Academic Press, New York, 1976, 289.
35. **Faulkner, P., Brown, M., and Potter, K. N.,** Utilization of tissue culture techniques to clone an insect cell line and to characterize strains of baculovirus, in *Invertebrate Tissue Culture, Applications in Medicine, Biology and Agriculture*, Kurstak, E. and Maramorosch, K., Eds., Academic Press, New York, 1976, 347.
36. **Sato, T.,** Establishment of eight cell lines from newborn larvae of Tortricids (Lepidoptera) and characteristics including susceptibility to insect viruses, in *Bull. Fruit Tree Res. Stat.*, Ser. A., No. 18, Ministry of Agriculture, Forestry and Fisheries, Tokyo, 1989, 115.
37. **Goodwin, R. H., Tompkins, G. J., Gettig, R. R., and Adams, J. R.,** Characterization and culture of virus replicating continuous insect cell lines from the bollworm, *Heliothis zea* (Boddie), *In Vitro*, 18, 843, 1982.
38. **McIntosh, A. H. and Ignoffo, C. M.,** Replication and infectivity of the single-embedded nuclear polyhedrosis virus, *Baculovirus heliothis*, in homologous cell lines, *J. Invert. Pathol.*, 37, 258, 1981.
39. **Granados, R. R. and Naughton, M.,** Replication of *Amsacta moorei* entomopoxvirus and *Autographa californica* nuclear polyhedrosis virus in hemocyte cell lines from *Estigmene acrea*, in *Invertebrate Tissue Culture, Applications in Medicine, Biology and Agriculture*, Kurstak, E. and Maramorosch, K., Eds., Academic Press, New York, 1976, 379.
40. **Goodwin, R. H., Tompkins, G. J., and McCawley, P.,** Gypsy moth cell lines divergent in viral susceptibility, *In Vitro*, 14, 485, 1978.
41. **Sohi, S. S., Percy, J., Arif, B. M., and Cunningham, J. C.,** Replication and serial passage of a singly enveloped baculovirus of *Orgyia leucostigma* in homologous cell lines, *Intervirology*, 21, 50, 1984.
42. **Dwyer, K. G., Webb, S. E., Shelton, A. M., and Granados, R. R.,** Establishment of cell lines from *Pieris rapae* embryos: characterization and susceptibility to baculoviruses, *J. Invert. Pathol.*, 52, 268, 1988.
43. **Granados, R. R., Derksen, A. C. G., and Dwyer, K. G.,** Replication of the *Trichoplusia ni* granulosis and nuclear polyhedrosis viruses in cell cultures, *Virology*, 152, 472, 1986.
44. **Miltenburger, H. G., Naser, W. L., Harvey, J. P., Huber, J., and Huger, A. M.,** The cellular substrate: a very important requirement for baculovirus *in vitro* replication, *Z. Naturforsch.*, 39C, 993, 1984.
45. **Lynn, D. E., Dougherty, E. M., McClintock, J. T., and Shapiro, M.,** Comparative replication of *Lymantria dispar* nuclear polyhedrosis virus strains in three continuous-culture cell lines, *Appl. Environ. Microbiol.*, 55, 1049, 1989.
46. **Lynn, D. E., Dougherty, E. M., McClintock, J. T., and Loeb, M.,** Development of cell lines from various tissues of Lepidoptera, in *Invertebrate and Fish Tissue Culture*, Kuroda, Y., Kurstack, E., and Maramorosch, K., Eds., Japan Scientific Society Press, Tokyo/Springer-Verlag, Berlin, 1988, 239.
47. **Danyluk, G. M. and Maruniak, J. E.,** *In vivo* and *in vitro* host range of *Autographa californica* nuclear polyhedrosis virus and *Spodoptera frugiperda* nuclear polyhedrosis virus, *J. Invert. Pathol.*, 50, 207, 1987.
48. **Crawford, A. M.,** Attempts to obtain *Oryctes baculovirus* replication in three insect cell cultures, *Virology*, 112, 625, 1981.
49. **Taniai, K. and Yamakawa, M.,** Characterization of a nuclear polyhedrosis virus from a Japanese oak silkworm, *Antheraea yamamai*, *J. Invert. Pathol.*, 60, 314, 1992.
50. **McIntosh, A. H., Barcenas, N., and Cate, J. R.,** Replication of *Autographa californica* baculovirus (AcMNPV) in a coleopteran cell line, *In Vitro Cell. Dev. Biol.*, 28, 557, 1992.
51. **McIntosh, A. H.,** *In vitro* infectivity of a clonal isolate of *Syngrapha falcifera* (celery looper) multiple nuclear polyhedrosis virus, *J. Invert. Pathol.*, 57, 441, 1991.
52. **Sieburth, P. J. and Maruniak, J. E.,** Susceptibility of an established cell line of *Anticarsia gemmatalis* (Lepidoptera: Noctuidae) to three nuclear polyhedrosis viruses, *J. Invert. Pathol.*, 52, 453, 1988.
53. **Lynn, D. E. and Hink, W. F.,** Comparison of nuclear *polyhedrosis virus* replication in five lepidopteran cell lines, *J. Invert. Pathol.*, 35, 234, 1980.
54. **McIntosh, A. H., Ignoffo, C. M., and Andrews, P. L.,** *In vitro* host range of five baculoviruses in lepidopteran cell lines, *Intervirology*, 23, 150, 1985.
55. **Kislev, N.,** Replication of two nuclear polyhedrosis viruses of *Spodoptera littoralis* in established insect lines, *J. Invert. Pathol.*, 47, 369, 1986.
56. **Roberts, P. L.,** Studies on *Autographa californica* nuclear polyhedrosis virus replication in *Spodoptera littoralis* cells including virus-induced protein synthesis, *Arch. Virol.*, 87, 49, 1986.
57. **Tsuda, K., Mizuki, E., Kawarabata, T., and Aizawa, K.,** Replication of *Xestia c-nigrum* (Lepidoptera: Noctuidae) polyhedrosis virus in continuous cell cultures, *Appl. Entomol. Zool.*, 19, 293, 1984.

58. **Lynn, D. E., Boucias, D. G., and Pendland, J. C.,** Nuclear polyhedrosis virus replication in epithelial cell cultures of Lepidoptera, *J. Invert. Pathol.*, 42, 424, 1983.
59. **Slavicek, J. M., Podgwaite, J., and Lanner-Herrera, C.,** Properties of two *Lymantria dispar* nuclear polyhedrosis virus isolates obtained from the microbial pesticide Gypchek, *J. Invert. Pathol.*, 59, 142, 1992.
60. **Cusack, T. and McCarthy, W. J.,** Effect of serial passage on genetic homogeneity of a plaque variant of *Lymantria dispar* nuclear polyhedrosis virus (Hamden LDP-67), *J. Gen. Virol.*, 70, 2963, 1989.
61. **McIntosh, A. H. and Ignoffo, C. M.,** Establishment of a persistent baculovirus infection in a lepidopteran cell line, *J. Invert. Pathol.*, 38, 395, 1981.
62. **Crawford, A. M. and Sheehan, C.,** Persistent baculovirus infections: *Spodoptera frugiperda* NPV and *Autographa californica* NPV in *Spodoptera frugiperda* cells, *Arch. Virol.*, 78, 65, 1983.
63. **Chao, Y.-C., Wood, H. A., Chang, C.-Y., Lee, H.-J., Shen, W.-C., and Lee, H.-T.,** Differential expression of Hz-1 baculovirus genes during productive and persistent viral infections, *J. Virol.*, 66, 1442, 1992.
64. **Zhong, J., Yue, Y.-X., and Su, D.-M.,** Productive and persistent infections of *Heliothis armigera* singly embedded nuclear polyhedrosis virus in its homologous cell line, in *Living with Insects, Proceedings XIX International Congress of Entomology, Abstracts*, Beijing, China, 1992, 274.
65. **Carpenter, W. M. and Bilimoria, S. L.,** A semipermissive nuclear polyhedrosis infection: characterization of infection kinetics and morphogenesis, *Virology*, 131, 227, 1983.
66. **McClintock, J. T., Dougherty, E. M., and Weiner, R. M.,** Semipermissive replication of a nuclear polyhedrosis virus of *Autographa californica* in a gypsy moth cell line, *J. Virol.*, 57, 197, 1986.
67. **Knudson, D. L. and Tinsley, T. W.,** Replication of a nuclear polyhedrosis virus in a continuous cell culture of *Spodoptera frugiperda:* purification, assay of infectivity, and growth characteristics of the virus, *J. Virol.*, 14, 934, 1974.
68. **Roberts, P. L.,** Infectivity assays for the *Autographa californica* nuclear polyhedrosis virus using *Spodoptera littoralis* cells, *J. Virol. Methods*, 10, 1, 1985.
69. **Scott, R. I., Blanchard, J. H., and Ferguson, C. H. R.,** Effects of oxygen on recombinant protein production by suspension cultures of *Spodoptera frugiperda* (Sf-9) insect cells, *Enzyme Microb. Technol.*, 14, 798, 1992.
70. **Neutra, R., Levi, B.-Z., and Shoham, Y.,** Optimization of protein-production by the baculovirus expression vector system in shake flasks, *Appl. Microbiol. Biotechnol.*, 37, 74, 1992.
71. **Licari, P. and Bailey, J. E.,** Factors influencing recombinant protein yields in an insect cell-baculovirus expression system: multiplicity of infection and intracellular protein degradation, *Biotechnol. Bioeng.*, 37, 238, 1991.
72. **Licari, P. and Bailey, J. E.,** Modeling the population dynamics of baculovirus-infected insect cells: optimizing infection strategies for enhanced recombinant protein yields, *Biotechnol. Bioeng.*, 39, 432, 1992.
73. **Wu, S.-C., Dale, B. E., and Liao, J. C.,** Kinetic characterization of baculovirus-induced cell death in insect cell cultures, *Biotechnol. Bioeng.*, 41, 104, 1993.
74. **King, G., Kuzio, J., Daugulis, A., Faulkner, P., Allen, B., Wu, J., and Gossen, M.,** Assessment of virus production and chloramphenicol acetyl transferase expression by insect cells in serum-free and serum-supplemented media using a temperature-sensitive baculovirus, *Biotechnol. Bioeng.*, 38, 1091, 1991.
75. **MacKinnon, E. A., Henderson, J. F., Stoltz, D. B., and Faulkner, P.,** Morphogenesis of nuclear polyhedrosis virus under conditions of prolonged passage *in vitro, J. Ultrastruct. Res.*, 49, 419, 1974.
76. **McIntosh, A. H., Shamy, R., and Ilsley, C.,** Interference with polyhedral inclusion body (PIB) production in *Trichoplusia ni* cells infected with a high passage strain of *Autographa californica* nuclear polyhedrosis virus (NPV), *Arch. Virol.*, 60, 353, 1979.
77. **Sherman, K. E.,** Considerations in the large-scale and commercial production of viral insecticides, in *Viral Insecticides for Biological Control*, Maramorosch, K., Sherman, K. E., Eds., Academic Press, New York, 1985, 757.
78. **Ramoska, W. A. and Hink, W. F.,** Electron microscope examination of two plaque variants from a nuclear polyhedrosis virus of the alfalfa looper, *Autographa californica, J. Invert. Pathol.*, 23, 197, 1974.
79. **Durand, J. P. and Summers, M. D.,** Alteration of *Autographa californica* nuclear polyhedrosis virus DNA upon serial passage in cell culture, *Virology*, 119, 223, 1982.
80. **Fraser, M. J., Smith, G. E., and Summers, M. D.,** Acquisition of host cell DNA sequences by baculoviruses: relationship between host DNA insertions and FP mutants of *Autographa californica* and *Galleria mellonella* nuclear polyhedrosis viruses, *J. Virol.*, 47, 287, 1983.
81. **Kumar, S. and Miller, L. K.,** Effects of serial passage of *Autographa californica* nuclear polyhedrosis virus in cell culture, *Virus Res.*, 7, 335, 1987.
82. **Kool, M., Voncken, J. W., van Lier, F. L. J., Tramper, J., and Vlak, J. M.,** Detection and analysis of *Autographa californica* nuclear polyhedrosis virus mutants with defective interfering properties, *Virology*, 183, 739, 1991.
83. **Wickham, T. J., Davis, T., Granados, R. R., Hammer, D. A., Shuler, M. L., and Wood, H. A.,** Baculovirus defective interfering particles are responsible for variations in recombinant protein production as a function of multiplicity of infection, *Biotechnol. Lett.*, 13, 483, 1991.

84. **Tompkins, G. J., Dougherty, E. M., Goodwin, R. H., and Adams, J. R.,** Maintenance of infectivity and virulence of nuclear polyhedrosis viruses during serial passage in noctuid (Lepidoptera: Noctuidae) cell lines, *J. Econ. Entomol.*, 84, 445, 1991.
85. **Stockdale, H. and Gardiner, G. R.,** The influence of the condition of cells and medium on production of polyhedra of *Autographa californica* nuclear polyhedrosis virus *in vitro, J. Invert. Pathol.*, 30, 330, 1977.
86. **Hink, W. E. and Strauss, E.,** Replication of alfalfa looper nuclear polyhedrosis virus in the *Trichoplusia ni* (TN-368) cell line, in *Invertebrate Tissue Culture, Applications in Medicine, Biology and Agriculture*, Kurstak, E. and Maramorosch, K., Eds., Academic Press, New York, 1976, 369.
87. **Wood, H. A., Johnston, L. B., and Burand, J. B.,** Inhibition of AcNPV replication in high-density *Trichoplusia ni* cell cultures, *Virology*, 119, 245, 1982.
88. **Caron, A. W., Archambault, J., and Massie, B.,** High-level recombinant protein production in bioreactors using the baculovirus-insect cell expression system, *Biotechnol. Bioeng.*, 36, 1133, 1990.
89. **Lynn, D. E. and Hink, W. F.,** Infection of synchronized TN-368 cell cultures with alfalfa looper nuclear polyhedrosis virus, *J. Invert. Pathol.*, 32, 1, 1978.
90. **Knittel, M. D. and Fairbrother, A.,** Effects of temperature and pH on survival of free nuclear polyhedrosis virus of *Autographa californica, Environ. Microbiol.*, 53, 2771, 1987.
91. **Wood, H. A. and Speyer, G.,** *In vitro* infectivity of the *Autographa californica* nuclear polyhedrosis virus under various conditions, *J. Gen. Virol.*, 40, 1, 1978.
92. **Ignoffo, C. M. and McIntosh, A. H.,** Low temperature stability of baculovirus virions from tissue culture and from inclusion bodies, *J. Invert. Pathol.*, 51, 304, 1988.
93. **Hink, W. F. and Bezanson, D. R.,** Invertebrate cell culture media and cell lines, in *Techniques in the Life Sciences, Cell Biology, C1, Techniques in Setting Up and Maintenance of Tissue and Cell Cultures, C111*, Kurstak, E., Ed., Elsevier Scientific Publishers, Ireland, Ltd., County Claire, Ireland, 1985, 1.
94. **Koval, T. M. and Suppes, D. L.,** pH dependency of cell attachment and growth at both clonal and subculture densities of cultured lepidopteran cells, *In Vitro. Cell. Dev. Biol.*, 26, 665, 1990.
95. **Kurtti, T. J., Chaudhary, S. P. S., and Brooks, M. A.,** Influence of physical factors on the growth of insect cells *in vitro, In Vitro*, 10, 3, 1974.
96. **Lindsay, D. A. and Betenbaugh, M. J.,** Quantification of cell culture factors affecting recombinant protein yields in baculovirus-infected insect cells, *Biotechnol. Bioeng.*, 39, 614, 1992.
97. **Goodwin, R. H.,** Growth of insect cells in serum-free media. *Techniques in the Life Sciences, Cell Biology, C1, Techniques in Setting Up and Maintenance of Tissue and Cell Cultures, C109*, Elsevier Scientific Publishers, Ireland, Ltd., County Claire, Ireland, 1985, 1.
98. **Vaughn, J. L.,** Low cost invertebrate cell culture media, *In Vitro Cell. Dev. Biol.*, 28, 43A, 1992.
99. **McIntosh, A. H. and Ignoffo, C. M.,** Notes on a low-cost medium for growth of insect cell lines and replication of insect viruses, *J. Kans. Entomol. Soc.*, 55, 354, 1982.
100. **Broussard, D. R. and Summers, M. D.,** Effects of serum concentration and media composition on the level of polyhedrin and foreign gene expression by baculovirus vectors, *J. Invert. Pathol.*, 54, 144, 1989.
101. **Zhang, J., Kalogerakis, N., and Behie, L. A.,** Investigation of reduced serum and serum-free media for the cultivation of insect cells (Bm5) and the production of baculovirus (BmNPV), *Biotechnol. Bioeng.*, 40, 1165, 1992.
102. **Sohi, S. S., Caputo, G. F., and Lalouette, W.,** Adaptation of *Orgyia leucostigma* (Lepidoptera: Lymantriidae) IPRI-OL-12 cells to serum-free media, *In Vitro Cell. Dev. Biol.*, 28, 87A, 1992.
103. **Vaughn, J. L., Fan, F., Dougherty, E. M., Adams, J. R., Guzo, D., and McClintock, J. T.,** The use of commercial serum replacements in media for the *in vitro* replication of nuclear polyhedrosis virus, *J. Invert. Pathol.*, 58, 297, 1991.
104. *Cell Culture Reagents Catalogue*, Sigma Chemical Co., St. Louis, Missouri, 1993, 40.
105. **Inoue, H., Shimizu, K., and Kuwabara, N.,** *In vitro* effect of insect hemolymph on the production of polyhedra of a nuclear polyhedrosis virus, *Appl. Entomol. Zool.*, 24, 486, 1989.
106. **Röder, A.,** Development of serum-free medium for cultivation of insect cells, *Naturwissenschaft*, 69, 92, 1982.
107. **Léry, X. and Fédière, G.,** A new serum-free medium for lepidopteran cell culture, *J. Invert. Pathol.*, 55, 342, 1990.
108. **Hink, W. F.,** A serum-free medium for the culture of insect cells and production of recombinant proteins, *In Vitro Cell. Dev. Biol.*, 27A, 397A, 1991.
109. **Goodwin, R. H.,** Replacement of vertebrate serum with lipids and other factors in the culture of invertebrate cells, tissues, parasites, and pathogens, *In Vitro Cell. Dev. Biol.*, 27A, 470A, 1991.
110. **Goodwin, R. H.,** Impediments to the culture of non-fibroblastoid cells from insects and other invertebrates, and their resolution, in *Eighth International Conference on Invertebrate and Fish Tissue Culture*, Fraser, M. J., Ed., Tissue Culture Association, Columbia, MD, 1991, 36.
111. **Goodwin, R. H. and Adams, J. R.,** Nutrient factors influencing viral replication in serum-free insect cell line culture, in *Invertebrate Systems In Vitro*, Kurstack, E., Maramorosch, K., Dübendorfer, A., Eds., Elsevier/North-Holland Biomedical Press, New York, 1980, 493.

112. **Inlow, D., Shauger, A., and Maiorella, B.,** Insect cell culture and baculovirus propagation in protein-free medium, *J. Tissue Culture Methods,* 12, 13, 1989.
113. **Reuveny, S., Kemp, C. W., Eppstein, L., and Shiloach, J.,** Carbohydrate metabolism in insect cell cultures during cell growth and recombinant protein production, *Ann. N. Y. Acad. Sci.,* 665, 230, 1992.
114. **Goodwin, R. H., Tompkins, G. J., Gettig, R. R., and Adams, J. R.,** Characterization and culture of virus replicating continuous insect cell lines from the bollworm, *Heliothis zea* (Boddie), *In Vitro,* 18, 843, 1982.
115. **Weiss, S. A., Smith, G. C., Vaughn, J. L., Dougherty, E. M., and Tompkins, G. J.,** Effect of aluminum chloride and zinc sulfate on *Autographa californica* nuclear polyhedrosis virus (AcNPV) replication in cell culture, *In Vitro,* 18, 937, 1982.
116. **Ogonah, O., Shuler, M. L., and Granados, R. R.,** Protein production (β-galactosidase) from a baculovirus vector in *Spodoptera frugiperda* and *Trichoplusia ni* cells in suspension culture, *Biotechnol. Lett.,* 13, 265, 1991.
117. **Wang, P., Granados, R. R., and Shuler, M. L.,** Studies on serum-free culture of insect cells for virus propagation and recombinant protein production, *J. Invert. Pathol.,* 59, 46, 1992.
118. **Chen, Q., McIntosh, A. H., Yu, F., Hong, H., Goodman, C. L., Grasela, J. J., and Ignoffo, C. M.,** The replication of *Autographa californica* baculovirus (AcMNPV) in two lepidopteran cell lines grown in serum-free media, *J. Invert. Pathol.,* in press.
119. **McIntosh, A. M., Grasela, J. J., and Lenz, C. L.,** Biological properties of lepidopteran cells in serum-free media, *In Vitro Cell. Dev. Biol.,* 27A, 105A, 1991.
120. **Schuler, M. L., Cho, T., Wickham, T., Ogonah, O., Kool, M., Hammer, D. A., Granados, R. R., and Wood, H. A.,** Bioreactor development for production of viral pesticides or heterologous proteins in insect cell cultures, *Ann. N. Y. Acad. Sci.,* 589, 399, 1990.
121. **Donofrio, D. M.,** Use of hollow fiber bioreactors to facilitate manufacture of cell products, *Am. Biotechnol. Lab.,* 9, 33, 1989.
122. **Weiss, S. A., Smith, G. C., Kalter, S. S., Vaughn, J. L., and Dougherty, E.,** Improved replication of *Autographa californica* nuclear polyhedrosis virus in roller bottles: characterization of the progeny virus, *Intervirology,* 15, 213, 1981.
123. **Kompier, R., Kislev, N., Segel, I., and Kadouri, A.,** Use of a stationary bed reactor and serum-free medium for the production of recombinant proteins in insect cells, *Enzymol. Microb. Technol.,* 13, 822, 1991.
124. **King, G. A., Daugulis, A. J., Faulkner, P., Bayly, D., and Goosen, M. F. A.,** Growth of baculovirus-infected insect cells in microcapsules to a high cell and virus density, *Biotechnol. Lett.,* 10, 683, 1988.
125. **de Gooijer, C. D., van Lier, F. L. J., van den End, E. J., Vlak, J. M., and Tramper, J.,** A model for baculovirus production with continuous insect cell cultures, *Appl. Microbiol. Biotechnol.,* 30, 497, 1989.
126. **Streett, D. A. and Hink, W. F.,** Oxygen consumption of *Trichoplusia ni* (TN-368) insect cell line infected with *Autographa californica* nuclear polyhedrosis virus, *J. Invert. Pathol.,* 32, 112, 1978.
127. **Garcia-Briones, M. and Chalmers, J. J.,** Cell-bubble interactions, mechanisms of suspended cell damage, *Ann. N.Y. Acad. Sci.,* 665, 219, 1992.
128. **Tramper, J., van den End, E. J., de Gooijer, C. D., Kompier, R., van Lier, F. L. J., Usmany, M., and Vlak, J. M.,** Production of baculoviruses in a continuous insect-cell culture: bioreactor design, operation, and modeling, *Ann. N.Y. Acad. Sci.,* 589, 423, 1990.
129. **Kompier, R., Tramper, J., and Vlak, J. M.,** A continuous process for the production of baculovirus using insect cell cultures, *Biotechnol. Lett.,* 10, 849, 1988.
130. **Luckow, V. A. and Summer, M. D.,** Trends in the development of baculovirus expression vectors, *Bio/Technology,* 6, 47, 1988.
131. **Fraser, M. J.,** Expression of eucaryotic genes in insect cell cultures, *In Vitro Cell. Dev. Biol.,* 25, 225, 1989.
132. **Maeda, S.,** Expression of foreign genes in insects using baculovirus vectors, *Annu. Rev. Entomol.,* 34, 351, 1989.
133. **Emory, V. C.,** Baculovirus expression vectors: choice of expression vector, in *Methods in Molecular Biology, Vol. 8: Practical Molecular Virology, Viral Vectors for Gene Expression,* Collins, M., Ed., Humana Press, Clifton, NJ, 1991, 287.
134. **Bishop, D. H. L., Entwistle, P. F., Cameron, I. R., Allen, C. J., and Possee, R. D.,** Field trials of genetically engineered baculovirus insecticides, in *The Release of Beneficially Engineered Micro-Organism,* Sussman, M., Collins, C. H., Skinner, F. A., and Stewart-Tull, D. E. S., Eds., Academic Press, New York, 1988, 143.
135. **Drahos, D. J.,** Field testing of genetically engineered microorganisms, *Biotechnol. Adv.,* 9, 157, 1991.
136. **Possee, R. D., Merryweather, A. T., Weyer, U., Harris, M. P. G., Hirst, M., and Bishop, D. H. L.,** The development and release of genetically engineered viral insecticides: a progress report 1986–1989, in *Molecular Insect Science,* Hagedorn, H. H., Ed., Plenum Press, New York, 1990, 113.
137. **Possee, R. D., King, L. A., and Weitzman, M. D.,** Progress in the genetic modification and field-release of baculovirus insecticides, in *The Release of Genetically Modified Microorganisms — REGEM2,* Stewart-Tull, D. E. S. and Sussman, M., Eds., Plenum Press, New York, London, 1992, 47.

138. **Hollander, A. K.,** Environmental impacts of genetically engineered microbial and viral biocontrol agents, in *Biotechnology for Biological Control of Pests and Vectors,* Maramorosch, K., Ed., CRC Press, Boca Raton, FL, 1991, 251.
139. **Wood, H. A.,** Development of genetically enhanced baculovirus pesticides, in *Biotechnology for Biological Control of Pests and Vectors,* Maramorosch, K., Ed., CRC Press, Boca Raton, FL, 1991, 69.
140. **Martens, J. W. M., Honée, Zuidema, D., van Lent, J. W. M., Visser, B., and Vlak, J. M.,** Insecticidal activity of a bacterial crystal protein expressed by a recombinant baculovirus in insect cells, *Appl. Environ. Microbiol.,* 56, 2764, 1990.
141. **Merryweather, A. T., Weyer, U., Harris, M. P. G., Hirst, M., Booth, T., and Possee, R. D.,** Construction of genetically engineered baculovirus insecticides containing the *Bacillus thuringiensis* subsp. *kurstaki* HD-73 delta endotoxin, *J. Gen. Virol.,* 71, 1535, 1990.
142. **Federici, B. A.,** Properties and analysis of insecticidal bacterial proteins produced in lepidopteran cell cultures using baculovirus vectors, *In Vitro Cell. Dev. Biol.,* 28, 50A, 1992.
143. **Maeda, S.,** Increased insecticidal effect by a recombinant baculovirus carrying a synthetic diuretic hormone gene, *Biochem. Biophys. Res. Commun.,* 165, 1177, 1989.
144. **Maeda, S.,** Baculovirus expression vectors expressing neuropeptide genes for insect control, in *Living with Insects, Proceedings XIX International Congress of Entomology, Abstracts,* Beijing, China, 1992, 96.
145. **Eldridge, R., Horodyski, F. M., Morton, D. B., O'Reilly, D. R., Truman, J. W., Riddiford, L. M., and Miller, L. K.,** Expression of an eclosion hormone gene in insect cells using baculovirus vectors, *Insect Biochem.,* 21, 341, 1991.
146. **Bonning, B. C., Hirst, M., Possee, R. D., and Hammock, B. D.,** Further development of a recombinant baculovirus insecticide expressing the enzyme juvenile hormone esterase from *Heliothis virescens, Insect Biochem. Mol. Biol.,* 22, 453, 1992.
147. **Hammock, B. D., McCutchen, B. F., Beetham, J., Choudary, P. V., Fowler, E., Ichonise, R., Ward, V. K., Vickers, J. M., Bonning, B. C., Harshman, L. G., Grant, D., Uematsu, T., and Maeda, S.,** Development of recombinant viral insecticides by expression of an insect-specific toxin and insect-specific enzyme in nuclear polyhedrosis viruses, *Arch. Insect Biochem. Physiol.,* 22, 315, 1993.
148. **Tomalski, M. D., King, T. P., and Miller, L. K.,** Expression of hornet genes encoding venom allergen antigen 5 in insects, *Arch. Insect Biochem. Physiol.,* 22, 303, 1993.
149. **Tomalski, M. D. and Miller, L. K.,** Expression of a paralytic neurotoxin gene to improve insect baculoviruses as biopesticides, *Bio/Technology,* 10, 545, 1992.
150. **Maeda, S., Volrath, S. L., Hanzlik, T. N., Harper, S. A., Majima, K., Maddox, D. W., Hammock, B. D., and Fowler, E.,** Insecticidal effects of an insect-specific neurotoxin expressed by a recombinant baculovirus, *Virology,* 184, 777, 1991.
151. **McCutchen, B. F., Choudary, P. V., Crenshaw, R., Maddox, D., Kamita, S. G., Palekar, N., Volrath, S., Fowler, E., Hammock, B. D., and Maeda, S.,** Development of a recombinant baculovirus expressing an insect-selective neurotoxin: potential for pest control, *Bio/Technology,* 9, 848, 1991.
152. **Stewart, L. M. D., Hirst, M., Ferber, M. L., Merryweather, A. T., Cayley, P. J., and Possee, R. D.,** Construction of an improved baculovirus insecticide containing an insect-specific toxin gene, *Nature,* 352, 85, 1991.
153. **Vlak, J. M.,** Engineering of biosafe baculovirus recombinants for insect control, in *Living with Insects, Proceedings, XIX International Congress of Entomology, Abstracts,* Beijing, China, 1992, 135.
154. **O'Reilly, D. R. and Miller, L. K.,** Improvement of a baculovirus pesticide by deletion of the EGT gene, *Bio/Technology,* 9, 1086, 1991.
155. **Kondo, A. and Maeda, S.,** Host range expansion by recombination of the baculoviruses *Bombyx mori* nuclear polyhedrosis and *Autographa californica* nuclear polyhedrosis virus, *J. Virol.,* 65, 3625, 1991.
156. **Mori, H., Nakazawa, H., Shirai, N., Shibata, N., Sumida, M., and Matsubara, F.,** Foreign gene expression by a baculovirus vector with an expanded host range, *J. Gen. Virol.,* 73, 1877, 1992.
157. **Ichinose, R., Kamita, S. G., Maeda, S., and Hammock, B. D.,** Pharmacokinetic studies of the recombinant juvenile hormone esterase in *Manduca sexta, Pestic. Biochem.,* 42, 13, 1992.
158. **Booth, T. F., Bonning, B. C., and Hammock, B. D.,** Localization of juvenile hormone esterase during development in normal and in recombinant baculovirus-infected larvae of the moth, *Trichoplusia ni, Tiss. Cell,* 24, 267, 1992.
159. **Morris, T. D., and Miller, L. K.,** Promoter influence on baculovirus-mediated gene expression in permissive and nonpermissive insect cell lines, *J. Virol.,* 66, 7397, 1992.
160. **Goodman,** unpublished data.

Chapter 3

SPECIFICITY OF BACULOVIRUSES*

Arthur H. McIntosh and James J. Grasela

TABLE OF CONTENTS

I. Introduction 59

II. General Properties 59

III. Nuclear Polyhedrosis Viruses 60
   A. MNPVs 60
   B. SNPVs 64

IV. Granulosis Viruses 64

V. Nonoccluded Baculoviruses 64

VI. Baculovirus Recombinants as Biological Control Agents 64

VII. Conclusion 65

References 65

* This chapter reports the results of research only. Mention of a proprietary product in this paper does not constitute a recommendation for use by the U.S. Department of Agriculture.

Arthur H. McIntosh, Sc.D., is a Research Microbiologist with the United States Department of Agriculture, Agricultural Research Services, and Research Leader of the Biological Control of Insects Research Laboratory (USDA-ARS-BCIRL), Columbia, Missouri. He received his B.A. from McMaster University, Hamilton, Canada (1959); his M.S. in Microbiology from the University of Guelph, Guelph, Canada (1962); and his Sc.D. in Microbiology from Harvard University (1969). He has held research positions at Boyce Thompson Institute (1971–1974), Waksman Institute of Microbiology (1974–1979), and USDA-ARS-BCIRL (1979 to present). He is a member of the Tissue Culture Association, Society of Invertebrate Pathology, Society of Applied Entomology and Zoology, and Sigma Xi. His research involves: biological control of *Helicoverpa-Heliothis* species; production of baculoviruses in lepidopteran cells; *in vitro* host range of baculoviruses.

I. INTRODUCTION

The topic of baculovirus specificity is both significant and pertinent at this juncture in time because of the potential use of baculoviruses as biological control agents aimed at reducing the usage of the more toxic chemical insecticides and for the expression of foreign genes.[1-14] Of particular interest is the generation of recombinants through cloning of toxin-coding genes into baculoviruses to increase their speed of action in the host.[15-17] The purported release of such recombinants into the environment brings up the question of risk assessment. In particular, host-range testing is of considerable importance for determining whether or not genetic manipulation alters virus host range as well as for assessing what effect the recombinant virus has on nontarget hosts. Some information is available on baculovirus host range for both *in vivo* and *in vitro* systems.[18-23] In general, the multiple nuclear polyhedrosis viruses (MNPVs) appear to have the widest host range for both larvae and cell cultures. Two MNPVs, *Autographa californica* (AcMNPV) and *Anagrapha falcifera* (AfMNPV) have been reported to infect members from 12 and 10 families of Lepidoptera, respectively.[24,25] Both of these baculoviruses infect over 30 species of Lepidoptera, of which only a partial list is shown in Table 2. A third baculovirus, *Mamestra brassicae* (MbMNPV), has been demonstrated to be cross-infective for 32 species of Lepidoptera spanning 4 families.[18]

One of the major advantages in employing baculoviruses for biological control of insect pests is their specificity of action, proving harmless to beneficial insects and noninfectious to vertebrates. It is currently accepted that baculoviruses from lepidopterous hosts are noninfectious for members of other orders. However, there have been recent reports on infection of members of Orthoptera and Isoptera by lepidopterous baculoviruses.[26,27] With regard to *in vitro* systems, McIntosh et al.[28] have recently demonstrated the replication of AcMNPV in a Coleopteran cell line from the cotton bollweevil, *Anthonomus grandis*. In view of these findings, further investigations are warranted in this area of "cross" specificity whereby baculoviruses formerly known only to infect members of one order are being reported to infect members of different orders within the Arthropoda.

In this review we will present information on insect hosts, their homologous cell lines, and their susceptibility to some of the more commonly employed baculoviruses. In addition, the host range of genetically engineered baculoviruses will be discussed in light of their potential usage as biocontrol agents for insect pests. For additional information on the subjects of baculovirus specificity, safety, and cell culture systems, the reader is referred to earlier reviews.[20,23,29-37]

II. GENERAL PROPERTIES

More than 600 baculoviruses have been isolated from arthropods, mainly insects,[38] and some have been successfully used for the control of many lepidopteran, hymenopteran, and coleopteran pests.[39] In this review we have not attempted to cover all these isolates, but rather to select some of the more commonly studied baculoviruses. Table 1 illustrates the most recent classification of representative type members of the Baculoviridae.[40] They are characterized by enveloped rod-shaped particles measuring approximately 50 × 250 nm and containing double-stranded circular DNA genomes of 100 to 170 kb. Two genera, the nuclear polyhedrosis viruses (NPVs) and the granulosis viruses (GVs), have been established within the subfamily Eubaculovirinae and the nonoccluded baculoviruses (NOB) within the subfamily Nudibaculovirinae. The NPVs occur in nature as multiple nucleocapsids (MNPVs), in which several nucleocapsids are packaged per envelope or as a single nucleocapsid (SNPV) in which there is only one nucleocapsid per envelope.[40] Many virions produced in the course of infection are occluded into proteinacious occlusion bodies (OB). The GV usually has only one enveloped nucleocapsid occluded per OB. The NOB do not produce OB (Table 1).

TABLE 1
Members of the Family Baculoviridae

Genus	Subgenus	Subfamily	Type/Species	OB
NPV	MNPV	Eubaculovirinae	AcMNPV	+
	SNPV	Eubaculovirinae	BmSNPV	+
GV	—	Eubaculovirinae	PiGV	+
NOB	—	Nudibaculovirinae	HzNOB	−

Note: AcMNPV = *Autographa californica* multiple nucleocapsid nuclear polyhedrosis virus; BmSNPV = *Bombyx mori* single nucleocapsid nuclear polyhedrosis virus; PiGV = *Plodia interpunctella* granulosis virus; HzNOB = *Helicoverpa zea* nonoccluded baculovirus.

Baculoviruses replicate in the nuclei of infected cells and bud from the cytoplasmic membrane as extracellular virus (ECV) particles as well as being incorporated into OB. The OB is the main means of virus spread in the environment. Insect larvae consume plant surfaces contaminated with OB, which are then degraded in the midgut as a result of alkali conditions and proteolytic action. The freed viral particles infect midgut cells and progeny virus as ECVs produce systemic infections. ECV is the form that is commonly used to initiate infections of cell cultures from infectious hemolymph. Both ECV and OB are produced in susceptible insect cell lines. In this review positive (+) susceptibility as indicated in Table 2 refers to the following: the production of OB in cell lines or larvae leading to the destruction of cells or death of larvae, the destruction of cells or death of larvae without OB production, or the detection of baculoviruses in the host cells by immunological means. The term cytopathic effect (CPE) is used to describe these various types of infection, and the reader should consult the specific reference to determine which criterion was used. It is also recognized that cell lines and/or larvae may be semipermissive for some baculovirus infections, but this subject will not be dealt with in this chapter.

III. NUCLEAR POLYHEDROSIS VIRUSES

A. MNPVs

The MNPVs are generally considered to have a wider host range than the SNPVs[41] and other baculoviruses in the family Baculoviridae, as evidenced in both *in vivo* and *in vitro* systems. Exceptions are the restricted host range of SfMNPV, which infects only species within the genus *Spodoptera* (Table 2), and the nonoccluded virus HZ-1, which has a wide *in vitro* host spectrum. Of the MNPVs, AcMNPV has been most widely studied because of its wide host range and its use as a vector for the expression of foreign genes. However, the AfMNPV that has been recently patented by the USDA is receiving recognition because of its wide host range as well as its ability to infect the major crop pests *Helicoverpa zea* and *Heliothis virescens*.[25] AcMNPV produces an aberrant infection in *H. zea* larvae, requiring high doses of OB to produce mortalities.[50] Since AfMNPV can be grown in a number of insect cell lines,[68] it should be amenable to genetic manipulation in much the same manner as has been done for AcMNPV. Inspection of Table 2 shows that *in vivo* results of successful infections with MNPVs closely correlate with *in vitro* infections of corresponding homologous cell lines. It has been shown that some cell lines from a particular host support virus replication, whereas others are refractile to the baculovirus in question.[67,71] Thus it is necessary to test more than one cell line from a particular host before concluding that the *in vitro* system does not support baculovirus replication.

TABLE 2
In Vivo and *In Vitro* Host Range of Selected Baculoviruses

Baculovirus	Insect host	Susceptibility spectrum: Larval host	Susceptibility spectrum: Cell line	References
AcMNPV	*Trichoplusia ni*	+	+	21,42,45,46
(Autographa californica)	*Spodoptera frugiperda*	+	+	21,22,44,47
	Plutella xylostella	+	+	48,49
	Helicoverpa zea	+	–	21,43,50
	Helicoverpa armigera	+	–	21,34
	Helicoverpa virescens	+	+	21,48,50
	Heliothis subflexa	+	+	25,51
	Anticarsia gemmatalis	U	+	48,52
	Bombyx mori	U	±	20,53,54
	Spodoptera exigua	+	+	42,43,55,56
	Spodoptera littoralis	–	+	54,57,58
	Spodoptera litura	+	U	54
	Estigmene acrea	+	+	43,44,59
	Mamestra brassicae	+	+	42,44,60–64
	Lymantria dispar	U	+	44,65,66
	Malacosoma disstria	U	+	20
	Manduca sexta	U	+	66,67
	Choristoneura fumiferana	U	±	67
	Anthonomus grandis	–	+	28
AfMNPV	*Anagrapha falcifera*	+	O	25
(Anagrapha falcifera)	*Trichoplusia ni*	+	+	25,68
	Spodoptera frugiperda	+	+	25,68
	Spodoptera ornithogalli	+	NT	25
	Helicoverpa zea	+	–	25
	Heliothis virescens	+	+	25,68
	Heliothis subflexa	+	+	25,68
	Helicoverpa armigera	NT	–	25
	Plutella xylostella	+	+	25,68
	Anticarsia gemmatalis	+	+	25,68
	Pieris rapae	+	NT	25
	Plodia interpunctella	+	NT	25
	Cydia pomonella	+	NT	25
	Agrotis ipsilon	+	O	25
	Manduca sexta	+	NT	25
	Amyelois transitella	+	NT	25
	Diatraea grandiosella	+	O	25
	Ostrinia nubilalis	+	O	25
MbMNPV	*Mamestra brassicae*	+	±	18,42,69–71
(Mamestra brassicae)	*Trichoplusia ni*	+	U	42
	Plutella xylostella	±	U	18,34
	Spodoptera exigua	+	U	42
	Spodoptera littoralis	±	U	18,42
	Helicoverpa armigera	+	U	18
	Galleria mellonella	–	O	34
LdMNPV	*Lymantria dispar*	+	+	65,72
(Lymantria dispar)	*Estigmene acrea*	+	NT	73
	Trichoplusia ni	U	–	36
	Spodoptera frugiperda	U	–	36
	Spodoptera littoralis	U	NT	36
	Manduca sexta	U	–	36
	Helicoverpa zea	U	–	36
	Bombyx mori	U	U	
SfMNPV	*Spodoptera frugiperda*	+	+	21,22,74,75
(Spodoptera frugiperda)	*Trichoplusia ni*	–	–	21,22

TABLE 2 (continued)
***In Vivo* and *In Vitro* Host Range of Selected Baculoviruses**

Baculovirus	Insect host	Susceptibility spectrum: Larval host	Susceptibility spectrum: Cell line	References
	Helicoverpa zea	U	–	21
	Helicoverpa armigera	U	–	21
	Heliothis virescens	U	–	21
	Spodoptera littoralis	+	U	76
AgMNPV	*Anticarsia gemmatalis*	+	+	52,77
(Anticarsia gemmatalis)	*Trichoplusia ni*	+	+	78,79
	Spodoptera frugiperda	–	+	78–80
	Helicoverpa zea	+	+	78,79
	Heliothis virescens	+	+	78,81
	Plodia interpunctella	U	+	79
	Spodoptera ornithogalli	+	NT	78
	Plathypena scabra	–	U	78
	Pseudoplusia includens	+	U	78
TnMNPV	*Trichoplusia ni*	+	+	21,82,83
(Trichoplusia ni)	*Spodoptera frugiperda*	U	+	21,36,84
	Heliothis virescens	U	+	21
	Helicoverpa zea	U	–	21
	Helicoverpa armigera	U	–	21
	Manduca sexta	U	+	36
GmNPV	*Trichoplusia ni*	+	+	21,85,86
(Galleria mellonella)	*Spodoptera frugiperda*	+	+	21
	Heliothis virescens	U	+	21
	Helicoverpa zea	+	–	21,85
	Helicoverpa armigera	U	–	21
	Manduca sexta	+	U	85
	Galleria mellonella	+	O	85,86
CfMNPV	*Choristoneura fumiferana*	+	+	87,88
(Choristoneura fumiferana)	*Malacosoma disstria*	U	+	87
	Estigmene acrea	+	U	73
	Orgyia leucostigma	U	–	67
OpMNPV	*Orgyia leucostigma*	U	+	89
(Orgyia leucostigma)	*Estigmene acrea*	+	U	73
SeMNPV	*Spodoptera exigua*	+	+	55,90
(Spodoptera exigua)	*Heliothis virescens*	–	U	90
	Trichoplusia ni	–	U	90
	Spodoptera frugiperda	–	+	55,90
	Spodoptera littoralis	–	U	91
	Mamestra brassicae	–	U	91
	Plutella xylostella	–	U	91
	Galleria mellonella	–	O	91
SlMNPV	*Spodoptera frugiperda*	+	+	54,76,92
(Spodoptera littoralis)	*Spodoptera littoralis*	+	+	54,57,92
	Helicoverpa armigera	–	U	93
	Spodoptera litura	+	U	93
	Spodoptera exigua	+	U	93
	Trichoplusia ni	U	–	54
	Bombyx mori	U	–	54
SuMNPV	*Spodoptera litura*	+	O	94
(Spodoptera litura)	*Spodoptera littoralis*	U	+	54
	Spodoptera frugiperda	U	+	54,95
	Trichoplusia ni	U	–	54
	Bombyx mori	U	–	54
HzSNPV	*Helicoverpa zea*	+	+	21,96–98
(Helicoverpa zea)	*Helicoverpa armigera*	+	–	21,99

TABLE 2 (continued)
In Vivo and *In Vitro* Host Range of Selected Baculoviruses

Baculovirus	Insect host	Susceptibility spectrum: Larval host	Susceptibility spectrum: Cell line	References
	Heliothis virescens	+	+	21,100
	Heliothis subflexa	+	–	51,101
	Spodoptera frugiperda	U	–	21
	Trichoplusia ni	U	–	21
	Heliothis paradoxa	+	O	102
	Heliothis peltigera	+	O	102
	Heliothis phloxiphaga	+	O	102
	Helicoverpa punctigera	+	O	99
HaSNPV	*Helicoverpa zea*	+	U	103
(Helicoverpa armigera)	*Helicoverpa armigera*	+	+	104–106
TnSNPV	*Trichoplusia ni*	+	+	107,108,41
(Trichoplusia ni)	*Rachiplusia ni*	–	O	41
	Pseudoplusia includens	–	O	41
	Autographa californica	+	O	41
OlSNPV	*Orgyia leucostigma*	U	+	109
(Orgyia leucostigma)				
BmSNPV	*Bombyx mori*	+	+	53,110
(Bombyx mori)	*Trichoplusia ni*	U	–	53
	Spodoptera frugiperda	U	–	53
	Spodoptera litura	–	O	53
	Spodoptera littoralis	–	U	111
	Pieris rapae	–	NT	111
	Lymantria dispar japonica	+	U	111
CpGV	*Cydia pomonella*	+	+	112,113
(Cydia pomonella)	*Helicoverpa zea*	–	U	113
	Mamestra brassicae	–	U	114
	Plutella xylostella	–	NT	114
	Amyelois transitella	–	U	113
	Antheraea pernyi	–	U	114
TnGV	*Trichoplusia ni*	+	+	108
(Trichoplusia ni)				
HaGV	*Helicoverpa armigera*	+	–	115,106
(Helicoverpa armigera)	*Helicoverpa zea*	+	–	115,116
	Heliothis virescens	+	NT	115
	Spodoptera frugiperda	+	U	116
	Spodoptera exigua	+	U	116
	Trichoplusia ni	+	U	116
	Helicoverpa zea	–	+	117,119
	Trichoplusia ni	U	+	117–119
	Spodoptera frugiperda	U	+	117–119
	Heliothis virescens	U	+	119
	Helicoverpa armigera	U	+	119
	Heliothis subflexa	U	+	119
Hz-1	*Anticarsia gemmatalis*	U	+	119
(nonoccluded)	*Plutella xylostella*	U	+	119
(Helicoverpa zea)	*Lymantria dispar*	U	+	117
	Mamestra brassicae	U	+	118
Oryctes	*Heteronychus arator*	+	+	120,123
(nonoccluded)	*Oryctes rhinoceros*	+	+	120–122
(Oryctes rhinoceros)	*Spodoptera frugiperda*	U	–	120

Note: + = susceptible; – = nonsusceptible; ± = reports of both susceptible and nonsusceptible species and/or cell lines; O = cell line not available; U = unknown; NT = not tested.

B. SNPVs

There are fewer reports of successful replication of SNPVs in cell cultures as compared with MNPVs (Table 2). HzSNPV is the most widely studied and has been shown to replicate in both *H. zea* and *H. virescens* cell lines, but failed to replicate in an *H. armigera* cell line. As more cell lines from *H. armigera* become available, it will be possible to determine if all cell lines from *H. armigera* are refractile, since *H. armigera* and *H. zea* are considered to be closely related species. The former occurs in the Old World and the latter in the New World. *H. punctigera*, a major pest of cotton and other crops in Australia, is susceptible to HzSNPV,[99] but there have been no reports of cell lines from this insect.

Granados et al.[108] reported on the establishment of 36 new *T. ni* cell lines from embryonic tissues of *T. ni* and demonstrated that 29 of these lines supported replication of TnSNPV.

Undoubtedly, further cell culture work on SNPVs will be largely dependent on the discovery and need for SNPVs. Because of the ease of cultivation and the wider host range of MNPVs, they are the preferred choice for cell culture studies. Continued research is advocated on SNPVs so that further knowledge can be acquired regarding the relationship between SNPVs and MNPVs. In this report, the baculovirus from *B. mori* is considered as an SNPV as reported by Raghow and Grace[110] and Maeda[124] and is the type species of the SNPVs.[40] Maeda et al.[54] have demonstrated a high degree of homology between BmSNPV and AcMNPV.

IV. GRANULOSIS VIRUSES

GVs tend to be more specific than the NPVs, and only 6 of 52 attempts to cross transmit were successful.[30] *In vitro* research on GVs has been restricted in the past because of failure to propagate them in cell lines. However, recent successes of the codling moth and the cabbage looper[108,112] GVs in cell lines should stimulate further research on GVs. In both cases susceptible cell lines were derived from embryonic tissue.

V. NONOCCLUDED BACULOVIRUSES

Nonoccluded baculoviruses have been reported to occur in a number of arthropods, including insects, mites, and crustacea[125] and do not produce OB in their host as previously mentioned. *Oryctes rhinoceros* NOB is well known for its successful control of the rhinoceros beetle in palm-growing areas[121] and replicates in an *H. arator* cell line but not in a *Spodoptera frugiperda* cell line.[120] A second NOB that has been successfully cultured in cell lines is the HZ-1 virus, which persistently infects the IMC-HZ-1 cell line of *H. zea*.[117] The HZ-1 appears to have the widest *in vitro* host range of baculoviruses, replicating in 10 cell lines including some lines that failed to support replication of several MNPVs (Table 2). Intrahemocoelic injections and virus administered per os failed to cause infection of *H. zea* larvae.[119]

VI. BACULOVIRUS RECOMBINANTS AS BIOLOGICAL CONTROL AGENTS

Baculovirus infections are restricted to the Arthropoda and are innocuous for vertebrates and plants. This high degree of specificity together with their pathogenic properties have made them the agent of choice for the biological control of insect pests. Some of the factors limiting the widespread use of baculoviruses as biological control agents include cost of production, their high specificity, lack of persistence in the environment (sensitivity to ultraviolet radiation), and their slow rate of kill as compared with the more toxic chemical insecticides. Toxin-encoding genes[16,17] and the juvenile hormone esterase[15] have been engineered into AcMNPV to produce recombinants that are faster acting or that disrupt the feeding pattern of the insect pests. Because of the extended host range of these recombinants, there is much concern for possible adverse effects on arthropod biodiversity.[126,127] It should be realized in actuality that it is the toxin that is expressed by the recombinant that results in the wider host range; the baculovirus is just acting as a vector for the expression of the toxin gene. In contrast to this approach, Kondo and Maeda[53] have produced genetic recombinants with wider host ranges by coinfection of cell lines with AcMNPV and BmSNPV. There is no doubt that there are both benefits and risks in employing recombinant baculoviruses for biological control; these two factors have to be properly assessed. This can best be accomplished by extensive investigations into the effect of these agents on nontarget organisms, including beneficials in the environment.

VII. CONCLUSION

Information on the specificity and safety of baculoviruses has been amassed over the years. Baculoviruses infect only members of the Arthropoda, and no deleterious effects on vertebrates have been reported. However, with possible modifications of the host range through genetic engineering of toxin-coding genes, there is need for additional research to assure the safe use of these agents in the environment.

As Table 2 illustrates, there is far more information on the *in vivo* specificity of baculoviruses than *in vitro* studies. With the ever-increasing establishment of cell lines, it is anticipated that more information will become available. Cell culture is a useful system for assessing specificity of baculoviruses and shows a high degree of correlation with *in vivo* tests. In both systems, it is of paramount importance to establish the identity of the viral isolate both before and following infection, to properly identify cell line or host, and to use the appropriate sensitive tests[20] for quantification.

REFERENCES

1. **Granados, R. R.**, Replication phenomena of insect viruses *in vivo* and *in vitro*, in *Safety Aspects of Baculoviruses as Biological Pesticides*, Miltenburger, H. G., Ed., Bundesministerium für Forschung und Technologie, Bonn, 1978, 163.
2. **Tinsley, T. W.**, Viruses and the biological control of insect pests, *BioScience*, 27, 659, 1977.
3. **Kawanishi, C. Y.**, Review of safety tests and methods for evaluating infectivity, in *Viral Pesticides: Present Knowledge and Potential Effects on Public and Environmental Health*, Summers, M. D. and Kawanishi, C. Y., Eds., U.S. Environmental Protection Agency, Research Triangle Park, NC, 1978, 199.
4. **Luckow, V. A. and Summers, M. D.**, Trends in the development of baculovirus expression vectors, *Bio/Technology*, 6, 47, 1988.
5. **Smith, G. E., Summers, M. D., and Fraser, M. J.**, Production of human β interferon in insect cells infected with a baculovirus expression vector, *Mol. Cell. Biol.*, 3, 2156, 1983.
6. **Pennock, G. D., Shoemaker, C., and Miller, L. K.**, Strong and regulated expression of *Escherichia coli* β-galactosidase in insect cells with a baculovirus vector, *Mol. Cell. Biol.*, 4, 399, 1984.
7. **Bishop, D. H. L.**, *Planned Release of Genetically Engineered Organisms, (TIBTECH/TREE Special Publication)*, Hodgson, J. and Sugden, A. M., Eds., Elsevier, 1988, S12.
8. **Miller, L. K.**, Baculoviruses as gene expression vectors, *Annu. Rev. Microbiol.*, 42, 177, 1988.
9. **Miller, L. K.**, Baculoviruses: high-level expression in insect cells, *Curr. Opinion Genet. Dev.*, 3, 97, 1993.
10. **Maeda, S.**, Expression of foreign genes in insects using baculovirus vectors, *Annu. Rev. Entomol.*, 34, 351, 1989.
11. **Fraser, M.**, Expression of eucaryoteic genes in insect cell cultures, *In Vitro Cell. Dev. Biol.*, 225, 1989.
12. **Bishop, D. H. L.**, Baculovirus expression vectors, *Semin. Virol.*, 3, 253, 1992.
13. **O'Reilly, D. R., Miller, L. K., and Luckow, V. A.**, *Baculovirus Expression Vectors: A Laboratory Manual*, W. H. Freeman and Company, New York, 1992.
14. **Possee, R. D. and King, L. A.**, *The Baculovirus Expression System: A Laboratory Guide*, Chapman and Hall, 1992.
15. **Hammock, B. D., Bonning, B. C., Possee, R. D., Hanzlik, T. N., and Maeda, S.**, Expression and effects of the juvenile hormone esterase in a baculovirus vector, *Nature*, 344, 458, 1990.

16. **Tomalski, M. D. and Miller, L. K.,** Insect paralysis by baculovirus-mediated expression of a mite neurotoxin gene, *Nature,* 352, 82, 1991.
17. **Stewart, L. M. D., Hirst, M., Ferber, M. L., Merryweather, A. T., Cayley, P. J., and Possee, R. D.,** Construction of an improved baculovirus insecticide containing an insect-specific toxin gene, *Nature,* 352, 85, 1991.
18. **Doyle, J. D., Hirst, M. L., Cory, J. S., and Entwistle, P. F.,** Risk assessment studies: detailed host range testing of wild-type cabbage moth *Mamestra brassicae* (Lepidoptera: Noctuidae) nuclear polyhedrosis virus, *Appl. Environ. Entomol.,* 56, 2704, 1990.
19. **Belloncik, S., Lavallee, C., and Hamelin, C.,** Relative pathogenicity of nuclear polyhedrosis viruses from *Mamestra brassicae, Euxoa messoria, Agrotis segetum, Autographa californica,* and *Heliothis* spp. for larvae of *Euxoa scandens* (Lepidoptera: Noctuidae), *J. Invert. Pathol.,* 47, 8, 1986.
20. **Knudson, D. L.,** Specificity of baculoviruses, in *Safety Aspects of Baculoviruses as Biological Insecticides: Symposium 13 — 15 November 1978,* Miltenburger, H. G., Ed., Bundesministerum für Forschung und Technologie, Bonn, 241, 1980.
21. **McIntosh, A. H., Ignoffo, C. M., and Andrews, P. L.,** *In vitro* host range of five baculoviruses in lepidopteran cell lines, *Intervirology,* 23, 150, 1985.
22. **Danyluk, G. M. and Maruniak, J. E.,** *In vivo* and *in vitro* host range of *Autographa californica* nuclear polyhedrosis virus and *Spodoptera frugiperda* nuclear polyhedrosis virus, *J. Invert. Pathol.,* 50, 207, 1987.
23. **Granados, R. R. and Hashimoto, Y.,** Infectivity of baculoviruses to cultured cells, in *Invertebrate Cell System Applications, Vol. II,* Mitsuhashi, J., Ed., CRC Press, Boca Raton, FL, 1989, 3.
24. **Granados, R. R. and Williams, K. A.,** *In vivo* infection and replication of baculoviruses, in *The Biology of Baculoviruses,* Vol. 1., Granados, R. R. and Federici, B. A., Eds., CRC Press, Boca Raton, FL, 1986, 177.
25. **Hostetter, D. L. and Puttler, B.,** A new broad host spectrum nuclear polyhedrosis virus isolated from a celery looper, *Anagrapha falcifera* (Kirby), (Lepidoptera: Noctuidae), *Environ. Entomol.,* 20, 1480, 1991.
26. **Al Fazairy, A. A. and Hassan, F. A.,** Infection of termites by *Spodoptera littoralis* nuclear polyhedrosis virus, *Insect Sci. Appl.,* 9, 37, 1988.
27. **Bensimon, A., Zinger, S., Gerassi, E., Hauschner, A., Harpaz, I., and Sela, I.,** Dark cheeks, a lethal disease of locusts provoked by a lepidopterous baculovirus, *J. Invert. Pathol.,* 50, 254, 1987.
28. **McIntosh, A. H., Barcenas, N., and Cate, J. R.,** Replication of *Autographa californica* baculovirus (AcMNPV) in a coleopteran cell line, *In Vitro Cell. Dev. Biol.,* 28A, 557, 1992.
29. **Summers, M., Engler, R., Falcon, L. A., and Vail, P. V.,** Eds., *Baculoviruses for Insect Pest Control: Safety Considerations,* American Society for Microbiology, Washington, D.C., 1975.
30. **Ignoffo, C. M.,** Specificity of insect viruses, *Bull. Entomol. Soc. Am.,* 14, 265, 1968.
31. **Ignoffo, C. M.,** Effects of entomopathogens on vertebrates, *Ann. N.Y. Acad. Sci.,* 217, 141, 1973.
32. **Krieg, A.,** Granulose- und Kernpolyeder-Viren: Hygienische Gesichtspunkte bei ihrer Produktion und Anwendung, *Z. Angew. Entomol.,* 82, 129, 1976.
33. **Tinsley, T. W.,** The potential of insect pathogenic viruses as pesticidal agents, *Annu. Rev. Entomol.,* 24, 63, 1979.
34. **Gröner, A.,** Specificity and safety of baculoviruses, in *The Biology of Baculoviruses, Vol. 1.,* Granados, R. R. and Federici, B. A., Eds., CRC Press, Boca Raton, FL, 1986, 177.
35. **Volkman, L. E. and Knudson, D. L.,** *In vitro* replication of baculoviruses, in *The Biology of Baculoviruses,* Vol. 1, Granados, R. R. and Federici, B. A., Eds., CRC Press, Boca Raton, FL, 1986, 109.
36. **Knudson, D. L. and Buckley, S. M.,** Invertebrate cell culture methods for the study of invertebrate associated animal viruses, in *Methods in Virology,* 6, Maramorosch, K. and Koaprowski, H., Eds., Academic Press, New York, 1977, 323.
37. **Vaughn, J. L.,** Insect cells for insect virus production, *Adv. Cell Cult.,* 1, 281, 1981.
38. **Martignoni, M. E. and Iwai, P. J.,** A catalog of viral diseases of insects, mites, and ticks, *U.S. Dept. Agric. for Serv. Publ. PNW-195,* Washington, D.C., 1986.
39. **Entwistle, P. F. and H. F. Evans,** Viral control, in *Comprehensive Insect Physiology, Biochemistry and Pharmacology,* Kerkut, G. and Gilbert, L. I., Eds., Pergamon Press, Elmsford, NY, 1985.
40. **Francki, R. I. B., Fauquet, C. M., Knudson, D. L., and Brown, F.,** Classification and nomenclature of viruses, *Fifth Report of the International Committee on Taxonomy of Viruses, Arch. Virol.* 2, (Supp.), 1991.
41. **Harper, J. D.,** Cross-infectivity of six plusiine nuclear polyhedrosis virus isolates in plusiine hosts, *J. Invert. Pathol.,* 27, 275, 1976.
42. **Burgerjon, A., Biache, G., and Chaufaux, J.,** Recherches sur la specificité de trois virus et polyedres nucleairs vis-a-vis de *Mamestra brassicae, Scotia segetum, Trichoplusia ni* et *Spodoptera exigua, Entomophaga,* 20, 153, 1975.
43. **Vail, P. V. and Jay, D. L.,** Pathology of nuclear polyhedrosis virus on the alfalfa looper in alternate hosts, *J. Invert. Pathol.,* 21, 198, 1973.
44. **Lynn, D. E. and Hink, W. F.,** Comparison of nuclear polyhedrosis virus replication in five lepidopteran cell lines, *J. Invert. Pathol.,* 35, 234, 1980.

45. **Hink, W. F. and Vail, P. V.**, A plaque assay for titration of alfalfa looper nuclear polyhedrosis virus in a cabbage looper (TN-368) cell line, *J. Invert. Pathol.*, 22, 168, 1973.
46. **Rockford, R., Dougherty, E. M., and Lynn, D. E.**, Establishment of a cell line from embryos of the cabbage looper, *Trichoplusia ni*, (Hübner), *In Vitro*, 20, 823, 1984.
47. **Vaughn, J. L., Goodwin, R. H., Tompkins, G. J., and McCawley, P.**, Establishment of two cell lines from the insect *Spodoptera frugiperda* (Lepidoptera: Noctuidae), *In Vitro*, 13, 213, 1977.
48. **McIntosh, A. H. and Ignoffo, C. M.**, Replication of *Autographa californica* nuclear polyhedrosis virus in five lepidopteran cell lines, *J. Invert. Pathol.*, 54, 97, 1989.
49. **Vail, P. V., Hunter, D. K., and Jay, D. L.**, Nuclear polyhedrosis of the diamond back moth, *Plutella xylostella*, *J. Invert. Pathol.*, 20, 216, 1972.
50. **Vail, P. V., Jay, D. L., Stewart, F. D., Martinez, A. J., and Dulmage, H. T.**, Comparative susceptibility of *Heliothis virescens and H. zea* to the nuclear polyhedrosis virus isolated from *Autographa californica*, *J. Econ. Entomol.*, 71, 293, 1978.
51. **McIntosh, A. H.**, unpublished data, 1987.
52. **Sieburth, P. J. and Maruniak, J. E.**, Growth characteristics of a continuous cell line from the velvetbean caterpillar, *Anticarsia gemmatalis*, Hübner (Lepidoptera: Noctuidae), *In Vitro Cell. Dev. Biol.*, 24, 195, 1988.
53. **Kondo, A. and Maeda, S.**, Host range expansion by recombination of the baculoviruses *Bombyx mori* nuclear polyhedrosis virus and *Autographa californica* nuclear polyhedrosis virus, *J. Virol.*, 65, 3625, 1991.
54. **Maeda, S., Mukohara, J., and Kondo, A.**, Characteristically distinct isolates of the nuclear polyhedrosis virus from *Spodoptera litura*, *J. Gen. Virol.*, 71, 2631, 1990.
55. **Gelernter, W. D. and Federici, B. A.**, Continuous cell line from *Spodoptera exigua* (Lepidoptera: Noctuidae) that supports replication of nuclear polyhedrosis viruses from *Spodoptera exigua* and *Autograph californica*, *J. Invert. Pathol.*, 48, 199, 1986.
56. **Smits, P. H. and Vlak, J. M.**, Selection of nuclear polyhedrosis viruses as biological control agents of *Spodoptera exigua* (Lep.: Noctuidae), *Entomophaga*, 33, 299, 1988.
57. **Knudson, D. L., Lescott, T., and Tinsley, T. W.**, New cell line, establishment of a continuous cell line of *Spodoptera littoralis* (Lepidoptera: Noctuidae), *In Vitro*, 16, 369, 1980.
58. **Mialhe, E., Quiot, J. M., and Paradis, S.**, Etablissement de deux lignées cellulaires de *Spodoptera littoralis* (Lep.: Noctuidae), Permissives pour des virus suseptibles d'etre utilises en lutte microbiologique, *Entomophaga*, 29, 347, 1984.
59. **Granados, R. R. and Naughton, M.**, Replication of the *Amscata moorei* entomopoxvirus and *Autographa californica* nuclear polyhedrosis virus in hemocyte cell lines from *Estigmene acrea*, in *Invertebrate Tissue Culture: Applications in Medicine, Biology and Agriculture*, Kurstak, E. and Maramorosch, K., Eds., Academic Press, New York, 1976, 379.
60. **Shapiro, M.**, *In vivo* mass production of insect viruses, in *Microbial and Viral Pesticides*, Kurstak, E., Ed., Marcel Dekker, New York, 1982, 463.
61. **Miltenburger, H. G., David, P., Mahr, U., and Zipp, W.**, Über die Erstellung von Lepidoptera-Dauerzellinien and die *in vitro*-Replikation von insektenpathogenen Viren. I. *Mamestra brassicae* L. (Kohleule)-Zellinien und NPV-Replikation, *Z. Angew. Entomol.*, 82, 306, 1977.
62. **Inoue, H. and Mitsuhashi, J.**, Further establishment of continuous cell lines from larval fat bodies of cabbage armyworm *Mamestra brassicae* (Lepidoptera: Noctuidae), *Appl. Entomol. Zool.*, 20, 496, 1985.
63. **Miltenburger, H. G. and David, P.**, Nuclear polyhedrosis virus replication in permanent cell lines of cabbage moth *(Mamestra brassicae L.)*, *Naturwissenschaften*, 63, 197, 1976.
64. **Mitsuhashi, J.**, Establishment and some characteristics of a continuous cell line derived from fat bodies of the cabbage armyworm (Lepidoptera: Noctuidae), *Dev. Growth Differ.*, 23, 63, 1981.
65. **Goodwin, R. H., Tompkins, G. J., and McCawley, P.**, Gypsy moth cell lines divergent in viral susceptibility. 1. Culture and identification, *In Vitro*, 14, 485, 1978.
66. **Volkman, L. E. and Goldsmith, P. A.**, Generalized immunoassay for *Autographa californica* nuclear polyhedrosis virus infectivity *in vitro*, *Appl. Environ. Microbiol.*, 44, 227, 1982.
67. **Hink, W. F. and Hall, R. L.**, Recently established invertebrate cell lines, in *Invertebrate Cell System Applications*, Mitsuhashi, J., Ed., CRC Press, Boca Raton, FL, 1989, 11, 269.
68. **McIntosh, A. H.**, *In vitro* infectivity at a cloned isolate of *Anagrapha (Syngrapha) falcifera* (celery looper) multiple nuclear polyhedrosis virus, *J. Invert. Pathol.*, 57, 441, 1991.
69. **Vlak, J. M. and Gröner, A.**, Identification of two nuclear polyhedrosis viruses from the cabbage moth, *Mamestra brassicae*, (Lepidoptera: Noctuidae), *J. Invert. Pathol.*, 35, 269, 1980.
70. **Evans, H. F., Lomer, C. J., and Kelly, D. C.**, Growth of nuclear polyhedrosis virus in larvae of the cabbage moth, *Mamestra brassicae* L., *Arch. Virol.*, 70, 207, 1981.
71. **Mitsuhashi, J. and Shozawa, A.**, Continuous cell lines from larval hemocytes of the cabbage armyworm, *Mamestra brassicae*, *Dev. Growth Differ.*, 27, 599, 1985.
72. **Magnoler, A.**, Susceptibility of gypsy moth larvae to *Lymantria* spp. nuclear- and cytoplasmic polyhedrosis viruses, *Entomophaga*, 15, 407, 1970.

73. **Shapiro, M., Martignoni, M. E., Cunningham, J. C., and Goodwin, R. H.,** Potential use of the saltmarsh caterpillar as a production host for nucleopolyhedrosis viruses, *J. Econ. Entomol.*, 75, 69, 1982.
74. **Goodwin, R. H., Vaughn, J. L., Adams, J. R., and Louloudes, S. J.,** Replication of a nuclear polyhedrosis virus in an established insect cell line, *J. Invert. Pathol.*, 16, 284, 1970.
75. **McIntosh, A. H. and Ignoffo, C. M.,** Establishment of a persistent baculovirus infection in a lepidopteran cell line, *J. Invert. Pathol.*, 38, 395, 1981.
76. **McKinley, D. L., Brown, D. A., Payne, C. C., and Harrap, K. A.,** Cross-infectivity and activation studies with four baculoviruses, *Entomophage*, 26, 79, 1981.
77. **Allen, G. E. and Knell, J. D.,** A nuclear polyhedrosis virus of *Anticarsia gemmatalis*. I. Ultrastructure, replication, and pathogenicity, *Fla. Entomol.*, 60, 233, 1977.
78. **Carner, G. R., Hudson, J. S., and Barnett, O. W.,** The infectivity of a nuclear polyhedrosis virus of the velvetbean caterpillar for eight noctuid hosts, *J. Invert. Pathol.*, 33, 211, 1979.
79. **Maruniak, J. E., Johnson, D., and Gowan, D.,** AgNPV replication in cell cultures and physical map of its DNA, *J. Cell. Biochem.*, 10c, 49, 1986.
80. **Pavan, O. H., Boucias, D. G., and Pendland, J. C.,** The effects of serial passage of a nucleopolyhedrosis virus through an alternate host system., *Entomophaga*, 26, 99, 1981.
81. **Grasela, J. J. and McIntosh A. H.,** unpublished data, 1992.
82. **Tompkins, G. J., Adams, J. R., and Heimpel, A. M.,** Cross infection studies with *Heliothis zea* using nuclear polyhedrosis viruses from *Trichoplusia ni*, *J. Invert. Pathol.*, 14, 342, 1969.
83. **Faulkner, P. and Henderson, J. F.,** Serial passage of a nuclear polyhedrosis virus of the cabbage looper *Trichoplusia ni* in a continuous tissue culture cell line, *Virology*, 50, 920, 1972.
84. **Goodwin, R. H., Vaughn, J. L., Adam, J. R., and Louloudes, S. J.,** The influence of insect cell lines and tissue-culture media on baculovirus polyhedra production, *Misc. Publ. Entomol. Soc. Am.*, 9, 66, 1973.
85. **Fraser, M. J. and Stairs, G. R.,** Susceptibility of *Trichoplusia ni, Heliothis zea* (Noctuidae), and *Manduca sexta* (Sphingidae) to a nuclear polyhedrosis virus from *Galleria mellonella* (Pyralidae), *J. Invert. Pathol.*, 40, 255, 1982.
86. **Witt, D. L. and Janus, C. A.,** Replication of *Galleria mellonella* nuclear polyhedrosis virus in cultured cells and in larvae of *Trichoplusia ni*, *J. Invert. Pathol.*, 29, 222, 1977.
87. **Sohi, S. S. and Bird, F. T.,** Replication of a nuclear polyhedrosis virus of *Choristoneura fumiferana* (Lepidoptera: Tortricidae) in *Malacosoma distria* (Lepidoptera: Lasiocampidae) hemocyte cultures, in *Invertebrate Tissue Culture, Applications in Medicine, Biology, and Agriculture*, Kurstak, E. and Maramorosch, K., Eds., Academic Press, New York, 1977, 361.
88. **Bilimoria, S. L. and Sohi, S. S.,** Development of an attached strain from a continuous insect cell line, *In Vitro*, 13, 466, 1977.
89. **Sohi, S. S., Percy, J., Cunningham, J. C., and Arif, B. M.,** Replication and serial passage of a multicapsid nuclear polyhedrosis virus of *Orgyia pseudotsugata* (Lepidoptera: Lymantridae) in continuous insect cell lines, *Can. J. Microbiol.*, 27, 1133, 1981.
90. **Gelernter, W. D. and Federici, B. A.,** Isolation, identification, and determination of a nuclear polyhedrosis virus from the beet armyworm, *Spodoptera exigua* (Lepidoptera: Noctuidae), *Environ. Entomol.*, 15, 240, 1986.
91. **Vlak, J. M., van Frankenhuyzen, K., Peters, D., and Gröner, A.,** Identification of a new nuclear polyhedrosis virus from *Spodoptera exigua.*, *J. Invert. Pathol.*, 38, 297, 1981.
92. **Kislev, N. and Edelman, M.,** DNA restriction pattern differences from geographic isolates of *Spodoptera littoralis* nuclear polyhedrosis virus, *Virology*, 119, 219, 1982.
93. **Harpaz, I. and Raccah, B.,** Nucleopolyhedrosis virus (NPV) of the Egyptian cotton worm, *Spodoptera littoralis* (Lepidoptera: Noctuidae): temperature and pH relations, host range, and synergism, *J. Invert. Pathol.*, 32, 368, 1978.
94. **Okada, T.,** Studies on the utilization and mass production of *Spodoptera litura* nuclear polyhedrosis virus for control of the tobacco cutworm, *Spodoptera litura* Fabricius, *Rev. Plant Prot. Res.*, 10, 102, 1977.
95. **Tsuda, K., Mizuki, E., Kawarabata, T., and Aizawa, K.,** Comparative neutralization of nuclear polyhedrosis viruses infectious to the tobacco cutworm, *Spodoptera litura* (Lepidoptera: Noctuidae) in a cell line of *Spodoptera frugiperda*, *J. Invert. Pathol.*, 52, 248, 1988.
96. **Ignoffo, C. M.,** The nuclear-polyhedrosis virus of *Heliothis zea* (Boddie) and *Heliothis virescens* (Fabricius). IV. Bioassay of virus activity, *J. Invert. Pathol.*, 7, 315, 1965.
97. **Ignoffo, C. M., Shapiro, M., and Hink, W. F.,** Replication and serial passage of infectious *Heliothis* nucleopolyhedrosis virus in an established line of *Heliothis zea* cells, *J. Invert. Pathol.*, 18, 131, 1971.
98. **McIntosh, A. H. and Ignoffo, C. M.,** Replication and infectivity of the single-embedded nuclear polyhedrosis virus, baculovirus *Heliothis*, in homologous cell lines, *J. Invert. Pathol.*, 37, 258, 1981.
99. **Teakle, R. E.,** Relative pathogenicity of nuclear polyhedrosis viruses from *Heliothis punctigera* and *Heliothis zea* for larvae of *Heliothis armigera* and *Heliothis punctigera*, *J. Invert. Pathol.*, 34, 231, 1979.

100. **Ignoffo, C. M.,** Susceptibility of the first-instar of the bollworm, *Heliothis zea,* and the tobacco budworm, *Heliothis virescens,* to *Heliothis* nuclear polyhedrosis virus, *J. Invert. Pathol.,* 8, 531, 1966.
101. **Ignoffo, C. M., McIntosh, A. H., Garcia, C., Huettel, M. D., and Hill, A. J.,** Relative resistance of *Heliothis subflexa* to a single-embedded nucleopolyhedrosis virus of *Heliothis* species, *J. Invert. Pathol.,* 42, 282, 1983.
102. **Ignoffo, C. M. and Couch, T. L.,** The nucleopolyhedrosis virus of *Heliothis* species as a microbial insecticide, in *Microbial Control of Pest and Plant Diseases,* 1970 — 1980, Burges, H. D., Ed., Academic Press, New York, 329, 1981.
103. **Hughes, P. R., Gettig, R. R., and McCarthy, W. J.,** Comparison of the time-mortality response of *Heliothis zea* to 14 isolates of *Heliothis* nuclear polyhedrosis virus, *J. Invert. Pathol.,* 41, 256, 1983.
104. **Zhu, G.-K. and Zhang, H.-J.,** The multiplication characterization of *Heliothis armigera* baculovirus in the established cell lines, in *Proc. 3rd Int. Cell Culture Congr.,* Mitsuhashi, J., Kuroda, Y., and Maramorosch, K., Eds., Sendai, Japan, 66, 1985.
105. **Teakle, R. E.,** Records of virus diseases in insects in Queensland, *Queensl. J. Agric. Anim. Sci.,* 30, 1991, 1973.
106. **Whitlock, V. H.,** Symptomatology of two viruses infecting *Heliothis armigera, J. Invert. Pathol.,* 23, 70, 1974.
107. **Heimpel, A. M. and Adams, J. R.,** A new nuclear polyhedrosis of the cabbage looper, *Trichoplusia ni, J. Invert. Pathol.,* 8, 340, 1966.
108. **Granados, R. R., Derksen, A. C. G., and Dwyer, K. G.,** Replication of the *Trichoplusia ni* granulosis and nuclear polyhedrosis viruses in cell cultures, *Virology,* 152, 472, 1986.
109. **Sohi, S. S., Percy, J., Arif, B. M., and Cunningham, J. C.,** Replication and serial passage of a singly enveloped baculovirus of *Orgyia leucostigma* in homologous cell lines, *Intervirology,* 21, 50, 1984.
110. **Raghow, R. and Grace, T. D. C.,** Studies on a nuclear polyhedrosis virus in *Bombyx mori* cells *in vitro,* 1. Multiplication kinetics and ultrastructural studies, *J. Ultrastruct. Res.,* 47, 384, 1974.
111. **Aratake, Y. and Kayamura, T.,** Pathogenicity of a nuclear-polyhedrosis virus of the silkworm, *Bombyx mori,* for a number of lepidopterous insects, *Jpn. J. Appl. Entomol. Zool.,* 17, 121, 1973.
112. **Naser, W. L., Miltenburger, H. G., Harvey, J. P., Huber, J., and Huger, A. M.,** *In vitro* replication of the *Cydia pomonella* (codling moth) granulosis virus, *FEMS Microbiol. Lett.,* 24, 117, 1984.
113. **Falcon, L. A., Kane, W. R., and Bethell, R. S.,** Preliminary evaluation of a granulosis virus for the control of codling moth, *J. Econ. Entomol.,* 61, 1208, 1968.
114. **Huber, J.,** About the host spectrum of the codling moth granulosis virus, in *Safety Aspects of Baculoviruses as Biological Insecticides,* Miltenburger, H. G., Ed., Bundesministerium für Forschung und Technologie, Projektträger Biotechnologie in der Kernforschungsanlage Jülich GmbH, Jülich, 1978, 75.
115. **Ignoffo, C. M., McIntosh, A. H., and Garcia, C.,** Susceptibility of *Heliothis zea, H. virescens,* and *H. armigera* (Lep.; Noctuidae) to 3 baculoviruses, *Entomophaga,* 28, 1, 1983.
116. **Hamm, J. J.,** Extension of the host range for a granulosis virus from *Heliothis armigera* from South Africa, *Environ. Entomol.,* 11, 159, 1982.
117. **Granados, R. R., Nguyen, T., and Cato, B.,** An insect cell line persistently infected with a baculovirus-like particle, *Intervirology,* 10, 309, 1978.
118. **Ralston, A. L., Huang, Y-S., and Kawanishi, C. Y.,** Cell culture studies with the IMC-Hz-1 nonoccluded virus, *Virology,* 115, 33, 1981.
119. **McIntosh, A. H. and Ignoffo, C. M.,** unpublished data, 1990.
120. **Crawford, A. M.,** Attempts to obtain *Oryctes* baculovirus replication in three insect cell cultures, *Virology,* 112, 625, 1981.
121. **Huger, A. M.,** A virus disease of the Indian rhinoceros beetle *Oryctes rhinoceros* L. caused by a new type of insect virus, *Rhabdinovirus oryctes, J. Invert. Pathol.,* 8, 38, 1966.
122. **Quiot, J., Monsarratt, P., Meynadier, G., Crozier, G., and Vago, C.,** Infection des cultures cellulaires de coleopteres par le "virus Oryctes," *C. R. Acad. Sci. Ser. D,* 276, 3229, 1973.
123. **Crawford, A. M. and Sheehan, C.,** Replication of *Oryctes* baculovirus in cell culture: viral morphogenesis, infectivity, and protein synthesis, *J. Gen. Virol.,* 66, 529, 1985.
124. **Maeda, S.,** A plaque assay and cloning of *Bombyx mori* nuclear polyhedrosis virus, *J. Seric. Sci. Jpn.,* 53, 547, 1984.
125. **Crawford, A. M. and Granados, R. R.,** Nonoccluded baculoviruses, in *Invertebrate Pathology and Microbial Conf. Proc. 3rd Int. Colloq. on Invertebrate Pathology,* September 6–10, University of Sussex, Brighton, 1982.
126. **Altmen, M.,** "Biopesticides" turning into new pests?, *Trends Ecol. Evol.,* 7, 65, 1992.
127. **Williamson, M.,** Biocontrol risks, *Nature,* 353, 394, 1991.

Chapter 4

APPLICATION OF INSECT CELL CULTURE TO THE STUDY OF *BACILLUS THURINGIENSIS* TOXINS

William J. McCarthy

TABLE OF CONTENTS

I. Introduction 73

II. Studies with Toxins Exhibiting Lepidopteran Specificity 73

III. Studies with Toxins Exhibiting Dipteran Specificity 84

IV. Closing Remarks 86

References 87

0-8493-4597-9/94/$0.00+$.50
© 1994 by CRC Press Inc.

William J. McCarthy, Ph.D., is an Associate Professor of Molecular Entomology at The Pennsylvania State University, University Park, PA. He received his B.A. in Biology-Chemistry 1963 at the University of Delaware, M.S. 1968 and Ph.D. 1971 at New York University. Post-doctoral traineeship, 1971–1974 at the Boyce Thompson Institute. Research Scientist, 1974–1979. Assistant Professor 1979–1983 and Associate Professor 1983–present at Penn State University. AAAS, ESA, TCA, ASV. Biochemistry of *Bacillus thuringiensis* toxicology, genetics of baculoviruses, biology of insect cells: DNA repair, DNA salvage pathways, mutation studies.

I. INTRODUCTION

Historically, investigations leading to an understanding of the molecular mechanisms of biological phenomena have progressed from *in vivo* studies and observations to the separation of the system of interest and study *in vitro.* The *in vitro* approach allows for controlled experiments and freedom from possible obscuring reactions of the host. The evolution of these types of studies reflects the precept that "only by studying discrete substances and their interactions can more complex phenomena be understood" (Arthur Kornberg). Many of the biological systems studied involve the interaction of diverse molecules or microorganisms with cells. In order to focus on the interaction of interest under controlled conditions, discrete populations of cells are necessary. However, biological systems studied separate from the host do not, in most systems, reflect the *in vivo* situation. Indeed, most cell studies *in vitro* involve immortal dedifferentiated cell lines derived from various host tissues or groups of tissues. Use of these cell lines offers the advantages of (1) control of environment, (2) cell homogeneity, and (3) repeatability of experiments. What is lost, however, are the specific cell interactions characteristic of the histology of the tissue and those systemic components involved in regulation *in vivo.* Nevertheless, although these fundamental differences exist, many functions persist *in vitro,* and as long as the limits of the model are appreciated, cell studies *in vitro* are a valuable tool for understanding complex biological phenomena.

Studies attempting to resolve the mechanism(s) of lepidopteran-specific *Bacillus thuringiensis* toxicity initially were hampered by a lack of *in vitro* systems, which are necessary to define the reaction in molecular terms. The early history of this biorational pesticide as documented in the literature presented the results of *in vivo* studies detailing the pathological events of toxicity, the various specificities of the numerous strains of this bacterium, identification of the proteinaceous parasporal crystal as the toxic moiety, and larval midgut cytopathology. Additional work on the insecticidal crystal protein identified the steps required for the solubilization and activation of the crystal protoxin, which included solubilization by the alkaline fluid in the larval midgut followed by proteolytic cleavage of approximately one-half of the protoxin molecule by midgut proteases. With the perfection of the methodology for establishment and maintenance of insect cell lines in the 1960s, new lines of study using the techniques of mammalian cell biology could be pursued in determining the molecular events of toxin-cell interaction.

There have been two excellent reviews up to and including work published in 1988 documenting *in vitro* studies with *Bacillus thuringiensis* toxic proteins using cells in culture.[1,2] It is not the purpose of this chapter to reiterate the information presented in those papers, but to review the research findings of the past five years; however, a brief overview of the significant research involving toxin-cell interaction is necessary for the continuity of this chapter.

II. STUDIES WITH TOXINS EXHIBITING LEPIDOPTERAN SPECIFICITY

Insect cell lysis *in vitro* was first documented by Murphy et al.,[3] using *Trichoplusia ni (T. ni)* gut fluid-activated protoxin from *Bacillus thuringiensis* (Bt) subsp. *kurstaki.* Cells from several lepidopteran lines (*Malacosoma disstria, Choristoneura fumiferana,* and *Trichoplusia ni*) were lysed to varying extents and exhibited a similar cytopathology of swelling and lysis as observed for midgut cells *in vivo.* In addition, Fast et al.[4] provided evidence suggesting that the protein toxin initially acts at the cell surface. Subsequent studies have documented the sensitivities of different lepidopteran cell lines to a variety of activated toxins (Table 1). Tables 1 and 2 comprise a comprehensive tabulation of the susceptibilities of different lepidopteran and dipteran cell lines to various toxin preparations. Other work documented the dynamics of

TABLE 1
Bt Toxin-Mediated Lysis of Lepidopteran Cells *In Vitro*

Insect	Cell line	Bt Strain	Solubilization and activation	Lysis	Reference
Choristoneura fumiferana	CF1	*kurstaki*-HD1	0.0135 *N* NaOH; chymotrypsin	+	37
	CF1	*israelensis*-HD500	0.0135 *N* NaOH	+	37
	CF1	*kurstaki*-HD1	50 m*M* Na_2CO_3/HCl pH 9.5, 10 m*M* DTT; *P. brassicae* gut juice	+	8
	CF1	*kurstaki*	10 m*M* Tris/HCl pH 8.5, 0.1 *M* KSCN, 1 m*M* Na-thioglycollate; chymotrypsinogen	+	13
	CF1	*kurstaki*-HD1	50 m*M* Na_2CO_3/HCl pH 9.5, 10 m*M* 2-mercapto EtOH; *Bombyx mori* gut juice	+	18
	CF1	*kurstaki*-NRD12	50 m*M* Na_2CO_3/HCl pH 9.5, 10 m*M* 2-mercapto EtOH; *Bombyx mori* gut juice	+	18
	CF1	*israelensis*	50 m*M* Na_2CO_3 pH 10.5, 10 m*M* DTT; *A. aegypti* gut juice	–	30
	CF1	*israelensis*	50 m*M* Na_2CO_3 pH 10.5, 10 m*M* DTT; trypsin	–	30
	CF1	*kurstaki*-HD73	0.1 *M* Caps pH 10.5; TPCK trypsin	+	21
	CF1	*kurstaki*-HD1	Thioglycollate-biocarbonate buffer pH 9.5; chymotrypsin	+	38
	CF1	*israelensis*-HD14	Thioglycollate-biocarbonate buffer pH 9.5; chymotrypsin	+	38
	CF1	*kurstaki*-HD73	50 m*M* KCO_3/HCl pH 9.5, 10 m*M* DTT; *P. brassicae* gut juice	+	19
	CF1	*thuringiensis*-HD350	50 m*M* Na_2CO_3/HCl pH 9.5, 10 m*M* DTT; *P. brassicae* gut juice	+	7
	CF1	*aizawai*-HD249	50 m*M* Na_2CO_3/HCl pH 9.5, 10 m*M* DTT; *P. brassicae* gut juice	+	7
	CF1	*israelensis*-IPS78	50 m*M* Na_2CO_3/HCl pH 10.5; *P. brassicae* gut juice	+	7
	CF1	*israelensis*	50 m*M* Na_2CO_3/HCl pH 10.5	+	17
	CF1	*colmeri*-HD847	50 m*M* Na_2CO_3/HCl pH 9.5, 10 m*M* DTT; *P. brassicae* gut juice	+	39
	CF1	*colmeri*-HD847	50 m*M* Na_2CO_3/HCl pH 9.5, 10 m*M* DTT, 1 *M* NaCl; *A. aegypti* gut juice	–	39
	CF1	*thuringiensis*-B4039	20 m*M* Tris/HCl pH 8.5, 50 m*M* KSCN, 1 m*M* Na-thioglycollate	+	40
	CF1	*thuringiensis*-B4039	20 m*M* Tris/HCl pH 8.5, 50 m*M* KSCN, 1 m*M* Na-thioglycollate; chymotrypsin	+	40
	CF1	*finitimus*-B4308	20 m*M* Tris/HCl pH 8.5, 50 m*M* KSCN, 1 m*M* Na-thioglycollate	–	40
	CF1	*kurstaki*-HD1	20 m*M* Tris/HCl pH 8.5, 50 m*M* KSCN, 1 m*M* Na-thioglycollate	+	40
	CF1	*kurstaki*-HD1	20 m*M* Tris/HCl pH 8.5, 50 m*M* KSCN, 1 m*M* Na-thioglycollate; chymotrypsin	+	40
	CF1	*kurstaki*-HD73	20 m*M* Tris/HCl pH 8.5, 50 m*M* KSCN, 1 m*M* Na-thioglycollate	+	40
	CF1	*kurstaki*-HD73	20 m*M* Tris/HCl pH 8.5, 50 m*M* KSCN, 1 m*M* Na-thioglycollate; chymotrypsin	+	40
	CF1	*dendrolimus*-B4311	20 m*M* Tris/HCl pH 8.5, 50 m*M* KSCN, 1 m*M* Na-thioglycollate	–	40

TABLE 1 (continued)
Bt Toxin-Mediated Lysis of Lepidopteran Cells *In Vitro*

Insect	Cell line	Bt Strain	Solubilization and activation	Lysis	Reference
	CF1	*sotto*-B4310	20 m*M* Tris/HCl pH 8.5, 50 m*M* KSCN, 1 m*M* Na-thioglycollate	–	40
	CF1	*sotto*-B4310	20 m*M* Tris/HCl pH 8.5, 50 m*M* KSCN, 1 m*M* Na-thioglycollate; chymotrypsin	+	40
	CF1	*kenyae*-B4312	20 m*M* Tris/HCl pH 8.5, 50 m*M* KSCN, 1 m*M* Na-thioglycollate	+	40
	CF1	*kenyae*-B4312	20 m*M* Tris/HCl pH 8.5, 50 m*M* KSCN, 1 m*M* Na-thioglycollate; chymotrypsin	+	40
	CF1	*galleriae*-B4313	20 m*M* Tris/HCl pH 8.5, 50 m*M* KSCN, 1 m*M* Na-thioglycollate	–	40
	CF1	*canadensis*-B4056	20 m*M* Tris/HCl pH 8.5, 50 m*M* KSCN, 1 m*M* Na-thioglycollate	–	40
	CF1	*subtoxicus*-B4057	20 m*M* Tris/HCl pH 8.5, 50 m*M* KSCN, 1 m*M* Na-thioglycollate	–	40
	CF1	*subtoxicus*-B4057	20 m*M* Tris/HCl pH 8.5, 50 m*M* KSCN, 1 m*M* Na-thioglycollate; chymotrypsin	+	40
	CF1	*aizawai*-B4314	20 m*M* Tris/HCl pH 8.5, 50 m*M* KSCN, 1 m*M* Na-thioglycollate	–	40
	CF1	*aizawai*-B4314	20 m*M* Tris/HCl pH 8.5, 50 m*M* KSCN, 1 m*M* Na-thioglycollate; chymotrypsin	+	40
	CF1	*morrisoni*-B4315	20 m*M* Tris/HCl pH 8.5, 50 m*M* KSCN, 1 m*M* Na-thioglycollate	–	40
	CF1	*morrisoni*-B4315	20 m*M* Tris/HCl pH 8.5, 50 m*M* KSCN, 1 m*M* Na-thioglycollate; chymotrypsin	+	40
	CF1	*tolworthi*-HD124	20 m*M* Tris/HCl pH 8.5, 50 m*M* KSCN, 1 m*M* Na-thioglycollate	–	40
	CF1	*tolworthi*-HD124	20 m*M* Tris/HCl pH 8.5, 50 m*M* KSCN, 1 m*M* Na-thioglycollate; chymotrypsin	+	40
	CF1	*darmstadiensis*-HD146	20 m*M* Tris/HCl pH 8.5, 50 m*M* KSCN, 1 m*M* Na-thioglycollate	–	40
	CF1	*darmstadiensis*-HD146	20 m*M* Tris/HCl pH 8.5, 50 m*M* KSCN, 1 m*M* Na-thioglycollate; chymotrypsin	+	40
	CF1	*toumanoffi*-B4059	20 m*M* Tris/HCl pH 8.5, 50 m*M* KSCN, 1 m*M* Na-thioglycollate	+	40
	CF1	*toumanoffi*-B4059	20 m*M* Tris/HCl pH 8.5, 50 m*M* KSCN, 1 m*M* Na-thioglycollate; chymotrypsin	+	40
	CF1	*israelensis*-HD500	20 m*M* Tris/HCl pH 8.5, 50 m*M* KSCN, 1 m*M* Na-thioglycollate	–	40
	CF1	*israelensis*-HD500	20 m*M* Tris/HCl pH 8.5, 50 m*M* KSCN, 1 m*M* Na-thioglycollate; chymotrypsin	+	40
	CF1	*aizawai*-HD249	50 m*M* Na_2CO_3/HCl pH 9.5, 10 m*M* DTT; *P. brassicae* gut juice	+	41
	CF1	*thuringiensis*-HD350	50 m*M* Na_2CO_3/HCl pH 9.5, 10 m*M* DTT; *P. brassicae* gut juice	+	41
	CF1	*kurstaki*	1 *M* KSCN, 100 m*M* MOPS, 100 m*M* DTT pH 7.8	+	42

TABLE 1 (continued)
Bt Toxin-Mediated Lysis of Lepidopteran Cells *In Vitro*

Insect	Cell line	Bt Strain	Solubilization and activation	Lysis	Reference
	CF1	*galleriae*	1 *M* KSCN, 100 m*M* MOPS, 100 m*M* DTT pH 7.8	+	42
	CF124	*kurstaki*	0.05 *M* CAPS buffer pH 10.5; *T. ni* gut juice	+	3
	CF124	*kurstaki*-HD1	10 m*M* Tris/HCl pH 8.5, 0.1 *M* KSCN, 1 m*M* Na-thioglycollate	+	43
	CF124	*kurstaki*-HD1	10 m*M* Tris/HCl pH 8.5, 0.1 *M* KSCN, 1 m*M* Na-thioglycollate; chymotrypsinogen	+	43
	CF124	*kurstaki*-HD1	10 m*M* Tris/HCl pH 8.5, 0.1 *M* KSCN, 1 m*M* Na-thioglycollate; Streptomyces griseus	+	43
	CF124	*kurstaki*-HD1	10 m*M* Tris/HCl pH 8.5, 0.1 *M* KSCN, 1 m*M* Na-thioglycollate; papain	+	43
	CF1	*darmstadiensis*-73E102	50 m*M* Na_2CO_3 pH 10.0, 10 m*M* DTT; *P. brassicae* gut juice	+	32
Heliothis virescens	HVAM1	*kurstaki*-HD263	20 m*M* Na_2CO_3 pH 9.5; trypsin	+	12
	HVAM1	*aizawai*-HD112	20 m*M* Na_2CO_3 pH 9.5; trypsin	+	12
	HVAM1	*kurstaki*-HD1	Thioglycollate-biocarbonate buffer pH 9.5; chymotrypsin	–	44
	HVAM1	*israelensis*-HD500	Thioglycollate-biocarbonate buffer pH 9.5; chymotrypsin	–	44
Heliothis zea	HZ1075	*kurstaki*-HD263	0.02 *M* $NaHCO_3$ pH 9.0; trypsin	+	11
	ND	*kurstaki*-HD1	50 m*M* Na_2CO_3/HCl pH 9.5, 10 m*M* DTT; *P. brassicae* gut juice	–	8
	ND	*kurstaki*-HD1	50 m*M* Na_2CO_3/HCl pH 9.5, 10 m*M* DTT; *P. brassicae* gut juice	–	41
	ND	*aizawai*-HD249	50 m*M* Na_2CO_3/HCl pH 9.5, 10 m*M* DTT; *P. brassicae* gut juice	+	41
	ND	*thuringiensis*-HD350	50 m*M* Na_2CO_3/HCl pH 9.5, 10 m*M* DTT; *P. brassicae* gut juice	+	41
	ND	*colmeri*-HD847	50 m*M* Na_2CO_3/HCl pH 9.5, 10 m*M* DTT; *P. brassicae* gut juice	+	39
	ND	*colmeri*-HD847	50 m*M* Na_2CO_3/HCl pH 9.5, 10 m*M* DTT, 1 *M* NaCl; *A. aegypti* gut juice	–	39
	ND	*darmstadiensis*-73E102	50 m*M* Na_2CO_3 pH 10.0, 10 m*M* DTT; *P. brassicae* gut juice	+	32
	HZ1075	*kurstaki*-HD263	20 m*M* Na_2CO_3 pH 9.5; trypsin	+	12
	HZ1075	*aizawai*-HD112	20 m*M* Na_2CO_3 pH 9.5; trypsin	+	12
Lymantria dispar	LD652Y	*kurstaki*-HD263	0.02 *M* $NaHCO_3$ pH 9.0; trypsin	+	11
	LD652Y	HD524	0.1 *M* Na_2CO_3 pH 10.2, 10 m*M* DTT, 2 *M* KSCN; trypsin	+	22
	LD652Y	*kurstaki*-HD263	20 m*M* Na_2CO_3 pH 9.5; trypsin	+	12
	LD652Y	*aizawai*-HD112	20 m*M* Na_2CO_3 pH 9.5; trypsin	+	12
Malacosoma disstria	Md-108	*kurstaki*	0.05 *M* CAPS buffer pH 10.5; T. ni gut juice	+	3
	Md-66	*kurstaki*	0.05 *M* CAPS buffer pH 10.5; T. ni gut juice	–	3
Mamestra brassicae	ND	*kurstaki*-HD1	50 m*M* Na_2CO_3/HCl pH 9.5, 10 m*M* DTT; *P. brassicae* gut juice	–	8
	ND	*colmeri*-HD847	50 m*M* Na_2CO_3/HCl pH 9.5, 10 m*M* DTT; *P. brassicae* gut juice	+	39
	ND	*colmeri*-HD847	50 m*M* Na_2CO_3/HCl pH 9.5, 10 m*M* DTT, 1 *M* NaCl; *A. aegypti* gut juice	–	39

TABLE 1 (continued)
Bt Toxin-Mediated Lysis of Lepidopteran Cells *In Vitro*

Insect	Cell line	Bt Strain	Solubilization and activation	Lysis	Reference
	ND	*kurstaki*-HD1	50 m*M* Na_2CO_3/HCl pH 9.5, 10 m*M* DTT; *P. brassicae* gut juice	–	41
	ND	*aizawai*-HD249	50 m*M* Na_2CO_3/HCl pH 9.5, 10 m*M* DTT; *P. brassicae* gut juice	+	41
	ND	*thuringiensis*-HD350	50 m*M* Na_2CO_3/HCl pH 9.5, 10 m*M* DTT; *P. brassicae* gut juice	–	41
Manduca sexta	CHE-20	*kurstaki*-HD1	0.0135 *N* NaOH; chymotrypsin	+	37
	CHE-20	*israelensis*-HD500	0.0135 *N* NaOH	+	37
	ND	*kurstaki*-HD73	50 m*M* KCO_3/HCl pH 9.5, 10 m*M* DTT; *P. brassicae* gut juice	+	19
	CHE-20	*kurstaki*-HD1	Thioglycollate-biocarbonate buffer pH 9.5; chymotrypsin	+	38
	CHE-20	*israelensis*-HD14	Thioglycollate-biocarbonate buffer pH 9.5; chymotrypsin	+	38
	CHE-20	*israelensis*-HD500	Thioglycollate-biocarbonate buffer pH 9.5; chymotrypsin	+	44
	CHE	*thuringiensis*-B4039	20 m*M* Tris/HCl pH 8.5, 50 m*M* KSCN, 1 m*M* Na-thioglycollate	+	40
	CHE	*thuringiensis*-B4039	20 m*M* Tris/HCl pH 8.5, 50 m*M* KSCN, 1 m*M* Na-thioglycollate; chymotrypsin	+	40
	CHE	*finitimus*-B4308	20 m*M* Tris/HCl pH 8.5, 50 m*M* KSCN, 1 m*M* Na-thioglycollate	+	40
	CHE	*finitimus*-B4308	20 m*M* Tris/HCl pH 8.5, 50 m*M* KSCN, 1 m*M* Na-thioglycollate; chymotrypsin	+	40
	CHE	*kurstaki*-HD1	20 m*M* Tris/HCl pH 8.5, 50 m*M* KSCN, 1 m*M* Na-thioglycollate	+	40
	CHE	*kurstaki*-HD1	20 m*M* Tris/HCl pH 8.5, 50 m*M* KSCN, 1 m*M* Na-thioglycollate; chymotrypsin	+	40
	CHE	*kurstaki*-HD73	20 m*M* Tris/HCl pH 8.5, 50 m*M* KSCN, 1 m*M* Na-thioglycollate	+	40
	CHE	*kurstaki*-HD73	20 m*M* Tris/HCl pH 8.5, 50 m*M* KSCN, 1 m*M* Na-thioglycollate; chymotrypsin	+	40
	CHE	*dendrolimus*-B4311	20 m*M* Tris/HCl pH 8.5, 50 m*M* KSCN, 1 m*M* Na-thioglycollate	+	40
	CHE	*dendrolimus*-B4311	20 m*M* Tris/HCl pH 8.5, 50 m*M* KSCN, 1 m*M* Na-thioglycollate; chymotrypsin	+	40
	CHE	*sotto*-B4310	20 m*M* Tris/HCl pH 8.5, 50 m*M* KSCN, 1 m*M* Na-thioglycollate	+	40
	CHE	*sotto*-B4310	20 m*M* Tris/HCl pH 8.5, 50 m*M* KSCN, 1 m*M* Na-thioglycollate; chymotrypsin	+	40
	CHE	*kenyae*-B4312	20 m*M* Tris/HCl pH 8.5, 50 m*M* KSCN, 1 m*M* Na-thioglycollate	+	40
	CHE	*kenyae*-B4312	20 m*M* Tris/HCl pH 8.5, 50 m*M* KSCN, 1 m*M* Na-thioglycollate; chymotrypsin	+	40
	CHE	*galleriae*-B4313	20 m*M* Tris/HCl pH 8.5, 50 m*M* KSCN, 1 m*M* Na-thioglycollate	+	40
	CHE	*galleriae*-B4313	20 m*M* Tris/HCl pH 8.5, 50 m*M* KSCN, 1 m*M* Na-thioglycollate; chymotrypsin	+	40

TABLE 1 (continued)
Bt Toxin-Mediated Lysis of Lepidopteran Cells *In Vitro*

Insect	Cell line	Bt Strain	Solubilization and activation	Lysis	Reference
	CHE	*canadensis*-B4056	20 m*M* Tris/HCl pH 8.5, 50 m*M* KSCN, 1 m*M* Na-thioglycollate	+	40
	CHE	*canadensis*-B4056	20 m*M* Tris/HCl pH 8.5, 50 m*M* KSCN, 1 m*M* Na-thioglycollate; chymotrypsin	+	40
	CHE	*subtoxicus*-B4057	20 m*M* Tris/HCl pH 8.5, 50 m*M* KSCN, 1 m*M* Na-thioglycollate	+	40
	CHE	*subtoxicus*-B4057	20 m*M* Tris/HCl pH 8.5, 50 m*M* KSCN, 1 m*M* Na-thioglycollate; chymotrypsin	+	40
	CHE	*aizawai*-B4314	20 m*M* Tris/HCl pH 8.5, 50 m*M* KSCN, 1 m*M* Na-thioglycollate	+	40
	CHE	*aizawai*-B4314	20 m*M* Tris/HCl pH 8.5, 50 m*M* KSCN, 1 m*M* Na-thioglycollate; chymotrypsin	+	40
	CHE	*morrisoni*-B4315	20 m*M* Tris/HCl pH 8.5, 50 m*M* KSCN, 1 m*M* Na-thioglycollate	+	40
	CHE	*morrisoni*-B4315	20 m*M* Tris/HCl pH 8.5, 50 m*M* KSCN, 1 m*M* Na-thioglycollate; chymotrypsin	+	40
	CHE	*tolworthi*-HD124	20 m*M* Tris/HCl pH 8.5, 50 m*M* KSCN, 1 m*M* Na-thioglycollate	+	40
	CHE	*tolworthi*-HD124	20 m*M* Tris/HCl pH 8.5, 50 m*M* KSCN, 1 m*M* Na-thioglycollate; chymotrypsin	+	40
	CHE	*darmstadiensis*-HD146	20 m*M* Tris/HCl pH 8.5, 50 m*M* KSCN, 1 m*M* Na-thioglycollate	+	40
	CHE	*darmstadiensis*-HD146	20 m*M* Tris/HCl pH 8.5, 50 m*M* KSCN, 1 m*M* Na-thioglycollate; chymotrypsin	+	40
	CHE	*toumanoffi*-B4059	20 m*M* Tris/HCl pH 8.5, 50 m*M* KSCN, 1 m*M* Na-thioglycollate	+	40
	CHE	*toumanoffi*-B4059	20 m*M* Tris/HCl pH 8.5, 50 m*M* KSCN, 1 m*M* Na-thioglycollate; chymotrypsin	+	40
	CHE	*israelensis*-HD500	20 m*M* Tris/HCl pH 8.5, 50 m*M* KSCN, 1 m*M* Na-thioglycollate	–	40
	CHE	*israelensis*-HD500	20 m*M* Tris/HCl pH 8.5, 50 m*M* KSCN, 1 m*M* Na-thioglycollate; chymotrypsin	+	40
Pieris brassicae	ND	*thuringiensis*-HD4412	50 m*M* Na_2CO_3 pH 9.5, 1 m*M* EDTA, 10 m*M* DTT; trypsin	–	45
	ND	*israelensis*-HD4444	50 m*M* Na_2CO_3 pH 9.5, 1 m*M* EDTA, 10 m*M* DTT	+	45
Plodia interpunctella	PID2	*kurstaki*-HD1	0.0135 *N* NaOH; chymotrypsin	–	37
	PID2	*israelensis*-HD500	0.0135 *N* NaOH	–	37
	PID2	*kurstaki*-HD1	Thioglycollate-biocarbonate buffer pH 9.5; chymotrypsin	–	38
	PID2	*israelensis*-HD14	Thioglycollate-biocarbonate buffer pH 9.5; chymotrypsin	–	38
	PID2	*kurstaki*-HD1	Thioglycollate-biocarbonate buffer pH 9.5; chymotrypsin	–	44
	PID2	*israelensis*-HD500	Thioglycollate-biocarbonate buffer pH 9.5; chymotrypsin	+	44

TABLE 1 (continued)
Bt Toxin-Mediated Lysis of Lepidopteran Cells *In Vitro*

Insect	Cell line	Bt Strain	Solubilization and activation	Lysis	Reference
	PI5.3	*kurstaki*-HD1	Thioglycollate-biocarbonate buffer pH 9.5; chymotrypsin	–	44
	PI5.3	*israelensis*-HD500	Thioglycollate-biocarbonate buffer pH 9.5; chymotrypsin	+	44
	PI5.4	*kurstaki*-HD1	Thioglycollate-biocarbonate buffer pH 9.5; chymotrypsin	–	44
	PI5.4	*israelensis*-HD500	Thioglycollate-biocarbonate buffer pH 9.5; chymotrypsin	+	44
	PI8.5	*kurstaki*-HD1	Thioglycollate-biocarbonate buffer pH 9.5; chymotrypsin	–	44
	PI8.5	*israelensis*-HD500	Thioglycollate-biocarbonate buffer pH 9.5; chymotrypsin	+	44
Spodoptera exigua	SEUCR1A	*kurstaki*-HD263	20 m*M* Na_2CO_3 pH 9.5; trypsin	+	12
	SEUCR1A	*aizawai*-HD112	20 m*M* Na_2CO_3 pH 9.5; trypsin	+	12
	SE1	*kurstaki*	1 *M* KSCN, 100 m*M* MOPS, 100 m*M* DTT pH 7.8	–	42
	SE1	*galleria*	1 *M* KSCN, 100 m*M* MOPS, 100 m*M* DTT pH 7.8	+	42
Spodoptera frugiperda	SF21AE	*kurstaki*-HD263	0.02 *M* $NaHCO_3$ pH 9.0; trypsin	+	11
	SF21AE	HD524	0.1 *M* Na_2CO_3 pH 10.2, 10 m*M* DTT, 2 *M* KSCN; trypsin	+	22
	SF	*israelensis*	0.1 *M* glycine-NaOH pH 10.0, 0.1 *M* EDTA, 1 m*M* PMSF	+	26
	SF	*israelensis*	0.05 *M* NaOH, 10 m*M* EDTA pH 11.7	+	26
	SF	*israelensis*	0.05 Na_2CO_3/HCl pH 10.5	+	26
	ND	*kurstaki*-HD1	50 m*M* Na_2CO_3/HCl pH 9.5, 10 m*M* DTT; *P. brassicae* gut juice	–	8
	SF21AE	*kurstaki*-HD263	20 m*M* Na_2CO_3 pH 9.5; trypsin	+	12
	SF21AE	*aizawai*-HD112	20 m*M* Na_2CO_3 pH 9.5; trypsin	+	12
	ND	*israelensis*-IPS78	50 m*M* Na_2CO_3/HCl pH 10.5	+	17
	SF21AE	*kurstaki*-HD1	Thioglycollate-biocarbonate buffer pH 9.5; chymotrypsin	–	44
	SF21AE	*israelensis*-HD500	Thioglycollate-biocarbonate buffer pH 9.5; chymotrypsin	+	44
	ND	*colmeri*-HD847	50 m*M* Na_2CO_3/HCl pH 9.5, 10 m*M* DTT; *P. brassicae* gut juice	+	39
	ND	*colmeri*-HD847	50 m*M* Na_2CO_3/HCl pH 9.5, 10 m*M* DTT, 1 *M* NaCl; *A. aegypti* gut juice	+	39
	ND	*kurstaki*-HD1	50 m*M* Na_2CO_3/HCl pH 9.5, 10 m*M* DTT; *P. brassicae* gut juice	–	41
	ND	*aizawai*-HD249	50 m*M* Na_2CO_3/HCl pH 9.5, 10 m*M* DTT; *P. brassicae* gut juice	+	41
	ND	*thuringiensis*-HD350	50 m*M* Na_2CO_3/HCl pH 9.5, 10 m*M* DTT; *P. brassicae* gut juice	–	41
Spodoptera litteralis	SL575	*kurstaki*-HD263	20 m*M* Na_2CO_3 pH 9.5; trypsin	+	12
	SL575	*aizawai*-HD112	20 m*M* Na_2CO_3 pH 9.5; trypsin	+	12
Trichoplusia ni	TN368	*aizawai*-I45	Silkworm gut juice	+	46
	TN368	*kurstaki*-HD263	0.02 *M* $NaHCO_3$ pH 9.0; trypsin	+	11
	TN368	HD524	0.1 *M* Na_2CO_3 pH 10.2, 10 m*M* DTT, 2 *M* KSCN; trypsin	+	22
	ND	*kurstaki*-HD1	50 m*M* Na_2CO_3/HCl pH 9.5, 10 m*M* DTT; *P. brassicae* gut juice	–	8
	TN368	*kurstaki*-HD263	20 m*M* Na_2CO_3 pH 9.5; trypsin	+	12
	TN368	*aizawai*-HD112	20 m*M* Na_2CO_3 pH 9.5; trypsin	+	12

TABLE 1 (continued)
Bt Toxin-Mediated Lysis of Lepidopteran Cells *In Vitro*

Insect	Cell line	Bt Strain	Solubilization and activation	Lysis	Reference
	ND	*israelensis*-IPS78	50 m*M* Na_2CO_3/HCl pH 10.5	+	17
	TN368	*kurstaki*-HD1	Thioglycollate-biocarbonate buffer pH 9.5; chymotrypsin	+	44
	TN368	*israelensis* HD500	Thioglycollate-biocarbonate buffer pH 9.5; chymotrypsin	+	44
	TN368	*kurstaki*	0.05 *M* CAPS buffer pH 10.5; T. ni gut juice	+	3
	TN368	*aizawai*-I45	PBS pH 7.0, 0.01% Triton X-100; *Bombyx mori* juice	+	47
	TN368	*aizawai*-I45	PBS pH 7.0, 0.01% Triton X-100; *Bombyx mori* juice	+	5

Note: ND = no cell line designation; + = evidence of cell lysis; – = no cell lysis.

TABLE 2
Bt Toxin-Mediated Lysis of Dipteran Cells *In Vitro*

Insect	Cell line	Bt Strain	Solubilization and activation	Lysis	Reference
Aedes aegypti	ND	*kurstaki*-HD1	0.0135 *N* NaOH; chymotrypsin	–	37
	ND	*israelensis*-HD500	0.0135 *N* NaOH	+	37
	Aa(s)	*kurstaki*-HD1	50 m*M* Na_2CO_3/HCl pH 9.5, 10 m*M* DTT; *P. brassicae* gut juice	–	8
	20A	*kurstaki*-HD1	50 m*M* Na_2CO_3/HCl pH 9.5, 10 m*M* DTT; *P. brassicae* gut juice	–	8
	ND	*israelensis*	50 m*M* Na_2CO_3 pH 10.5, 10 m*M* DTT; *A. aegypti*	+	30
	ND	*israelensis*	50 m*M* Na_2CO_3 pH 10.5, 10 m*M* DTT; trypsin	–	30
	ND	*kurstaki*-HD1	Thioglycollate-biocarbonate buffer pH 9.5; chymotrypsin	–	38
	ND	*israelensis*-HD14	Thioglycollate-biocarbonate buffer pH 9.5; chymotrypsin	+	38
	ND	*kurstaki*-HD1	Thioglycollate-biocarbonate buffer pH 9.5; chymotrypsin	–	44
	ND	*israelensis*-HD500	Thioglycollate-biocarbonate buffer pH 9.5; chymotrypsin	+	44
	ND	*colmeri*-HD847	50 m*M* Na_2CO_3/HCl pH 9.5, 10 m*M* DTT; *P. brassicae* gut juice	–	39
	ND	*colmeri*-HD847	50 m*M* Na_2CO_3/HCl pH 9.5, 10 m*M* DTT, 1 *M* NaCl; *A. aegypti* gut juice	+	39
	ND	*kurstaki*-HD1	50 m*M* Na_2CO_3/HCl pH 9.5, 10 m*M* DTT; *P. brassicae* gut juice	–	41
	ND	*aizawai*-HD249	50 m*M* Na_2CO_3/HCl pH 9.5, 10 m*M* DTT; *P. brassicae* gut juice	+	41
	ND	*thuringiensis*-HD350	50 m*M* Na_2CO_3/HCl pH 9.5, 10 m*M* DTT; *P. brassicae* gut juice	–	41
	ND	*darmstadiensis*-73E102	50 m*M* Na_2CO_3 pH 10.0, 10 m*M* DTT; *P. brassicae* gut juice	+	32
Aedes albopictus	ND	*kurstaki*-HD1	50 m*M* Na_2CO_3/HCl pH 9.5, 10 m*M* DTT; *P. brassicae* gut juice	–	8
	C6/36	*israelensis*	0.1 *M* glycine/NaOH pH 10.0, 0.1 *M* EDTA, 1 m*M* PMSF	+	26

TABLE 2 (continued)
Bt Toxin-Mediated Lysis of Dipteran Cells *In Vitro*

Insect	Cell line	Bt Strain	Solubilization and activation	Lysis	Reference
	C6/36	*israelensis*	0.05 *M* NaOH with 10 m*M* EDTA pH 11.7	+	26
	C6/36	*israelensis*	0.05 Na_2CO_3/HCl pH 10.5	+	26
	ND	*israelensis*-IPS78	50 m*M* Na_2CO_3/HCl pH 10.5	+	17
	ND	*colmeri*-HD847	50 m*M* Na_2CO_3/HCl pH 9.5, 10 m*M* DTT; *P. brassicae* gut juice	–	39
	ND	*colmeri*-HD847	50 m*M* Na_2CO_3/HCl pH 9.5, 10 m*M* DTT, 1 *M* NaCl; *A. aegypti* gut juice	+	39
	ND	*kurstaki*-HD1	50 m*M* Na_2CO_3/HCl pH 9.5, 10 m*M* DTT; *P. brassicae* gut juice	–	41
	ND	*aizawai*-HD249	50 m*M* Na_2CO_3/HCl pH 9.5, 10 m*M* DTT; *P. brassicae* gut juice	–	41
	ND	*thuringiensis*-HD350	50 m*M* Na_2CO_3/HCl pH 9.5, 10 m*M* DTT; *P. brassicae* gut juice	–	41
	ND	*darmstadiensis*-73E102	50 m*M* Na_2CO_3 pH 10.0, 10 m*M* DTT; *P. brassicae* gut juice	+	32
	ND	*darmstadiensis*-73E102	50 m*M* Na_2CO_3 pH 10.0; *P. brassicae* gut juice	+	32
Anopheles gambiae	ND	*kurstaki*-HD1	0.0135 *N* NaOH; chymotrypsin	–	37
	ND	*israelensis*-HD500	0.0135 *N* NaOH	+	37
	ND	*israelensis*	50 m*M* Na_2CO_3 pH 10.5, 10 m*M* DTT; *A. aegypti*	–	30
	ND	*israelensis*	50 m*M* Na_2CO_3 pH 10.5, 10 m*M* DTT; trypsin	–	30
	ND	*kurstaki*-HD1	Thioglycollate-biocarbonate buffer pH 9.5; chymotrypsin	–	38
	ND	*israelensis*-HD14	Thioglycollate-biocarbonate buffer pH 9.5; chymotrypsin	+	38
	ND	*kurstaki*-HD1	Thioglycollate-biocarbonate buffer pH 9.5; chymotrypsin	–	44
	ND	*israelensis*-HD500	Thioglycollate-biocarbonate buffer pH 9.5; chymotrypsin	+	44
	ND	*thuringiensis*-B4039	20 m*M* Tris/HCl pH 8.5, 50 m*M* KSCN, 1 m*M* Na-thioglycollate	–	40
	ND	*finitimus*-B4308	20 m*M* Tris/HCl pH 8.5, 50 m*M* KSCN, 1 m*M* Na-thioglycollate	–	40
	ND	*kurstaki*-HD1	20 m*M* Tris/HCl pH 8.5, 50 m*M* KSCN, 1 m*M* Na-thioglycollate	–	40
	ND	*kurstaki*-HD73	20 m*M* Tris/HCl pH 8.5, 50 m*M* KSCN, 1 m*M* Na-thioglycollate	–	40
	ND	*kurstaki*-HD73	20 m*M* Tris/HCl pH 8.5, 50 m*M* KSCN, 1 m*M* Na-thioglycollate; chymotrypsin	+	40
	ND	*dendrolimus*-B4311	20 m*M* Tris/HCl pH 8.5, 50 m*M* KSCN, 1 m*M* Na-thioglycollate	–	40
	ND	*sotto*-B4310	20 m*M* Tris/HCl pH 8.5, 50 m*M* KSCN, 1 m*M* Na-thioglycollate	–	40
Anopheles gambiae	ND	*kenyae*-B4312	20 m*M* Tris/HCl pH 8.5, 50 m*M* KSCN, 1 m*M* Na-thioglycollate	–	40
	ND	*galleriae*-B4313	20 m*M* Tris/HCl pH 8.5, 50 m*M* KSCN, 1 m*M* Na-thioglycollate	–	40
	ND	*canadensis*-B4056	20 m*M* Tris/HCl pH 8.5, 50 m*M* KSCN, 1 m*M* Na-thioglycollate	–	40
	ND	*subtoxicus*-B4057	20 m*M* Tris/HCl pH 8.5, 50 m*M* KSCN, 1 m*M* Na-thioglycollate	–	40

TABLE 2 (continued)
Bt Toxin-Mediated Lysis of Dipteran Cells *In Vitro*

Insect	Cell line	Bt Strain	Solubilization and activation	Lysis	Reference
	ND	*aizawai*-B4314	20 m*M* Tris/HCl pH 8.5, 50 m*M* KSCN, 1 m*M* Na-thioglycollate	–	40
	ND	*aizawai*-B4314	20 m*M* Tris/HCl pH 8.5, 50 m*M* KSCN, 1 m*M* Na-thioglycollate; chymotrypsin	+	40
	ND	*morrisoni*-B4315	20 m*M* Tris/HCl pH 8.5, 50 m*M* KSCN, 1 m*M* Na-thioglycollate	–	40
	ND	*tolworthi*-HD124	20 m*M* Tris/HCl pH 8.5, 50 m*M* KSCN, 1 m*M* Na-thioglycollate	–	40
	ND	*darmstadiensis*-HD146	20 m*M* Tris/HCl pH 8.5, 50 m*M* KSCN, 1 m*M* Na-thioglycollate	–	40
	ND	*darmstadiensis*-HD146	20 m*M* Tris/HCl pH 8.5, 50 m*M* KSCN, 1 m*M* Na-thioglycollate; chymotrypsin	+	40
	ND	*toumanoffi*-B4059	20 m*M* Tris/HCl pH 8.5, 50 m*M* KSCN, 1 m*M* Na-thioglycollate	–	40
	ND	*toumanoffi*-B4059	20 m*M* Tris/HCl pH 8.5, 50 m*M* KSCN, 1 m*M* Na-thioglycollate; chymotrypsin	+	40
	ND	*israelensis*-HD500	20 m*M* Tris/HCl pH 8.5, 50 m*M* KSCN, 1 m*M* Na-thioglycollate	+	40
	ND	*israelensis*-HD500	20 m*M* Tris/HCl pH 8.5, 50 m*M* KSCN, 1 m*M* Na-thioglycollate; chymotrypsin	+	40
	ND	*kurstaki*-HD1	50 m*M* Na_2CO_3 pH 9.5, 10 m*M* DTT; *P. brassicae* gut juice	–	41
	ND	*aizawai*-HD249	50 m*M* Na_2CO_3 pH 9.5, 10 m*M* DTT; *P. brassicae* gut juice	+	41
	ND	*thuringiensis*-HD350	50 m*M* Na_2CO_3 pH 9.5, 10 m*M* DTT; *P. brassicae* gut juice	–	41
	ND	*darmstadiensis*-73E102	50 m*M* Na_2CO_3 pH 10.0, 10 m*M* DTT; *P. brassicae* gut juice	+	32
	ND	*kurstaki*-HD1	50 m*M* Na_2CO_3 pH 9.5, 10 m*M* DTT; *P. brassicae* gut juice	–	8
	ND	*kurstaki*-HD1	50 m*M* Na_2CO_3 pH 9.5, 10 m*M* DTT; *P. brassicae* gut juice	–	41
	ND	*aizawai*-HD249	50 m*M* Na_2CO_3 pH 9.5, 10 m*M* DTT; *P. brassicae* gut juice	+	41
	ND	*thuringiensis*-350	50 m*M* Na_2CO_3 pH 9.5, 10 m*M* DTT;	–	41
Culex quinquefasciatus	C2	*kurstaki*-HD1	50 m*M* Na_2CO_3/HCl pH 9.5, 10 m*M* DTT; *P. brassicae* gut juice	–	8
	ND	*israelensis*	50 m*M* Na_2CO_3 pH 10.5, 10 m*M* DTT; *A. aegypti*	–	30
	ND	*israelensis*	50 m*M* Na_2CO_3 pH 10.5, 10 m*M* DTT; trypsin	–	30
	ND	*colmeri*-HD847	50 m*M* Na_2CO_3/HCl pH 9.5, 10 m*M* DTT; *P. brassicae* gut juice	–	39
	ND	*colmeri*-HD847	50 m*M* Na_2CO_3/HCl pH 9.5, 10 m*M* DTT, 1 *M* NaCl; *A. aegypti* gut juice	+	39
	ND	*kurstaki*-HD1	50 m*M* Na_2CO_3/HCl pH 9.5, 10 m*M* DTT; *P. brassicae* gut juice	–	41
	ND	*aizawai*-HD249	50 m*M* Na_2CO_3/HCl pH 9.5, 10 m*M* DTT; *P. brassicae* gut juice	–	41
	ND	*thuringiensis*-HD350	50 m*M* Na_2CO_3/HCl pH 9.5, 10 m*M* DTT; *P. brassicae* gut juice	–	41

TABLE 2 (continued)
Bt Toxin-Mediated Lysis of Dipteran Cells *In Vitro*

Insect	Cell line	Bt Strain	Solubilization and activation	Lysis	Reference
	ND	*darmstadiensis*-73E102	50 m*M* Na_2CO_3 pH 10.0, 10 m*M* DTT; *P. brassicae* gut juice	+	32
Culex tarlis	ND	*israelensis*	0.1 *M* glycine/NaOH pH 10.0, 0.1 *M* EDTA, 1 m*M* PMSF	+	26
	ND	*israelensis*	0.05 *M* NaOH, 10 m*M* EDTA pH 11.7	+	26
	ND	*israelensis*	0.05 Na_2CO_3/HCl pH 10.5	+	26
Drosophila melangaster	ND	*kurstaki*-HD1	50 m*M* Na_2CO_3/HCl pH 9.5, 10 m*M* DTT; *P. brassicae* gut juice	–	8
	ND	*colmeri*-HD847	50 m*M* Na_2CO_3/HCl pH 9.5, 10 m*M* DTT; *P. brassicae* gut juice	–	39
	ND	*colmeri*-HD847	50 m*M* Na_2CO_3/HCl pH 9.5, 10 m*M* DTT, 1 *M* NaCl; *A. aegypti* gut juice	–	39
	ND	*kurstaki*-HD1	50 m*M* Na_2CO_3/HCl pH 9.5, 10 m*M* DTT; *P. brassicae* gut juice	–	41
	ND	*aizawai*-HD249	50 m*M* Na_2CO_3/HCl pH 9.5, 10 m*M* DTT; *P. brassicae* gut juice	+	41
	ND	*thuringiensis*-HD350	50 m*M* Na_2CO_3/HCl pH 9.5, 10 m*M* DTT; *P. brassicae* gut juice	–	41
	D2	*thuringiensis*-HD4412	50 m*M* Na_2CO_3 pH 9.5, 1 m*M* EDTA, 10 m*M* DTT; trypsin	+	45
	D2	*israelensis*-HD4444	50 m*M* Na_2CO_3 pH 9.5, 1 m*M* EDTA, 10 m*M* DTT	+	45
	ND	*darmstadiensis*-73E102	50 m*M* Na_2CO_3 pH 10.0, 10 m*M* DTT; *P. brassicae* gut juice	+	32

Notes: ND = no cell line designation; + = evidence of cell lysis; – = no cell lysis.

cell lysis and initial ion fluxes occurring during toxin-cell interaction.[5,6] Results from analyses of toxin-mediated swelling and lysis combined with findings involving extracellular osmoprotectants lead to the hypothesis that a specific binding step followed by or in concert with formation of an ion channel (pore) of specific size (approximately 0.5 to 1.0 nm radius) defined lepidopteran Bt toxicity.[7] In a preliminary study, Knowles and Ellar[8] identified a 146 kDa toxin binding glycoprotein on CF-1 cell membranes using Bt subsp. *kurstaki* (HD-1) in a blotting technique. Using the same technique, CF-1 membrane proteins of 68 kDa and 120 kDa, and a 68 kDa protein in *Spodoptera frugiperda* cell membranes were bound by a Bt subsp. *aizawai* toxin.[9] Evidence that some of the binding proteins are glycoproteins was obtained from experiments using lectins and aminosugar inhibition of Bt subsp. *kurstaki* (HD-1) lysis of CF-1 cells.[10] Also, inhibition of TN 368 cell lysis to different extents was obtained by preincubation of the toxic proteins from either Bt subsp. *kurstaki* (HD-263) or subsp. *aizawai* (HD-112) with the amino- and *N*-acetylamino-derivatives of galactose, mannose, and glucose.[11,12]

There has been one *in vitro* study examining Bt resistance *in vitro*, and this experiment involved an attempt to select a population of CF-1 cells resistant to Bt subsp. *kurstaki* (HD-1) toxin.[13] Although an 89-fold increase in resistance to the toxin was obtained under constant selection pressure, when the selective agent was removed from the medium, the cells gradually returned to the level of resistance exhibited by the parental strain. The only reported instances of Bt resistance have occurred with native populations of lepidoptera.[14,15] However, the stability of this resistance has not been established as yet.

Recent developments using Bt toxins and lepidopteran cells have come from a number of diverse studies, including a new assay procedure (the lawn assay) involving cells suspended in an agar matrix.[16] Application of toxin solutions results in cell lysis visualized by subsequent treatment with trypan blue stain. The size of the blue areas is proportional to toxin concentration. This assay is rapid, sensitive to picogram levels of toxin, and only small volumes are needed. In addition, toxins can be applied at high pH; however, relatively large numbers of cells are required for the assay (10^8 per 8.5 cm diameter petri dish). Other assays used or in use to assay cytolytic damage by Bt toxins include trypan blue, ATP release, and quantification of cell lysis.[17,4]

An extension of earlier findings involving susceptibility of CF-1 cells to CryIA toxins has shown that CryIA(c) toxin from Bt subsp. *kurstaki* (HD-1) is the only CryIA protein toxic to CF-1 cells.[18] In addition, Knowles et al.[19] indicate that *N*-acetylgalactosamine may be a component of the CryIA(c) toxin receptor of CF-1 cells. Their conclusions resulted from experiments involving *N*-acetylgalactosamine inhibition of CryIA(c) toxicity to CF-1 cells; however, both galactose and *N*-acetylglucosamine also reduced lysis of CF-1 cells, although not to the extent of that with *N*-acetylgalactosamine.

An interesting interpretation of some data has been advanced by Swartz et al.[20] using a microspectrofluorescence technique for analysis of early events in the interaction of *Spodoptera frugiperda* (Sf-9) cells and CryIC toxin. Briefly, cells are loaded with the ester derivative of a Ca^{++} sensitive probe (FURA/2AM), the ester group of which is cleaved by esterases in metabolically active cells. Incubation of these cells with Bt toxins triggers a rise in intracellular Ca^{++} via Ca^{++} channels either endogenous to the cells or formed *de novo* prior to the formation of monovalent ion channels. Formation of these Ca^{++} channels leads to an increase in fluorescence. These studies have lead the authors to propose a novel model of Bt toxin cell lysis involving specific binding to a surface receptor, activation of Ca^{++} channels which facilitates toxin internalization, and toxin-mediated formation of chloride channels from the cytosol side of the cell membrane. Additional data is required for proof of this model.

Lepidopteran cell lines have also been used in recent studies with amino-acid modified toxins. Derivatization of the cysteine and lysine residues of Btk (HD-73) protoxin [CryIA(c)] yielded modification of all the cysteine residues by iodoacetic acid (carboxymethyl protoxin), iodoacetamide (cabaminomethyl protoxin), or *N*-(*B*-iodoethyl)trifluoroacetamide (aminoethyl protoxin), while only 35% of the lysine residues were modified with succinic anhydride.[21] Except for the aminoethyl derivative, which showed low alkaline solubility and low cytolytic activity with the 67 kDa CryIA(c) toxin in a CF-1 cell lawn assay, the other derivatives were not affected. All derivative protoxins were equally toxic *in vivo* against *Heliothis virescens* larvae. In another approach, modification of the trypsin-resistant toxins from Bt subsp. *thuringiensis* (HD-524), which by electrophoresis consist of 60 kDa (major band) and 69 kDa (minor band) proteins, by substitution of tyrosine residues with tetranitromethane reduced lysis of a *Trichoplusia ni* cell line (TN368) by approximately 60%.[22] Analysis of the absorbance data indicated that approximately 12 of 24 tyrosine residues were nitrated. Modification of accessible lysine residues by *O*-methylisourea did not affect lysis. Since there are only 3 to 6 residues in the molecule that are positioned at the C-terminal domain, alteration of these residues did not affect binding or channel formation.

III. STUDIES WITH TOXINS EXHIBITING DIPTERAN SPECIFICITY

A number of mosquitocidal and dual-specificity (dipteran and lepidopteran) strains of Bt have been studied using cells in culture (Table 2). Essentially all of the information involving toxin dipteran cell interaction has resulted from studies with cells in culture in contrast to the lepidopteran-specific toxins. Most of the data involving dipteran specificity, mechanism of action, ion channel formation, etc. have been obtained within the past ten years.

The first isolation of a Bt strain toxic to mosquitos was reported by Goldberg and Margalit,[23] and the initial observation that Bti solubilized toxin is cytolytic to mosquito cells in culture was reported by Thomas and Ellar.[17] Their preparation was also cytolytic to lepidopteran cells and hemolytic to mammalian red blood cells. The three most related isolates exhibiting dipteran toxicity are Bt subsp. *israeliensis* strain 1884 (Bti), Bt subsp. *darmstadiensis* strain 73-E10-2 (Btd), and Bt subsp. *morrisoni* PG-14 (Btm). The crystals of each are similar in size, shape, and protein composition except for the Btm and Btd strains, which in addition to the common proteins of 135, 128, 67, and 27 kDa contain an additional 144 kDa protein.[24] In general, the crystals contain several proteins, which can vary in number and size depending on the extraction conditions and presence of proteases.[25] Initially, it was found that not only could *Aedes albopictus* cells be lysed by Bti proteins, but *Spodoptera frugiperda* cells and the red blood cells of several mammals could be lysed as well.[26] It was subsequently found that Bti proteins in the range 22 to 28 kDa caused both cytolysis of mosquito cells and hemolysis without undergoing proteolytic processing.[27,28] Morphologic changes in cells undergoing lysis by Bti proteins included swelling and lysis similar to the cytopathology observed with lepidopteran cells and lepidopteran-specific toxins. These observations, combined with leakage studies of ions of varying atomic radii during treatment of *A. aegypti* cells with Bti proteins, lead to the concept of colloid-osmotic lysis as the toxic mechanism in dipteran cells, as well as in lepidopteran cells. No evidence could be developed that implicated the involvement of specific Bti toxin binding proteins on dipteran cells, although cytolytic activity was inhibited by *A. albopictus* cell lipids.[17,26] Additional work showed that specific lipids (phosphatidylcholine, sphingomyelin, and phosphatidylethanolamine) were bound by Bti toxin and that the 28 kDa Bti protein could be inserted into phosphatidylcholine liposomes.[17,29] In recent work with the cloned cryIVB gene from Bti trypsin, treatment of the 128 kDa protein yielded polypeptides of predominantly 46 to 48 kDa that were not lytic to either dipteran or lepidopteran cells; however, treatment of the 128 kDa protein with *A. aegypti* gut extract produced polypeptides of 60 to 65 kDa that were lytic to *A. aegypti* cells, but not to cells from the *A. gambiae, C. quinquefasciatus,* or CF-1 cell lines.[30] Apparently the less severe proteolytic processing of the 128 kDa molecule by the as-yet-undefined proteases of *A. aegypti* gut extract produces a polypeptide with cytolytic activity *in vitro.*

The alkaline-soluble proteins from the crystals of Btd (73-E10-2) did not exhibit hemolytic and insect cell cytolytic activities; however, after incubation with *P. brassicae* gut extracts, the proteins were cytolytic to lepidopteran cells (CF-1) as well as dipteran cells and were inhibited by *A. albopictus* membrane phospholipids.[31] A toxic mechanism similar to Bti has been postulated for this strain. In recent work with Btd (73-E10-2), it was found that the 28 kDa inactive protoxin is immunologically distinct from the crystal proteins from Bti, Btk (HD-1), Bt subsp. *aizawai* 1CI, and Btm PG-14.[32] Conversion to the hemolytic and cytotoxic 23 kDa protein was accomplished by alkaline solubilization and proteolytic processing. Cytolytic activity of this protein included the dipteran cell lines *A. aegypti, A albopictus,* and the lepidopteran cell line CF-1 and an *H. zea* line.

Two interesting strains of Bt subsp. *morrisioni* exhibiting toxicity to diptera have been studied. The PG-14 strain is not toxic to lepidoptera *in vivo* and contains an array of proteins similar to Bti except for an additional 144 kDa protein.[33,34] The alkaline-soluble crystal proteins are antigenically related to Bti proteins and are cytolytic for lepidopteran cells (CF-1) and the dipteran cell lines from *A. gambiae* and *C. quinquefasciatus,* but not for *A. aegypti* cells. In contrast, the HD-12 strain shares no plasmids with the PG-14 strain and is toxic to both lepidoptera and diptera *in vivo.* A group of proteins in the range of 126 to 140 kDa has been identified and, following proteolytic processing with *Pieris brassicae* gut extract, resulted in polypeptides cytolytic to lepidopteran cells (CF-1) and dipteran cells (*A. gambiae* and *C. quinquefasciatus*). However, no cytolysis occurred with lepidopteran cell lines from *S.*

frugiperda, M. brassicae, and *L. dispar* or dipteran cell lines from *A. aegypti, A. albopictus,* and *D. melanogaster.*

In a study of another dual specificity isolate designated Bt subsp. *aizawai* 1CI (originally classified as Bta subsp. *colmeri*-HD-847), a gene encoding a 130 kDa protoxin was expressed in a recombinant *E. coli.*[9] Processing of the solubilized protoxin by trypsin yielded a 55 kDa protein cytolytic to CF-1 cells, but not to *A. albopictus* cells. Conversely, incubation of the protoxin with *A. aegypti* gut proteases yielded a 53 kDa protein that was cytolytic to *A. albopictus* cells, but not to CF-1 cells. Therefore, different modes of proteolytic processing of certain protoxins are reflected in different toxicity specificities *in vitro.* An extension of this work by Haider, Smith, and Ellar[9] utilizing deletion mutants of the Bta 1CI cloned toxin gene showed that lepidopteran cell toxicity was retained in a polypeptide comprising the amino acid residues ile30-glu595. Yet removal of 37 residues from the C-terminal domain produced a toxin active only to mosquito larvae and *A. aegypti* cells. Cytolytic activity to CF-1 cells was lost.

In addition to the Bta strains, which are studied predominantly for their toxic properties to lepidoptera although exhibiting dipteran cell toxicity, several other strains toxic to lepidopteran larvae *in vivo* have been found to contain a mosquitocidal toxin. A toxin protein exhibiting dual toxicity is found in several Bt strains in crystals distinct from the accompanying crystals containing the lepidopteran-specific toxins. This crystal protein was initially identified in the Btk (HD-1) strain and was designated P2 as opposed to the lepidopteran-specific P1 (CryIA) proteins. The P2 protein has since been classified as the CryIIA protein expressed by the cryIIA gene. It is a 71 kDa protein sparingly soluble at pH<10 and requires proteolytic processing for activity.[35] Recent work by Nicholls et al.[36] documented the toxic specificity of five CryIIA proteins from the strains Btk (HD-1), Bt subsp. *thuringiensis* (HD-117) and (HD-270), and Bt subsp. *galleriae* (HD-29) and (HD-916). This family of proteins was related antigenically, but analysis of their molecular weights and polypeptide profiles indicated two distinct groups of proteins. Solubilization of the crystals at pH 12 or by *P. brassicae* gut fluid yielded proteins with no cytolytic activity; however, treatment of the solubilized proteins with chymotrypsin produced cytolytic proteins of 49 and 51 kDa. The CryIIA protein from HD-1 only lysed CF-1 cells, whereas the CryIIA proteins from the Bt subsp. *thuringiensis* strains lysed *A. aegypti* cells, but only after a long incubation (22 hr) while the CryIIA proteins from Bt subsp. *galleriae* strains lysed cells from the CF-1, *H. zea, S. frugiperda, A aegypti,* and *A. gambiae* cell lines.

IV. CLOSING REMARKS

Although important information on various aspects of Bt toxin biochemistry and mode of action has been obtained through the use of established insect cell lines, a review of the recent literature indicates a decline in studies with lepidopteran cell lines. The application of midgut brush border vesicles (BBMV) in the mid-1980s to studies of toxin-receptor interaction and the fact that BBMV more closely resemble the *in vivo* system obviated the need for lepidopteran cell lines for controlled *in vitro* studies. Yet the study of channel formation and lysis requires functioning cells, and for these studies lepidopteran cell lines are still being used. Recently, however, with the perfection of techniques for the preparation of primary cultures of viable midgut cells, it is probable that lepidopteran cell lines will be used even less frequently, although eventual establishment of midgut cell lines will provide additional latitude for toxin-cell research. Nevertheless, a still fruitful line of research will involve comparative studies of the toxicity mechanisms and binding proteins between lepidopteran cells in culture and primary cultures of midgut cells from the same species.

Studies with mosquitocidal toxins must still rely on dipteran cell lines for data development on toxin-cell interaction and for determining the molecular aspects of toxin specificity, since it is not feasible to obtain midgut tissue from dipteran larvae.

REFERENCES

1. **Davidson, E. W.,** Insect cell cultures as tools in the study of bacterial protein toxins, in *Advances in Cell Culture,* Vol. 7, Maramorosch, K. and Sato, G. H., Eds., Academic Press, San Diego, 1989, 125.
2. **Schwab, G. E. and Culver, P.,** *In vitro* analyses of *Bacillus thuringiensis* delta-endotoxin action, in *Analytical Chemistry of Bacillus Thuringiensis,* Hickle, L. A. and Fitch, W. L., Eds., American Chemical Society, Washington, D.C., 1990, 36.
3. **Murphy, D. W., Sohi, S. S., and Fast, P. G.,** *Bacillus thuringiensis* enzyme-digested delta-endotoxin: effect on cultured insect cells, *Science* 194, 954.
4. **Fast, P. G., Murphy, D. W., and Sohi, S. S.,** *Bacillus thuringiensis* delta-endotoxin: evidence that toxin acts at the surface of susceptible cells, *Experientia,* 15(6), 762, 1978.
5. **Himeno, M., Koyama, N., Funato, T., and Komano, T.,** Mechanism of action of *Bacillus thuringiensis* insecticidal delta-endotoxin on insect cells *in vitro, Agric. Biol. Chem.,* 49(5), 1461, 1985.
6. **English, L. H. and Cantley, L. C.,** Delta-endotoxin inhibits Rb^+ uptake, lowers cytoplasmic pH and inhibits a K^+-ATPase in *Manduca sexta* CHE cells, *J. Membrane Biol.,* 85, 199, 1985.
7. **Knowles, B. H. and Ellar, D. J.,** Colloid-somotic lysis is a general feature of the mechanism of action of *Bacillus thuringiensis* delta-endotoxins with different insect specificity, *Biochim. Biophys. Acta,* 924, 509, 1987.
8. **Knowles, B. H. and Ellar, D. J.,** Characterization and partial purification of a plasma membrane receptor for *Bacillus thuringiensis* var. *kurstaki* lepidopteran-specific delta-endotoxin, *J. Cell Sci.,* 83, 89, 1986.
9. **Haider, M. Z., Smith, G. P., and Ellar, D. J.,** Delineation of the toxin coding fragments and an insect-specificity region of a dual toxicity *Bacillus thuringiensis* crystal protein gene, *FEMS Microbiol. Lett.,* 58, 157, 1989.
10. **Knowles, B. H., Thomas, W. B., and Ellar, D. J.,** Lectin-like binding of *Bacillus thuringiensis* var. *kurstaki* lepidopteran delta-endotoxin is an initial step in insecticidal action, *FEBS Lett.,* 168, 197, 1984.
11. **McCarthy, W. J., Aronson, J. N., and Labenberg, J.,** A 63 kDa toxic polypeptide from *Bacillus thuringiensis* subsp. *kurstaki* (HD-263): effects on several lepidopteran cell lines, *In Vitro Cell. Dev. Biol.,* 24(1), 59, 1988.
12. **McCarthy, W. J.,** unpublished data, 1994.
13. **Johnson, D. E.,** Selection for resistance to *Bacillus thuringiensis* delta-endotoxin in an insect cell line *(Choristoneura fumiferana), Experientia,* 40, 274, 1984.
14. **McGaughey, W. H.,** Insect resistance to the biological insecticide *Bacillus thuringiensis, Science,* 229, 193, 1985.
15. **Tabashnik, B. E., Cushing, N. L., Finson, A., and Johnson, M. W.,** Field development of resistance to *Bacillus thuringiensis* in diamond back moth *(Lepidoptera: Plutellidae), J. Econ. Entomol.,* 85, 1671, 1990.
16. **Gringorten, J. L., Witt, D. P., Milne, R. E., Fast, P. G., Sohi, S. S., and van Frankenhuyzen, K.,** An *in vitro* system for testing *Bacillus thuringiensis* toxins: the lawn assay, *J. Invert. Pathol.,* 56, 237, 1990.
17. **Thomas, W. E. and Ellar, D. J.,** *Bacillus thuringiensis* var *israelensis* crystal delta-endotoxin: effects on insect and mammalian cells *in vitro* and *in vivo, J. Cell Sci.,* 60, 181, 1983.
18. **Masson, L., Prefontaine, G., Peloquin, L., Lau, P. C. K., and Brousseau, R.,** Comparative analysis of the individual protoxin components in P1 crystals of *Bacillus thuringiensis* subsp. *kurstaki* isolates NRD-12 and HD-1, *Biochem. J.,* 269, 507, 1989.
19. **Knowles, B. H., Knight, P. J. K., and Ellar, D. J.,** N-acetyl galactosamine is part of the receptor in insect gut epithelia that recognizes an insecticidal protein from *Bacillus thuringiensis, Proc. R. Soc. Lond. B,* 245, 31, 1991.
20. **Schwartz, J., Garneau, L., Masson, L., and Brousseau, R.,** Early response of cultured lepidopteran cells to exposure to delta-endotoxin from *Bacillus thuringiensis:* involvement of calcium and anionic channels, *Biochim. Biophys. Acta,* 1065, 250, 1991.
21. **Choma, C. T. and Kaplan, H.,** *Bacillus thuringiensis* crystal protein: effect of chemical modification of the cysteine and lysine residues, *J. Invert. Pathol.,* 59, 75, 1992.
22. **Yan, X. J. and McCarthy, W. J.,** Chemical modification of *Bacillus thuringiensis* susp. *thuringiensis* (HD-524) trypsin-activated endotoxin: implication of tyrosine residues in lepidopteran cell lysis, *J. Invert. Pathol.,* 57(1), 101, 1991.
23. **Goldberg, L. J. and Margalit, J.,** A bacterial spore demonstrating rapid larvicidal activity against *Anopheles sergenti, Uranstaenia unguievlata, Culex univitattus, Aedes aegypti,* and *Culex pipiens, Mosq. News,* 37, 355, 1977.
24. **Federici, B. A., Ibarra, J. E., Padua, L. E., Galjart, N. J., and Sivasubramanian, N.,** Parasporal body of mosquitocidal subspecies of *Bacillus thuringiensis,* in *Biotechnology in Invertebrate Pathology and Cell Culture,* Maramorosch, K., Ed., Academic Press, San Diego, 1987, chap. 9.
25. **Hofte, H. and Whiteley, H. R.,** Insecticidal crystal proteins of *Bacillus thuringiensis, Microbiol. Rev.,* 53, 242, 1989.
26. **Gill, S. S. and Hornung, J. M.,** Cytolytic activity of *Bacillus thuringiensis* proteins to insect and mammalian cell lines, *J. Invert. Pathol.,* 50, 16, 1987.

27. **Yamamoto, T., Lizuka, T., and Aronson, J. N.,** Mosquitocidal protein of *Bacillus thuringiensis* subsp. *israeliensis;* identification and partial isolation of the protein, *Curr. Microbiol.,* 9, 279, 1983.
28. **Gill, S. S., Sungh, G. J. P., and Hornung, J. M.,** Cell membrane interaction of *Bacillus thuringiensis* subsp. *israelensis* cytolytic toxins, *Infect. Immunol.,* 55, 1300, 1987.
29. **Thomas, W. E. and Ellar, D. J.,** Mechanism of action of *Bacillus thuringiensis* var. *israelensis* insecticidal delta-endotoxin, in *Bacterial Protein Toxins,* Academic Press, London, 1984, 199.
30. **Angsuthanasombat, C., Crickmore, N., and Ellar, D. J.,** Cytotoxicity of a cloned *Bacillus thuringiensis* subsp. *israelensis* CryIVB toxin to an *Aedes aegypti* cell line, *FEMS Microbiol. Lett.,* 83, 273, 1991.
31. **Drobniewski, F. A. and Ellar, D. J.,** Identification of the principal active mosquitocidal polypeptide of *Bacillus thuringiensis* var. *darmstadiensis* strain 73-E10-2, in *Fundamental and Applied Aspects of Invertebrate Pathology,* Samson, R. A., Vlak, J. M., and Peters, D., Eds., Fourth Int. Colloq. Invert. Pathol., Wageningen, 1986, 38.
32. **Drobniewski, F. A. and Ellar, D. J.,** Purification and properties of a 28-kilodalton hemolytic and mosquitocidal protein toxin of *Bacillus thuringiensis* subsp. *darmstadiensis* 73-E10-2, *J. Bacteriol.,* 171(6), 3060, 1989.
33. **Ibarra, J. E. and Federici, B. A.,** Parasporal bodies of *Bacillus thuringiensis* subsp. *morrisoni* (PG-14) and *Bacillus thuringiensis* subsp. *israelensis* are similar in protein composition and toxicity, *FEMS Microbiol. Lett.,* 34, 79, 1986.
34. **Padua, L. E. and Federici, B. A.,** Development of mutants of the mosquitocidal bacterium *Bacillus thuringiensis* subspecies *morrisoni* (PG-14) toxic to lepidopterous or dipterous insects, *FEMS Microbiol. Lett.,* 66, 257, 1990.
35. **Yamamoto, T. and McLaughlin, R. E.,** Isolation of a protein from the parasporal crystal of *Bacillus thuringiensis* var. *kurstaki* toxic to the mosquito larva, *Aedes taeniorhynchus, Biochem. Biophys. Res. Commun.,* 103, 414, 1981.
36. **Nicholls, C. N., Ahmad, W., and Ellar, D. J.,** Evidence for two different types of insecticidal P2 toxins with dual specificity in *Bacillus thuringiensis* subspecies, *J. Bacteriol.,* 171(9), 5141, 1989.
37. **Johnson, D. E.,** Entomocidal activity of crystal proteins from *Bacillus thuringiensis* toward cultured insect cells, in *Biotechnology in Invertebrate Pathology and Cell Culture,* Maramorosch, K., Ed., Academic Press, San Diego, 1987, chap. 4.
38. **Johnson, D. E. and Davidson, L. I.,** Specificity of cultured insect tissue cells for bioassay of entomocidal protein from *Bacillus thuringiensis, In Vitro,* 20, 66, 1984.
39. **Haider, M. Z., Knowles, B. H., and Ellar, D. J.,** Specificity of *Bacillus thuringiensis* var. *colmeri* insecticidal delta-endotoxin is determined by differential proteolytic processing of the protoxin by larval gut proteases, *J. Biochem.,* 156, 531, 1986.
40. **Johnson, D. E.,** Incidence of insect cell cytolytic activity among *Bacillus thuringiensis* serotypes, *FEMS Microbiol. Lett.,* 43, 121, 1987.
41. **Knowles, B. H., Francis, P. H., and Ellar, D. J.,** Structurally related *Bacillus thuringiensis* delta-endotoxins display major differences in insecticidal activity *in vivo* and *in vitro, J. Cell Sci.,* 84, 221, 1986.
42. **Witt, D. P., Carson, H., and Hodgdon, J. C.,** Cytotoxicity of *Bacillus thuringiensis* delta-endotoxins to cultured Cf-1 cells does not correlate with *in vivo* activity toward spruce budworm larvae, in *Fundamental and Applied Aspects of Invertebrate Pathology,* Samson, R. A., Vlak, J. M., and Peters, D., Eds., Fourth Int. Colloq. Invert. Pathol., Wageningen, 1986, 3.
43. **Johnson, D. E.,** Toxicity of *Bacillus thuringiensis* entomocidal protein toward cultured insect tissue, *J. Invert. Pathol.,* 38, 94, 1981.
44. **Johnson, D. E.,** Specificity of cultured insect tissue cells for the bioassay of entomocidal protein of *Bacillus thuringiensis,* in *Invertebrate Cell System Applications,* Vol. 2, Mitsuhashi, J., Ed., CRC Press, Boca Raton, 1989, chap. 10.
45. **Hofmann, C. and Luthy, P.,** Binding and activity of *Bacillus thuringiensis* delta-endotoxin to invertebrate cells, *Arch. Microbiol.,* 146, 7, 1986.
46. **Himeno, M.,** Mechanism of *Bacillus thuringiensis* insecticidal delta-endotoxin action on insect cells *in vitro,* in *Biotechnology in Invertebrate Pathology and Cell Culture,* Maramorosch, K., Ed., Academic Press, San Diego, 1987, chap. 3.
47. **Nishiitsutsuji-Uwo, J. and Endo, Y.,** Mode of action of *Bacillus thuringiensis* delta-endotoxin: effect on TN-368 cells, *J. Invert. Pathol.,* 34, 267, 1979.

Chapter 5

LARGE SCALE INSECT CELL PRODUCTION

Spiros N. Agathos

TABLE OF CONTENTS

I. Introduction 91

II. Prerequisites of Large-Scale Insect Cell Culture: Choice of Insect Cell Lines and Expression Vectors 91

III. Large-Scale Insect Cell Culture in Bioreactors 96

IV. Development of Convenient Serum-Free Media 99

V. Concluding Remarks 100

References 101

0-8493-4597-9/94/$0.00+$.50
© 1994 by CRC Press Inc.

Spiros N. Agathos, Ph.D., is Professor of Bioengineering at the Catholic University of Louvain, Louvain-la-Neuve, Belgium. He received his Dipl.Eng. in Chemical Engineering in 1973 at the National Technical University of Athens, Greece, his M.Eng. in Chemical Engineering in 1976 at McGill University, Montreal, Canada, and his Ph.D. in Biochemical Engineering in 1983 at the Massachusetts Institute of Technology, Cambridge, MA. Lecturer and Assistant Professor, University of Western Ontario, London, Canada, 1982–1985, Assistant Professor, Rutgers University, New Brunswick, NJ, 1985–1993, Professor, Catholic University of Louvain, 1993-date. He holds memberships in the American Institute of Chemical Engineers, American Chemical Society, American Society for Microbiology, Society for Industrial Microbiology, American Association for the Advancement of Science, Sigma Xi, New York Academy of Sciences, etc. He is a member of the Editorial Boards of the scientific journals *Applied Biochemistry and Biotechnology,* and *Journal of Industrial Microbiology.* His research interests include biochemical and enzyme engineering, bioreactor design, fungal and insect cell culture and scale up, and bioreactor-based pollutant degradation.

I. INTRODUCTION

Insect cell culture is recognized as an emerging technology for the production of biologicals, including recombinant proteins and biological insecticides, primarily through the use of baculovirus expression vectors (BEV). In an exhaustive review,[1] the ever-expanding literature was examined in the field of cloning and expression of heterologous genes in lepidopteran insect cells with BEV, and there was noted an excess of 350 publications from 1983 till mid-1990, with most of them appearing since 1987. This increasing flow of information can be ascribed to several desirable characteristics of this technology. Among the chief advantages of insect cell-BEV systems are the very high levels of foreign gene expression compared with most mammalian and several microbial cell systems; their capacity to produce proteins that are correctly processed post-translationally in the insect host cell, and are, as a result, enzymatically, immunologically, and functionally similar to their authentic counterparts, in contrast with many proteins expressed in bacterial or yeast systems; their helper-independence and inability to activate dormant oncogenes, unlike several mammalian expression systems; and the easy manipulation of BEV, which allows the insertion of an astonishing variety of coding sequences from practically any cellular or tissue origin, including cytotoxic gene products. These biological features, coupled with the technological similarities between *in vitro* insect cell propagation and mammalian cell culture, suggest that the insect cell-BEV system is poised to grow in importance not only for production of biological materials useful in research and product development, but also for bulk manufacturing of therapeutics, diagnostics, and biological control agents for agriculture. Consequently, this means that the commercial use of insect cell systems will require the efficient cultivation of these cells *in vitro* using current and emerging developments in bioreactor engineering.

Currently, the lepidopteran cell-BEV system is by far the main driving force behind insect cell culture engineering. In this system, the standard cell line most widely used is line Sf9 from *Spodoptera frugiperda,* and the most common vector for protein expression consists of genetically engineered *Autographa californica* nuclear polyhedrosis virus (AcNPV).[2-4] The cultivation of dipteran cell lines is pursued much less than that of lepidopteran cell lines, since the former are not efficient hosts to BEV. However, new advances in dipteran cell use for regulated protein expression, as illustrated by the Schneider 2 cell line of *Drosophila melanogaster* stably integrating multiple copies of a foreign gene in the insect chromosome, promise to complement the already high potential of the lepidopteran cell-BEV system.[5]

The developing status of insect cell culture as an established biotechnological production-scale operation is influenced by a number of factors, including (a) progress in basic understanding of the biological functions of insect cells as vehicles for protein synthesis, (b) improvements in the design and operation of appropriate bioreactors, and (c) development of inexpensive and convenient culture media, especially serum-free and low-protein media of chemically defined composition. Trends in these areas are briefly outlined below.

II. PREREQUISITES OF LARGE-SCALE INSECT CELL CULTURE: CHOICE OF INSECT CELL LINES AND EXPRESSION VECTORS

Even though a large number of widely differing proteins have been cloned and expressed in insect cell culture using BEV (a recent compilation numbers over 130 different proteins),[1] their overwhelming majority has been produced in Sf9 cells infected with recombinant AcNPV. Nonetheless, there are alternative insect cell lines and alternative expression vectors suitable for production of recombinant proteins. In general, selection of a cell line and choice of serum-containing or serum-free medium must be determined empirically for a given recombinant protein. Among the dozens of established insect cell lines and strains available for *in vitro* cultivation,[6] many are susceptible to AcNPV or other nuclear polyhedrosis viruses that are specific to the particular insect species. For example, a widely studied cell line, TN 368 from the cabbage looper, *Trichoplusia ni,* is used for propagation of wild-type and recombinant AcNPV.[7,8] Wild-type AcNPV is known to infect cultured cell lines from *Estigmene acrea* (saltmarsh caterpillar), *Lymantria dispar* (gypsy moth), *Mamestra brassicae* (cabbage worm), *Heliothis virescens* (tobacco budworm) and others.[9-11] The *Heliothis zea* (cotton bollworm) nuclear polyhedrosis virus, HzNPV,[12] and the *Bombyx mori* (silkworm) nuclear polyhedrosis virus, BmNPV,[13,14] are used for BEV development and gene cloning in their respective insect cell lines.

There is a paucity of comparative studies of gene expression level and of cellular or volumetric protein productivity in diverse cell lines using exactly the same BEV construct under standardized conditions. Such empirical studies are essential in efforts to increase consistently the level of expression of recombinant genes in BEV systems. Our current quantitative understanding of the influence cultured insect cells have on protein expression is still inadequate, and the few reports comparing protein production in parallel among several cell lines are inconclusive. For instance, a dozen years ago, when the driving force behind insect cell culture technology was the development of ample quantities of viral insecticides, the replication of wild-type AcNPV was compared in cultured insect cell lines from *S. frugiperda, E. acrea, L. dispar, M. brassicae* and *T. ni.*[9] A line from *M. brassicae,* IZD-MB0503, was found superior in terms of susceptibility to infection, level of polyhedrin occlusion bodies (POB), and rapidity of extracellular virus production. More recent work[10] comparing cellular growth rate, infectivity, and productivity of wild-type AcNPV among five cell lines from *S. frugiperda, T. ni, H. virescens, Plutella xylostella* (diamond-back moth), and *Anticarsa gemmitalis* (soybean caterpillar) suggested that the highest viral titer producers were the *T. ni* and *H. virescens* cell lines, although the *P. xylostella* cell line was superior in POB yield. Given the still uncertain relationship between intact wild-type virus infectivity, polyhedrin gene expression, and recombinant (polyhedrin replacement) gene expression in the same host, it is not automatically evident that the number of POB produced per cell and the rate of viral replication are a reliable quantitative predictor of the level of foreign gene expression expected from a genetically engineered virus infecting the cell line in question. It is easy to see that this underlying uncertainty also precludes rational projections of protein productivity among diverse host cell lines on the basis of wild-type virus infection data in these cell lines. In fact, the progression of the infection and the cytopathic effects observed in Sf9 cells differ between wild-type AcNPV infection and infection with a recombinant β-galactosidase-encoding AcNPV.[15] In the paragraphs below, I will assess the few available data from comparative studies of expression of the same proteins in several cell lines, in order to emphasize the need to screen many alternative hosts for optimal expression of a given protein before going to production scale.

In their 1991 study, Hink and coworkers[16] examined the question of comparative yields and expression characteristics of the same recombinant proteins using BEV to infect many insect cell lines in small-scale, static suspension culture. These workers chose three different types of genes for insertion into AcNPV, and they used the resulting vectors to infect 23 different lepidopteran cell lines or strains. The genes cloned were a structural gene from a virus (coding for Gp50T, an envelope *O*-linked glycoprotein from pseudorabies virus), a bacterial gene (coding for the enzyme β-galactosidase from *E. coli*), and a mammalian gene (coding for the human pro-enzyme HPg, whose proteolytic cleavage generates plasmin). The 23 host cell lines or strains originated from 12 different species of lepidopteran insects, including the standard Sf9 cell line from *S. frugiperda.* For HPg, line IZD-MB0503 from *M. brassicae* had the highest yield, and four other cell lines produced higher levels than Sf9. Also for Gp50T, four different cell lines were superior to Sf9, although certain cell lines produced Gp50T with molecular mass about 1,000 daltons larger than that from Sf9 cells, which may indicate increased glycosylation. Lines Sf9, IZD-MB0503, and BCIRL-PX2-HNV3 (the latter from *P. xylostella*) produced equally high levels of β-galactosidase, the largest yield among all cell types tested. The authors concluded that no single cell line emerged as the highest producer for all three proteins. For two of the recombinant proteins, HPg and β-galactosidase, they reported a direct comparative relationship between wild-type AcNPV replication and the proteins' level of expression in line IZD-MB0503; they also indicated that for four of the cell lines (IZD-MB0503 plus the other three lines in the earlier study by Lynn and Hink[9]) the relative numbers of POB produced are directly related to the levels of expression of the above two (HPg and β-galactosidase) of the three heterologous genes. In the same work, Hink and coworkers[16] examined the foreign protein yields from cells cultivated in serum-free and serum-supplemented media, using cell lines Sf9, TN 368, EAA, and Sf 21AE, the parent line from which strain Sf9 was derived. They concluded that, in general, the production levels in serum-free medium were equal to or better than those in serum-supplemented medium. An examination of protein yields between parent cell lines and clonal strains derived from them was inconclusive, in that, for some proteins, yields were similar between parent and clone, whereas for others yields were higher or lower in the parent. A final conclusion from this comparative study was that medium composition had a much more pronounced effect on heterologous protein expression than on susceptibility of the insect cells to wild-type virus, which may indicate that, generally speaking, infectivity data with intact AcNPV may *not* be an adequate predictor of cloned gene expression in a given line.

The type of research exemplified by the work summarized above[16] is of high theoretical and, especially, practical significance, since new and exciting applications of BEV technology and its scale-up call for more efficient and more widely applicable insect cell lines. On the other hand, the difficulties of precise comparisons at all stages and levels of such studies often are considerable. For instance, it is known empirically that environmental factors affecting foreign gene expression in cultured insect cells using BEV include, but are not limited to, cell density, cell viability, growth stage, multiplicity of infection (MOI), time postinfection (for assessing protein yield), nutrition (media ingredients, serum, or substitutes), available dissolved oxygen concentration (DO), pH, and osmotic pressure.[17] Therefore, comparative studies among diverse cell lines should not only employ the exact same BEV construct, but also maintain the same standardized protocol of cell cultivation and infection that would imply constancy of all the above-mentioned environmental factors. Although some of these parameters can be maintained at substantially similar levels between cell lines (e.g., pH, DO, cell density, and MOI at the time of infection), others are harder to control at the desirable values or ranges. For instance, different cell lines have different growth characteristics (e.g., doubling time) in the same cultivation medium, and, thus, growth stage, time postinfection (for evaluating protein yield), and, perhaps more generally, "physiological competence" for gene expression may be much more elusive when comparing cells growing in different media. In the study by Hink and coworkers[16] an effort was made to maintain the same medium, pH, osmotic pressure, DO, cell density, growth stage, and MOI for each given cell line, and only the recombinant AcNPV vector differed, corresponding to each of the three different genes expressed. However different cell lines and strains were cultivated in different media and exhibited different growth characteristics from one another. Only the comparison among four cell lines (targeted at the influence of serum-free versus serum-supplemented media, as described above) involved a single serum-free medium (ISFM) for all four lines, but three different serum-containing media (IPL-41, TNM-FH, GTC-100). On the other hand, these workers[16] attempted to standardize the procedure for infecting the cell lines, by keeping the cell density constant at 7.5×10^5 cells per ml and in exponential growth phase when exposed to the virus, by maintaining the same volume of virus inoculum for all lines, and by not carrying any spent media over into the infected cell cultures. There were, however, no consecutive kinetic data of cell growth, baculovirus infection, and recombinant protein production in the 23 cell lines tested, and the protein yields were reported on a volumetric basis, with no indication of protein productivity on a cellular basis (it is known that smaller cells could reach higher cell densities but produce lower protein levels per cell than larger cells). Therefore, the measuring of protein productivities in different media, and the reporting of yields only at 2 and 3 days postinfection rather than over the whole period of the batch until complete cell lysis, make the above comparisons among the various cell lines somewhat problematic. Perhaps more importantly for large-scale applications, it is not known to what extent data from static cultivation (T-flasks or petri dishes) can be extrapolated to agitated suspension culture (spinner flasks and bioreactors). It is safe to suggest that data from spinner-based cultivation be used as a guide for scaling up in bioreactors, in particular of the stirred tank type.

In a comparative study between TN 368 and Sf9 cells grown in spinner flasks in the same medium, GTC 100 (with 10% serum), and infected with the same AcNPV vector encoding β-galactosidase, Ogonah et al.[8] found that TN 368 cells produced about 30% more β-galactosidase on a per cell basis, and twice as much enzyme per reactor volume, than Sf9 cells, while peak β-galactosidase concentration was reached in TN 368 cells one day earlier than in Sf9 cells (2.5 vs. 3.5 days postinfection). Nonetheless, these authors concluded that, because of their comparative ease of cultivation and higher resistance to shear, Sf9 cells may be more suitable for large-scale production of recombinant proteins than TN 368 cells.

A third comparative study of recombinant protein production was carried out using the same serum-free medium, Excell 400, with Sf9 cells and two alternative candidate cell lines, IPLB-LdElta (LdElta) from *L. dispar* and IPLB-HvT1 (HvT1) from *H. virescens.*[18] Both LdElta and HvT1 cells were reported to produce higher total activity levels of recombinant β-galactosidase in static monolayer culture than Sf9 cells (by more than 40%) after infection with the same AcNPV construct, but only LdElta had a higher product yield than Sf9 (by 25%) on an activity per total cell protein basis. When comparing LdElta with Sf9 for the production of a different recombinant product, rotaviral protein VP4, total production levels of this protein were almost double the levels of Sf9, and the VP4 yields (on a per total protein basis) achieved by LdElta were also higher (by 38%) than those in Sf9 monolayer cultures. The choice of total cell protein rather than cell number as the basis for calculating the recombinant protein yield was made by these researchers[18] because the three cell lines differ in cell size. The comparison on this basis was also facilitated by the absence of serum or other proteins in the medium. In suspension culture experiments performed in spinner flasks, the same workers compared Sf9 and LdElta cells in terms of growth characteristics and recombinant β-galactosidase production. While maximum viable cell densities for LdElta were about half the values achieved by Sf9, the LdElta cells were much larger than Sf9 cells, as were their total final protein levels. Additionally, LdElta cells grew as aggregates of 50 to 200 cells and at longer doubling times than Sf9 cells, which, in contrast, grew as single cells or small aggregates. The final β-galactosidase yields on a per total cell protein basis were similar to those observed in the monolayer cultures; i.e., LdElta cells were more productive (by 38%) than Sf9 cells, whereas total (volumetric basis) enzyme production in LdElta cells was double that of Sf9 cells. An examination of the time courses of recombinant protein production in the two cell lines showed that the pronounced difference in total productivity was partially due to the longer postinfection viability and sustained gene expression manifested by the LdElta cell line compared to Sf9.[18] Long viability of the cells during the course of the infection may not be as important a factor in the overall productivity of a lepidopteran cell-BEV system since, for instance, proteolytic degradation of the foreign protein product in the infection stage may adversely influence its yield.[19]

A careful comparison of recombinant β-galactosidase production on a specific (per cell and per microgram of uninfected cellular protein) and on a volumetric (per milliliter of culture) basis was the subject of a fourth study.[20] In this examination of eight cell lines derived from the lepidopteran insects *Spodoptera frugiperda, Trichoplusia ni, Mamestra brassicae,* and *Estigmene acrea,* cell density beyond a certain level seemed to exert inhibition of specific protein and of virus production, which was ascribed to cell-cell contact: a drop of 3- to 6-fold in specific production per cell was seen in cultures at confluent cell densities in five of the eight cell lines examined, cell lines Sf9 and Sf21 being the most resistant to this adverse density effect. On the other hand, the two adherent cell lines from *Trichoplusia ni,* BTI Tn 5B1-4 and BTI TnM, were found to be the most productive among all lines tested in recombinant enzyme production both on a specific and on a volumetric basis.[20] Cell line Tn 5B1-4 (the "high-five" cell line) also had the highest viability among all cell lines tested at 3 days postinfection and was adaptable to serum-free media. A consensus emerging from several laboratories at this time is that this particular cell line may be the insect cell line of choice when a glycosylated and secreted protein is to be cloned and overexpressed in the insect cell/baculovirus system. A final result of the study was the lack of any correlation between recombinant protein and nonoccluded virus (NOV) production, even though both were seen to decrease with increasing cell density.[20]

Trying to assess the qualitative and, wherever possible, quantitative trends in β-galactosidase production among various strains that were compared in the four characteristic studies reviewed here[16-18,20] may lead to contradictory conclusions. As pointed out above, scrupulously standardized conditions of cultivation and infection are crucial in establishing the host cell system of choice for the abundant expression of a given recombinant protein through insect cell culture technology. This empirical examination ought to be viewed as a necessary initial stage in any serious attempt to produce proteins or virus on a large scale. A more complete understanding of the various (and often interdependent) factors having an influence on the final titer, the integrity/function of the foreign protein or virus product, and the relative robustness of each cell line under bioreactor cultivation conditions will be crucial before researchers can select rationally the best host cell system for the expression of a given product in insect cells.

In addition to the influence of the host cell line, the construction of the baculovirus vector, i.e., the exact location in the viral genome where the recombinant gene is inserted, is clearly of fundamental significance in achieving high levels of protein expression. For instance, even among similar BEV constructs, differences in the point of insertion by a few base pairs can have a profound effect on the levels of expression, which can differ by as much as 20-fold.[21] There are continuing efforts in the area of vector construction for high-level expression of foreign genes that are not fused with the polyhedrin gene. Successful strategies in this area include the alteration of downstream sequences[22] or of the polyhedrin ATG start codon to ATT or ATC by site-directed mutagenesis.[23,24] Analysis of expressed proteins from genes cloned in-frame with various alterations in the polyhedrin initiator codon will not only result in further enhancement of the translational efficiency of foreign gene expression in the insect cell-BEV system, but may also offer new insights in our current understanding of the rules underlying alternative initiator codons in eukaryotic systems. There is increasing evidence that the secondary structure of the mRNA near the initiator codon influences decisively the initiation of translation at non-AUG codons.[25]

The expression of more than one foreign protein, which is now possible through the construction of new, multiple gene expression vectors,[26,27] is providing further impetus for practical applications. For instance, both surface and core antigens of the hepatitis B virus have been expressed,[28] each under the control of the polyhedrin promoter in opposite orientations. Additionally, the implementation of strong promoters other than the polyhedrin promoter, which drive the expression of other non-essential late-stage genes, offers new options in the quest for abundant foreign gene expression. One such possibility stems from the p10 promoter,[29] which drives the expression of a protein that seems nonessential in the replication and production of extracellular or occluded baculovirus, even though it may have some role in the cytopathic effects (loss of viability and eventual lysis) of the infected insect cells.[30] Obviously, the use of novel expression vectors under the control of a promoter such as the p10 must offer significant advantages over the ones based on the polyhedrin promoter if they are to be considered for extensive BEV construction and use in industrial applications.

A recent development that may extend the applicability of lepidopteran cell lines beyond the lytic baculovirus-based gene expression is the stable transformation of insect cells using baculoviruses with foreign genes under the control of the promoter IE1, an immediate early viral gene.[31] The complex human glycoprotein, tissue plasminogen activator (tPA), was recently shown to be produced more efficiently in stably transformed Sf9 cells compared to normally (i.e., lytically) infected cells of the same line.[31] In general, expression levels for secreted, glycosylated proteins using the standard lytic BEV systems in insect cells are relatively low, in comparison to other recombinant products.[1] Thus, the optimal expression of complex, posttranslationally modified proteins might be one area where the stable transfection of insect cells with genetically engineered baculovirus vectors under the control of early promoters could emerge as complementary to the standard lytic BEV approach.

III. LARGE-SCALE INSECT CELL CULTURE IN BIOREACTORS

In its most essential practice, insect cell cultivation is only a variant of mammalian cell culture, and this is true independent of propagation scale. Therefore, the process options for commercial-scale insect cell culture, in terms of hardware and operational mode, revolve around the same basic repertoire of bioreactor designs and functional strategies as those developed or under development for mammalian cell-based bioprocesses. A simple engineering analysis shows that, despite the numerous available reactor designs, including an abundance of materials for cell immobilization, the operational alternatives in mode of bioprocessing are limited. Cells are grown either in batch or (semi-) continuously, in suspension or attached form. It can be easily seen from the literature of the last dozen years or so, which illustrates the decisive advancement of the art and science of animal cell cultivation for industrial applications, that among the many alternative bioreactor designs,[32,33] no single system has proven to be universally superior in productivity and scalability. Indeed, a good case can be made that, with an adequate knowledge of the physiological requirements of the subject animal cell line, one can grow the cells in such simple equipment as shake flasks, large spinner flasks, and minimally modified microbial fermentors, with the same yields and productivities as those obtainable from most of the specialized animal cell reactor systems on the market today.[34] Therefore, trends in efficient bioreactor cultivation of insect cells most likely will be shaped by the evolving technology in other animal cell propagation. In terms of bioprocessing implications, it now appears that the differences between cultured insect and mammalian cells[35] are not as overwhelming as it was thought even a few years ago. For example, the myths of insect cells' excessive fragility to hydrodynamic shear and exceptionally high oxygen demand are not supported by experimental data.[36] In fact, it is becoming progressively clearer that resistance to shear is, to a very large extent, a function of the identity of the insect cell line and of the presence of media ingredients, just as is true for other animal cells.[37,38]

For insect cell cultivation, as for the *in vitro* propagation of other animal cells, the two most basic criteria for improved bioreactor design and scale-up are (a) the aerobic metabolism of these cells and the physical, mechanical, and geometric constraints imposed on the devices used for homogeneous mixing and oxygen supply to the culture, and (b) the desirability of high productivity per unit bioreactor volume, which, ideally, implies both higher rates of product formation per cell and high cell density per unit reactor volume.

In addressing the first set of design considerations, i.e., the satisfaction of the cellular oxygen demand both during the growth and during the infection stage, when oxygen uptake rates are higher,[15,39] the means of aeration should be compatible with the relative sensitivity to shear expected of the insect cells in comparison with microbial cells, based on the former cells' larger size and lack of cell wall. Additionally, the relative sensitivity to shear among alternative insect cell lines will be an important priority for investigation, since it may dictate different strategies for oxygenation of different cell lines. For instance, the higher fragility of *T. ni* cells compared to *S. frugiperda* cells, which may be due to their larger size and more spindle-like shape, has been independently confirmed by spinner growth experiments,[8] by viscometric experiments,[37] and by experiments aimed at adapting static cell cultures to shake-flask cultures.[40]

A wide variety of agitated reactors with sparging or surface aeration, of airlift reactors, and of bubble columns, spanning a broad range of working volumes (from less than 1 liter to over 40 liters) have already been documented in insect cell cultivation, as seen in several reviews.[7,35,41-43] The loss of insect cell viability in agitated and aerated reactors or bubble columns has been primarily attributed to shear forces exerted as a result of air sparging, which could cause cell death either in the area of bubble disengagement in the free liquid surface or in the area of air injection.[62,68,69] The supplementation of culture media with pluronic polyols (non-ionic surfactants) has been shown to be effective in preventing or greatly diminishing insect cell death in agitated and aerated bioreactors.[60,62] Although the mechanism of cell damage by hydrodynamic shear is not yet fully understood, it is becoming progressively clear that, in most well-documented cases, cell death actually occurs solely as a consequence of shear forces generated by film drainage around bubbles or by breakup during bubble disengagement at the culture surface.[50] Repeated failure of several research workers to cultivate insect cells in unsupplemented airlift reactors[54,78] has also implicated shear forces stemming from bubble disengagement. Controlling the diameter of air bubbles to avoid bubble coalescence, and adding pluronic surfactant to the serum-free medium has enabled Maiorella et al.[36] to cultivate successfully Sf9 cells in a 21-liter airlift reactor.

For agitated bioreactors, Kunas and Papoutsakis[57] have attempted to separate the detrimental effects of intense agitation on suspension-cultured hybridoma cells from those of air sparging. These workers have shown that two fluid-mechanical mechanisms are involved in cell injury and death: the first is present only when there is a gas phase, and is due to bubble breakup either because of vortex formation and gas entrainment or because of direct sparging, whereas the second mechanism causes cell injury in the absence of a gas phase (and, therefore, in the absence of bubbles) only at very high agitation rates (e.g., above 700 rpm in a 2-liter bioreactor with a 7-cm diameter impeller) by stresses in the bulk turbulent liquid. With the latter mechanism, cell damage correlates with Kolmogorov eddy sizes of the same magnitude or smaller than the cell size (9 to 15 μm). Since, in most agitated bioreactors of practical use for cell cultivation, agitation rates do not normally exceed 300 to 500 rpm, Papoutsakis[38] concludes that bubble breakup is the main mechanism of cell damage, as is true for bubble column and airlift bioreactors. It is reasonable to expect that the same mechanisms are responsible for the adverse effects seen in the cultivation of suspended insect cells in agitated and/or aerated bioreactors.

Although there are some indications that the protective effect of pluronic surfactants correlates well with the hydrophilic-lipophilic balance of these block copolymers and is perhaps acting by some kind of interaction with the cell membrane,[61] the preponderance of evidence lies mostly on the side of a physical mechanism of protection.[50,59] The physical mechanism appears to imply that the additives change the average bubble size and affect the draining and bursting properties of the bubbles at the point of disengagement. In a thoughtful review, Papoutsakis[38] has analyzed the available evidence for mechanisms of shear-induced cell death and the protective effect of media additives and has shown that, in many cases, both fluid-mechanical mechanisms and biological mechanisms (fast-acting cell membrane interactions and slower-acting metabolic effects) may mediate this apparent protection. Elegant current experimentation aimed at visualizing the processes associated with bubble bursting (e.g., film drainage, liquid jet ejection) through video microscopy and fluorescent dye activation (J. Chalmers, Ohio State University; D.I.C. Wang, Massachusetts Institute of Technology; private communications) promises to add significantly to our knowledge of the phenomenon of cultured animal cell death under realistic bioreactor conditions, including, perhaps, the eventual elucidation of the protective role played by media additives such as pluronic polyols, carboxymethylcelluloses, serum proteins, etc.[38]

Strategies aimed at outright avoiding of direct sparging or, more generally, of any extended liquid-gas interface, such as attempts to aerate the insect cell culture through semipermeable silicon or teflon tubing[46,49] are of limited practical use for scale-up. On the contrary, with better understanding of the physical and biological basis of cell death due to shear, standard agitated and airlift reactors can be utilized in the future, possibly with small modifications, for large-scale applications of insect cell culture. The retrofitting of such microbial cultivation vessels with low-shear agitators (marine propellers or hydrofoils), the introduction of simple ring spargers providing noncoalescing bubbles, the strategic placement of spargers with respect to the agitators, the complementation between efficient surface aeration and nonaggressive sparging, and the reasoned use of shear-protective media additives should make the scale-up of insect cell culture quite feasible, without having to resort to specialized designs.

The second set of considerations in insect cell bioreactor development, namely the attainment of high volumetric productivity, will certainly follow along the general lines of current developments in animal cell bioreactors. This goal can be approached not only by the judicious selection of most productive insect host cell-vector systems on a per cell (or, as seen above, on a per unit cell protein) basis, but also by establishing proven biochemical engineering methodologies, such as high-density suspension culture by cell immobilization, perfusion culture, continuous culture with recycle, or combinations of these concepts. For example, *S. frugiperda* cells could be successfully propagated in a 8-liter spin-filter bioreactor through continuous perfusion of medium and removal of waste by-products, leading to high cell densities.[74] A variation of this technique, in which continuous removal of the spent medium is effected by a hollow fiber filtration module, proved beneficial for high cell density and enhancement of polyhedra formation after infection with a wild-type AcNPV.[53] In implementing high-density suspension culture of insect cells, the question will arise whether the actual foreign protein productivity will increase with high cell concentration. There are reports that infection of high-density *T. ni* cell culture with AcNPV was inhibited, compared to low-density infection,[76] and that the per cell specific productivity of a recombinant rotaviral capsid protein in Sf9 cells at high cell densities dropped, compared to that obtained at lower densities.[47] In addition, after infection of cells of eight cell lines at low density, specific β-galactosidase production was found to fall several fold in five of the cell lines tested when plated on tissue culture plates at near-confluent and confluent cell densities.[20] Typically, "low" cell densities mean approximately one-half to one million cells per milliliter of culture, whereas "high" cell densities imply typically more than two million cells per milliliter. It is not yet clear from the available studies whether this apparent inhibition is real or artifactual, and, if real, whether the mechanism is associated with depletion of nutrients including oxygen, accumulation of metabolic inhibitors, or cell-to-cell contact preventing viral receptor binding. The timely elucidation of these points will lead to optimized bioreactor cultivation and infection protocols.

Cell density increases are also possible via immobilization or attachment to inert matrixes. Abundant production of recombinant β-galactosidase has been shown in high-density cultivation of an anchorage-dependent *T. ni* strain on glass beads in a packed bed bioreactor.[7] Also, Sf9 cells have attained very high cell densities and baculovirus titers per unit volume of support matrix when immobilized in alginate-polylysine microcapsules.[52] Ultimately, the commercial attractiveness of these innovations will rest squarely on the actual densities of viable cells attained per unit bioreactor volume, on the translation of these values to correspondingly high productivities of recombinant protein or of baculovirus titer, and on the ease of product recovery and purification, compared to simple suspension culture.

The mode of bioreactor operation will also influence the overall productivity of the process. Continuous insect cell cultivation has been proposed in order to attain high and uniform productivity. In this process, two[55] or three[71] bioreactors are run in series, the first for growth of uninfected Sf9 cells followed by one or two bioreactors for cell infection and virus production. By all measures (polyhedra per cell, polyhedra-containing cells, nonoccluded virus), virus productivity has been shown to drop with time, thus limiting the useful period of continuous operation to approximately one month.[70] The reason for this situation of diminishing returns has been recently attributed to the so called "passage effect," i.e., the well-known reduction of AcNPV infectivity upon prolonged passage through multiple infection cycles.[70] It has been recently shown that the basis for this effect is the progressive enrichment of the serially passaged virus in defective interfering particles, i.e., deletion mutant particles interfering with the production of fully active AcNPV.[56,75] Thus, low-passage viral inocula will need to be used in future developments of commercial continuous or semicontinuous bioprocessing schemes. On the other hand, the recently announced nonlytic system of stably transformed lepidopteran cells in culture[31] and the also newly developed system of stably transformed *Drosophila* cells in culture[5] lend themselves naturally to continuous bioreactor propagation, since the principle behind these systems is totally different from the lytic BEV-based system: the foreign gene is inserted in the insect cell chromosome, and, in the *Drosophila* case, it is highly amplified.[5] Because of this feature, high-rate perfusion culture of *Drosophila* cells will be an important tool in increased recombinant productivity, given the growth-associated production pattern of the recombinant product.[65]

IV. DEVELOPMENT OF CONVENIENT SERUM-FREE MEDIA

Further progress in the generalized use of insect cell culture for commercial applications will depend on the development of cheap, conveniently formulated culture media. The desirability of completely defined, serum-free, and low-protein or protein-free media has been amply recognized and is one of the clearest trends in insect cell technology today.[72,73] This trend will continue not only for the purpose of eliminating serum (which is expensive, of variable composition, unnecessary for insect cell growth, and often interferes with downstream processing), but also in order to improve the nutrition of the cell population and its susceptibility to infection, and to supply ingredients necessary for surviving the rigors of the bioreactor environment, as seen above.

Careful studies addressing the detailed nutritional and biosynthetic requirements of insect cells in culture, as well as the development of adaptation protocols, will be necessary for serum-free medium design and optimization, especially in the area of commercial serum replacements and shear protective agents.[51,72] An additional direction, especially for the mass production of commodity products like biopesticide baculoviruses, will be the design of "universal" or "totipotent" media, which could be easily formulated, autoclavable, and could support the growth of several insect cell lines.[54]

The selection of serum-free media, including the basal medium formulation or the partial elimination of serum, will be often influenced by the level of foreign protein production and by the degree or type of posttranslational processing, which might be compromised. For instance, there is evidence that high serum concentration in serum-supplemented TNM-FH medium (e.g., 10%), while beneficial for fast cell growth, may not be as conducive to high recombinant β-galactosidase expression in AcNPV-infected Sf9 cells as low levels of serum (e.g., 0.5%), as shown by Broussard and Summers.[45] Conversely, less than 4% serum in the same medium affected adversely the production of recombinant erythropoietin, whereas the best medium formulation for this product was TNM-FH medium containing 4% serum and only one-fifth of its normal concentrations of yeastolate and lactalbumin hydrolysate.[64] Also, the complete absence of serum brought about a decrease in the yield of a recombinant rotaviral protein, even though the IPL-41 serum-free medium used supported a high Sf9 cell growth rate, as shown by Caron et al.[47] These workers also noted an increased sensitivity of the cells to shear-induced injury under serum-free conditions, compared to serum-supplemented cultivation, as expected from the absence of protective additives (serum proteins). Thus, in most cases, serum-free media will require a well-calibrated complement of cell-protective additives (e.g., pluronic polyols) for use in agitated or aerated vessels. Additionally, careful studies will need to be conducted in addressing the oxygen requirements of the cell line in serum-free versus serum-supplemented media, as there is preliminary evidence indicating considerably higher oxygen uptake rates for Sf9 cells in serum-free than in serum-containing media.[79] Finally, in the current absence of adequate data, the influence of media formulation on the form and extent of posttranslational modification (e.g., glycosylation, folding, excretion, etc.) of relatively complex recombinant proteins will have to be addressed in the immediate future.

V. CONCLUDING REMARKS

The previous sections have addressed a number of fronts where new knowledge will continue being generated in the area of insect cell-based biotechnology. From all available indications it is safe to assume that insect cell culture engineering is here to stay, and that it promises to expand our horizons in both fundamental understanding of BEV-driven gene expression and in efficient production of valuable products. In a matter of only a few years, foreign gene expression in cultured insect cells has emerged as a major methodological approach in the arsenal of today's biotechnologist, and, if current research and development productivity in academia and in the private sector is any indicator, the trend will continue toward further expansion of this technology for commercially important applications.

The design of specialized as well as all-purpose baculovirus vectors will offer even more power to high-level insect cell-based gene expression. The choice of cell lines will be influenced by the level, as well as the type of protein expression, including possible posttranslational modifications. Even though our knowledge of this area is still in its infancy, it is known that insect cells glycosylate proteins somewhat differently from mammalian cells,[48,58] and there are indications that there are differences in the pattern of posttranslational processing of the same protein by different cell lines.[51] Answers from this type of research will influence to an important degree the future of insect cell culture engineering, since it is conceivable that, at least in some therapeutic or diagnostic applications, the absolute structural identity between insect cell-derived and human cell-derived protein may make a decisive difference in favor of the latter. Nonetheless, the preponderance of the currently available evidence points to the fact that the overwhelming advantage of the insect cell-BEV system lies in its very abundant expression level of biologicals that are *functionally* identical to their native form.[1] Hence, further development and commercialization of this tool for industrial applications, including appropriate bioreactor scale-up, will be amply justified, provided, of course, that the relevant regulatory agencies are cooperative in this area.

Despite the obvious attraction of therapeutic or other medically oriented proteins as possible high-value products targeted for expression in insect cell culture, we should not overlook the reemerging potential of this technology for the production of genetically engineered pest control agents.[4,77] Exciting innovations in this area include, among others, the development of a recombinant *egt* deletion mutant AcNPV with superior pesticide capacity against infected larvae,[63] and the construction of much more lethal AcNPV against lepidopteran larvae by incorporating toxin-producing genes from a mite[67] or from a scorpion species[66] into the viral genome. Indeed, with a progressively favorable regulatory climate and in view of encouraging preliminary field test results,[77] the use of massive quantities of viral insecticides as an environmentally desirable alternative to toxic synthetic pesticides may yet bring insect cell culture engineering to the forefront of the new industrial biotechnology revolution and transform it into a dominant industrial practice for the next century.

REFERENCES

1. **Luckow, V. A.,** Cloning and expression of heterologous genes in insect cells with baculovirus vectors, in *Recombinant DNA Technology and Applications*, Prokop, A., Bajpai, R. K., and Ho, C. S., Eds., McGraw-Hill, New York, 1991, 97.
2. **Luckow, V. A. and Summers, M. D.,** Trends in the development of baculovirus expression vectors, *Bio/Technology*, 6, 1988.
3. **Miller, L. K.,** Baculoviruses as gene expression vectors, *Annu. Rev. Microbiol.*, 42, 177, 1988.
4. **Cameron, I. R., Possee, R. D., and Bishop, D. H. L.,** Insect cell culture technology in baculovirus expression systems, *Trends Biotechnol.*, 7, 66, 1989.
5. **Culp, J. S., Johansen, H., Hellmig, B., Beck, J., Matthews, T. J., Delers, A., and Rosenberg, M.,** Regulated expression allows high level production and secretion of HIV-1 gp120 envelope glycoprotein in *Drosophila* Schneider cells, *Bio/Technology*, 9, 173, 1991.
6. **Hink, W. F. and Hall, R. L.,** Recently established invertebrate cell lines, in *Invertebrate Cell System Applications*, Vol. 2, Mitsuhashi, J., Ed., CRC Press, Boca Raton, FL, 1989, 269.
7. **Shuler, M. L., Cho, T., Wickham, T., Ogonah, O., Kool, M., Hammer, D. A., Granados, R. R., and Wood, H. A.,** Bioreactor development for production of viral pesticides or heterologous proteins in insect cell cultures, *Ann. N.Y. Acad. Sci.*, 589, 399, 1990.
8. **Ogonah, O., Shuler, M. L., and Granados, R. R.,** Protein production (β-galactosidase) from a baculovirus vector in *Spodoptera frugiperda* and *Trichoplusia ni* cells in suspension culture, *Biotechnol. Lett.*, 13, 265, 1991.
9. **Lynn, D. E. and Hink, W. F.,** Comparison of *Autographa californica* nuclear polyhedrosis virus replication in five lepidopteran cell lines, *J. Invert. Pathol.*, 35, 234, 1980.
10. **McIntosh, A. H. and Ignoffo, C. M.,** Replication of *Autographa californica* nuclear polyhedrosis virus in five lepidopteran cell lines, *J. Invert. Pathol.*, 54, 97, 1989.
11. **Granados, R. R. and Hashimoto, Y.,** Infectivity of baculoviruses to cultured cells, in *Invertebrate Cell System Applications*, Vol. 2, Mitsuhashi, J., Ed., CRC Press, Boca Raton, FL, 1989, 3.
12. **Fraser, M. J.,** Expression of eucaryotic genes in insect cell cultures, *In Vitro Cell. Devel. Biol.*, 25, 225, 1989.
13. **Maeda, S.,** Expression of foreign genes in insects using baculovirus vectors, *Annu. Rev. Entomol.*, 34, 351, 1989.
14. **Iatrou, K., Meidinger, R. G., and Goldsmith, R.,** Recombinant baculoviruses as vectors for identifying proteins encoded by intron-containing members of complex multigene families, *Proc. Nat. Acad. Sci. U.S.A.*, 86, 9129, 1989.
15. **Schopf, B., Howaldt, M. W., and Bailey, J. E.,** *J. Biotechnol.*, 15, 169, 1989.
16. **Hink, W. F., Thomsen, D. R., Davidson, D. J., Meyer, A. L., and Castellino, F. J.,** Expression of three recombinant proteins using baculovirus vectors in 23 insect cell lines, *Biotechnol. Prog.*, 7, 9, 1991.
17. **Volkman, L. E. and Knudsen, D. L.,** *In vitro* replication of baculoviruses, in *The Biology of Baculoviruses*, Vol. 1, Granados, R. R. and Federici, B. A., Eds., CRC Press, Boca Raton, FL, 1986, 109.
18. **Betenbaugh, M. J., Balog, L., and Lee, P.-S.,** Production of recombinant proteins by baculovirus-infected gypsy moth cells, *Biotechnol. Prog.*, 7, 462, 1991.
19. **Licari, P. and Bailey, J. E.,** Factors influencing recombinant protein yields in an insect cell-baculovirus expression system: multiplicity of infection and intracellular protein degradation, *Biotechnol. Bioeng.*, 37, 238, 1991.
20. **Wickham, T. J., Davis, T., Granados, R. R., Shuler, M. L., and Wood, H. A.,** Screening of insect cell lines for the production of recombinant proteins and infectious virus in the baculovirus expression system, *Biotechnol. Progr.*, 8, 391, 1992.
21. **Matsuura, Y., Possee, R. D., and Bishop, D. H. L.,** Expression of the S-coded genes of lymphocytic choriomeningitis arenavirus using a baculovirus vector, *J. Gen. Virol.*, 67, 1515, 1986.
22. **Ooi, B. G., Rankin, C., and Miller, L. K.,** Downstream sequences augment transcription from the essential initiation site of a baculovirus polyhedrin gene, *J. Mol. Biol.*, 210, 721, 1989.
23. **Luckow, V. A. and Summers, M. D.,** High level expression of nonfused foreign genes with *Autographa californica* nuclear polyhedrosis virus expression vector, *Virology*, 170, 31, 1989.
24. **Page, M. J., Hall, A., Rhodes, S., Skinner, R. H., Murphy, V., Sydenham, M., and Lowe, P. N.,** Expression and characterization of the Ha-*ras* p21 protein produced at high levels in the insect/baculovirus system, *J. Biol. Chem.*, 264, 19147, 1989.

25. **Kozak, M.**, Downstream secondary structure facilitates recognition of initiator codons by ekaryotic ribosomes, *Proc. Natl. Acad. Sci. U.S.A.*, 87, 8301, 1990.
26. **French, T. J. and Roy, P.**, Synthesis of bluetongue virus (BTV) corelike particles by a recombinant baculovirus expressing the two major structural core proteins of BTV, *J. Virol.*, 64, 1530, 1990.
27. **Darcy, P. A., Agathos, S. N., and Kumar, R.**, Production of human hemoglobin in insect cells using a baculovirus expression system, presented at Biochemical Engineering VIII, Engineering Foundation Conferences, Princeton, NJ, July 11 to 16, 1993.
28. **Takehara, K., Ireland, D., and Bishop, D. H. L.**, Co-expression of the hepatitis B surface and core antigens using baculovirus multiple expression vectors, *J. Gen. Virol.*, 69, 2763, 1988.
29. **Weyer, U. and Possee, R. D.**, Analysis of the promoter of the *Autographa californica* nuclear polyhedrosis virus p10 gene, *J. Gen. Virol.*, 70, 203, 1989.
30. **Williams, G. V., Rohel, D. Z., Kuzio, J., and Faulkner, P.**, A cytopathological investigation of *Autographa californica* nuclear polyhedrosis virus p10 gene function using insertion-deletion mutants, *J. Gen. Virol.*, 70, 187, 1989.
31. **Jarvis, D. L., Fleming, J. G. W., Kovacs, G. R., Summers, M. D., and Guarino, L. A.**, Use of early baculovirus promoters for continuous expression and efficient processing of foreign gene products in stably transformed lepidopteran insect cells, *Bio/Technology*, 8, 950, 1990.
32. **Griffiths, B.**, *Cytotechnology*, 3, 109, 1990.
33. **Hu, W.-S. and Peshwa, M. V.**, *Can. J. Chem. Eng.* 69, 409, 1991.
34. **Persson, B., Kierulff, J., and Emborg, C.**, in *Production of Biologicals from Animal Cells in Culture* (Proc. 10th Meet. European Society for Animal Cell Technology, Avignon, France 1990), Spier, R. E., Griffiths, J. B., and Meignier, B., Eds., Butterworth-Heinemann, Oxford/Boston, 1991, 381.
35. **Agathos, S. N.**, Production scale insect cell culture, *Biotechnol. Adv.*, 9, 51, 1991.
36. **Maiorella, B., Inlow, D., Shauger, A., and Harano, D.**, *Bio/Technology*, 6, 1406, 1988.
37. **Goldblum, S., Bae, Y.-K., Hink, W. F., and Chalmers, J.**, *Biotechnol. Prog.*, 6, 383, 1990.
38. **Papoutsakis, E. T.** *Trends Biotechnol.*, 9, 316, 1991.
39. **Kamen, A. A., Tom, R. L., Caron, A. W., Chavarie, C., Massie, B., and Archambault, J.**, *Biotechnol. Bioeng.*, 38, 619, 1991.
40. **Wu, J., King, G., Daugulis, A. J., Faulkner, P., Bone, D. H., and Goosen, M. F. A.**, *J. Ferment. Bioeng.*, 70, 1, 1990.
41. **Wu, J. Y., King, G., Daugulis, A. J., Faulkner, P., Bone, D. H., and Goosen, M. F. A.**, *Appl. Microbiol. Biotechnol.*, 32, 249, 1989.
42. **Agathos, S. N., Jeong, Y.-H., and Venkat, K.** *Ann. N.Y. Acad. Sci.*, 589, 372, 1990.
43. **Agathos, S. N.**, Mass production of viral insecticides, in *Biotechnology for Biological Control of Pests and Vectors*, Maramorosch, K., Ed., CRC Press, Boca Raton, FL, 1991, 217.
44. **Goosen, M. F. A.**, *Can. J. Chem. Eng.*, 69, 450, 1991.
45. **Broussard, D. A. and Summers, M. D.**, *J. Invertebr. Pathol.*, 54, 144, 1989.
46. **Buntemeyer, H., Bodeker, B. G. D., and Lehmann, J.**, in *Modern Approaches to Animal Cell Technology* (Proc. 8th Meet. European Society for Animal Cell Technology, Tiberias, Israel, 1987), Spier, R. E. and Griffiths, J. B., Eds., Butterworths, Kent, U. K., 411, 1987.
47. **Caron, A. W., Archambault, J., and Massie, B.**, *Biotechnol. Bioeng.*, 36, 1133, 1990.
48. **Davidson, D. J., Fraser, M. J., and Castellino, F. J.**, *Biochemistry*, 29, 5584, 1990.
49. **Eberhardt, U. and Schugerl, K.**, *Dev. Biol. Standard*, 66, 325, 1987.
50. **Handa-Corrigan, A., Emery, A. N., and Spier, R. E.**, *Enzyme Microb. Technol.*, 11, 230, 1989.
51. **Hink, W. F.**, *In Vitro Cell. Dev. Biol.*, 27A, 397, 1991.
52. **King, G. A., Daugulis, A. J., Faulkner, P., and Goosen, M. F. A.**, *Biotechnol. Lett.*, 10, 683, 1988.
53. **Kloppinger, M., Fertig, G., Fraune, E., and Miltenburger, H. G.**, *Cytotechnology*, 4, 271, 1990.
54. **Koike, M. and Sato, K.**, in *Invertebrate and Fish Tissue Culture*, Kuroda, Y., Kurstak, E., and Maramorosch, K., Eds., Springer-Verlag, New York, 1988, 7.
55. **Kompier, R., Tramper, J., and Vlak, J. M.**, *Biotechnol. Lett.*, 10, 849, 1988.
56. **Kool, M., van Lier, F. L. J., Vlak, J. M., and Tramper, J.**, in *Agricultural Biotechnology in Focus in the Netherlands*, Dekkers, J. J., van der Plas, H. C., and Vuijk, D. H., Eds., Pudoc, Wageningen, The Netherlands, 1990, 64.
57. **Kunas, K. T. and Papoutsakis, E. T.**, *Biotechnol. Bioeng.*, 36, 476, 1990.
58. **Kuroda, K., Geyer, H., Geyer, R., Doerfler, W., and Klenk, H.-D.**, *Virology* 174, 418, 1990.
59. **Michaels, J. D., Petersen, J. F., McIntire, L. V., and Papoutsakis, E. T.**, *Biotechnol. Bioeng.*, 38, 169, 1991.
60. **Murhammer, D. W. and Goochee, C. F.**, *Bio/Technology*, 6, 1411, 1988.
61. **Murhammer, D. W. and Goochee, C. F.**, *Biotechnol. Prog.*, 6, 142, 1990a.
62. **Murhammer, D. W. and Goochee, C. F.**, *Biotechnol. Prog.*, 6, 391, 1990b.
63. **O'Reilly, D. R. and Miller, L. K.**, *Bio/Technology*, 9, 1086, 1991.
64. **Quelle, F. W., Caslake, L. F., Burkert, R. E., and Wojchowski, D. M.**, *Blood*, 74, 652, 1989.

65. **Schütz, C., Jager, V., Driesel, A. J., and Wagner, R.**, in *Production of Biologicals from Animal Cells in Culture* (Proc. 10th Meet. European Society for Animal Cell Technology, Avignon, France, 1990), Spier, R. E., Griffiths, J. B., and Meignier, B., Eds., Butterworth-Heinemann, Oxford/Boston, 460–465, 1991.
66. **Stewart, L. M D., Hirst, M., Lopez Ferber, M., Merryweather, A. T., Cayley, P. J., and Possee, R. D.**, *Nature (London)*, 352, 85, 1991.
67. **Tomalski, M. D. and Miller, L. K.**, *Nature (London)*, 352, 82, 1991.
68. **Tramper, J., Williams, J. B., Joustra, D., and Vlak, J. M.**, *Enz. Microb. Technol.*, 8, 33, 1986.
69. **Tramper, J., Smit, D., Straatman, J., and Vlak, J. M.**, *Bioproc. Eng.*, 3, 37, 1988.
70. **Tramper, J., van den End, E. J., de Gooijer, C. D., Kompier, R., van Lier, F. L. J., Usmany, M., and Vlak, J. M.**, *Ann. N.Y. Acad. Sci.*, 589, 423, 1990.
71. **van Lier, F. L. J., van den End, E. J., de Gooijer, C. D., Vlak, J. M., and Tramper, J.**, *Appl. Microbiol. Biotechnol.*, 33, 43, 1990.
72. **Vaughn, J. L. and Fan, F.**, *In Vitro Cell. Dev. Biol.*, 25, 143, 1989.
73. **Vaughn, J. L. and Weiss, S. A.**, *BioPharm*, 4(Feb.), 16, 1991.
74. **Weiss, S. A., Belisle, B. W., DeGiovanni, A., Godwin, G., Kohler, J., and Summers, M. D.**, *Continuous Cell Lines as Substrates for Biologicals*, 70, 271, 1989.
75. **Wickham, T. J., Davis, T., Granados, R. R., Hammer, D. A., Shuler, M. L., and Wood, H. A.**, *Biotechnol. Lett.*, 13, 483, 1991.
76. **Wood, H. A., Johnston, L. B., and Burand, J. P.**, *Virology*, 119, 245, 1982.
77. **Wood, H. A. and Granados, R. R.**, *Annu. Rev. Microbiol.*, 45, 69, 1991.
78. **Wudtke, M. and Schugerl, K.**, In *Modern Approaches to Animal Cell Technology* (Proc. 8th Meet. European Society for Animal Cell Technology, Tiberias, Israel, 1987), Spier, R.E. and Griffiths, J.B., Eds., Butterworths, Kent, U.K., 1987, 297.
79. **Hensler, W., Singh, V., and Agathos, S. N.**, Sf9 insect cell growth and β-galactosidase production in serum and serum-free media, *Ann. N.Y. Acad. Sci.*, 1993, in press.

Chapter 6

GROWTH AND INTERACTIONS OF CELLS FROM THE INSECT NERVOUS SYSTEM *IN VITRO*

L. A. Oland and H. Oberlander

TABLE OF CONTENTS

I. Introduction 107

II. Organ Culture — The Brain and Neurosecretion 107

III. Primary Cell Culture 109
  A. Neurogenesis and Differentiation 109
  B. Development of Neurotransmitter Phenotype 110
  C. Ion Channel Expression 111
  D. Neurite Outgrowth 113
  E. Synaptogenesis 115
  F. Role of Glial Cells 116
  G. Trophic Interactions 117

IV. Conclusions and Prospects 119

Acknowledgments 121

References 121

0-8493-4597-9/94/$0.00+$.50
© 1994 by CRC Press Inc.

Lynne A. Oland, Ph.D., is an Assistant Research Scientist at the Arizona Research Laboratories, Division of Neurobiology, University of Arizona, Tucson, Arizona. She received her B.S. degree in Nursing in 1972, an M.S. degree in Nursing from The Catholic University of America in 1975, and her Ph.D. in Physiology from the University of North Carolina at Chapel Hill in 1985. NIH post-doctoral fellow, Georgetown University, 1985–86, and University of Arizona, 1987. Sigma Theta Tau, Society for Neuroscience, Association for Women in Science, AAAS. Neuronal-glial interactions in development, development of sensory systems, electron and confocal microscopy, physiology, tissue culture.

I. INTRODUCTION

The methodology of culturing tissues outside the organism under conditions that permit detailed morphological examination of growth and development was established by Ross Harrison in 1907 when he placed pieces of embryonic frog neural tube in drops of lymph and observed the outgrowth of nerve fibers. Only a few years later, also at Yale, Richard Goldschmidt performed the first organ culture experiments with insects. Surprisingly, the merging of these two historic approaches, i.e., insect and neuronal tissue culture, was decades away (see review by Beadle and Hicks).[1] Moreover, it wasn't until the breakthrough in the development of insect culture media and the establishment of the first insect cell lines that organ culture approaches became widely used on insect tissues (see Chapter 7).

Although a number of laboratories made occasional studies on cultures of the insect nervous system, it wasn't until the incisive work of Levi-Montalcini that substantial attention was brought to this subject. In an early review article Levi-Montalcini and Aloe[2] stated, "A new opening in this area came very recently with the discovery of the remarkable capacity of the insect nervous tissue and of other tissues from insect embryos to survive in simple chemically defined media for many months and to produce, under these conditions, vigorous nerve fiber outgrowth and cell migration." Levi-Montalcini had already achieved substantial fame for her role in the discovery of nerve growth factor, for which she used a tissue culture assay based on culturing chick ganglia in proximity to a tumor fragment (see discussion in Purves and Lichtman).[3] Thus, the attention of Levi-Montalcini's laboratory to the culture of insect nervous system brought substantial focus to this approach.

Meanwhile, rapid progress in the use of tissue culture methodology continued in research with vertebrate nervous systems; it included the establishment of cell lines, a feat yet to be accomplished from insect nervous tissue. These approaches have been so successful with the vertebrate nervous system that an extensive manual on cell culture in the *Methods in Neurosciences* series was recently published.[4] By contrast there have been few publications on tissue culture of the insect nervous system, and only one comprehensive review,[1] while separate chapters on cockroach and *Drosophila* neuronal culture have been published.[5,6]

In the present work we have chosen to focus on the developmental aspects of tissue culture studies with the insect nervous system, and especially with investigations of interactions among components of this system. *In vitro* technology promises to help us both ask and answer specific questions about the development of the nervous system in insects in a manner not possible *in vivo*. For example, organotypic tissue culture allows the establishment of the neuroendocrine system under defined conditions so that the role of specific neurosecretory cells can be established. Explants from the nervous system at various stages of development permit the evaluation of outgrowth of neurites under different conditions. The ability to culture dissociated cells allows the developmental capacities of neuroblasts to be examined, as well as the interactions of different elements of the nervous system both during embryogenesis and metamorphosis. The profound transformations that are part of insect metamorphosis provide a unique opportunity for determining the factors that control morphological and functional alterations in neuronal function, and tissue culture methodology is an important asset in such research. Thus, tissue culture of the nervous system allows the advantages of the direct tissue culture approach to be applied to the intriguing biological questions that can be asked of the developing nervous system.

II. ORGAN CULTURE — THE BRAIN AND NEUROSECRETION

One of the first applications of tissue culture to neurophysiology in insects was based on attempts to recreate the neuroendocrine control of molting in a culture chamber. Among the early approaches were efforts to determine whether insect brains remained endocrinologically active *in vitro.* Several reports were published that indicated that cultured brains from orthopteran and lepidopteran species maintained, produced, or released neurosecretory granules, based on histological and ultrastructural studies.[7-10] Apparently, there was some specificity to the capacities of insect brains to maintain function, as Borg and Granger[8] observed that neurosecretory granules were present in the medial M-2 cells of the brain but not in other neurosecretory cells. These results provided encouragement that the neurosecretory system could produce hormones *in vitro,* but this required direct testing of hormonal activity.

Among the first successes with assessing the endocrinological activity of cultured brains were experiments in saturniid moths. Active brains and inactive pupal prothoracic glands (the source of the molting hormone, ecdysone) were incubated together in hanging drop cultures, resulting in increased RNA synthesis in the prothoracic glands.[11] This approach was followed by the use of *in vitro* bioassays to determine the ecdysteroid activity from the incubation of prothoracic glands with endocrinologically active brains. For example, Kambysellis and Williams[12] cultured active brains with inactive prothoracic glands and pupal testes from *Cynthia* silkworm pupae. Under these conditions spermatogenesis in the testes was promoted. Another indirect bioassay was developed by Agui[13] in which *Mamestra brassicae* pupal brains were cocultured with prothoracic glands, and the stimulation of molting of isolated fragments of epidermis from the larval stage of another moth, *Chilo supressalis,* was observed. In these experiments brains, but not abdominal ganglia or fat body, stimulated molting of the target epidermis when prothoracic glands were present in the cultures.

One of the most clear-cut early demonstrations of neurohormonal release and action in organ culture was the assay of "hindgut stimulating neurohormone" (HSN) in the cockroach, *Leucophaea maderae.* The HSN was synthesized and stored by cultured brains, but if the corpora cardiaca-allatum complex was still attached with the nerves intact, the HSN did not accumulate in the brains, but was released into the medium. Although most of the released HSN was inactivated, there was a sufficient amount present to positively bioassay in the hindgut test.[14-15]

Another success in demonstrating neurosecretion *in vitro* was achieved with the cricket, *Locusta migratoria.* In this case neurosecretion in tissue culture was observed when either the entire cerebral neurosecretory system or isolated neurosecretory lobes of the corpora cardiaca were incubated *in vitro.* Release of gonadotropic neurohormones was assessed by the direct observation of release of radioactively labeled (^{35}S-cysteine) neurosecretory material. In addition, immature adult locusts were injected with medium conditioned by the neurosecretory sources and ovarian development was stimulated. Positive results for both release of neurosecretory material and stimulation of gonadogenesis were observed when the pars intercerebralis was electrically stimulated to initiate neurosecretion.[16]

The tissue culture of prothoracic glands combined with a sensitive radioimmune assay for secreted ecdysteroids provided the basis for a breakthrough in understanding endocrine interactions during diapause in insects. Bollenbacher et al.,[17] demonstrated that prothoracic glands from *Manduca sexta* pupae could be stimulated to produce ecdysteroids upon exposure to brain extracts with prothoracicotropic activity (PTTH). This approach was used to demonstrate that prothoracic glands from diapause pupae of *M. sexta* were refractory to PTTH stimulation.[18] Amazingly, brain-retrocerebral complexes from last instar larvae of *M. sexta* could be cultured *in vitro* under different photoperiodic regimes and be reprogrammed. Thus, brains that received a long-day photoperiod *in vitro* were implanted into recipient larvae destined for pupal diapause, and produced instead nondiapause pupae.[19]

Insect nervous systems have been exploited for developmental studies because they offer a number of special advantages to the investigator. These advantages include:

1. That they have a smaller number of cells in their nervous systems than vertebrates
2. That individual neurons, and in some cases glial cells, can be identified and followed readily through development

3. That the cells tend to have more restricted, less plastic, developmental programs
4. That there is often considerable understanding of the intricacies of neuronal networks underlying particular behaviors
5. In the case of *Drosophila,* that there is an increasing array of mutant phenotypes available with which to explore questions of function and regulation during development

Nevertheless, the much-applauded simplicity of these systems is only relative, and many experiments designed to examine particular developmental interactions or patterns of growth are confounded by the complexity of the *in vivo* environment. The jump into cell culture was inevitable and made with the expectation that the advantages of the insect nervous system could be pursued in a more precisely controlled experimental environment. Culture of cells from insect nervous systems holds considerable promise as an approach for exploration of many issues in development.

III. PRIMARY CELL CULTURE

For the portion that follows, we have chosen to review tissue culture research that addresses questions of general interest to developmental neurobiologists. While much of the *in vivo* work on insect nervous systems has focused upon exploration of particular neuronal circuits, the concern of these culture studies has not been the recreation of such circuits, but rather simplification of the system to permit examination of specific cellular processes or interactions that occur during development.

A. NEUROGENESIS AND DIFFERENTIATION

In vivo studies in insects have provided detailed information about the patterns of division and the differentiation of neurons within various lineages,[20-29] but it has been difficult *in vivo* to dissect out the nature and relative importance of exogenous and intrinsic influences that guide the generation and development of neuronal lineages. The effort to resolve some of these influences using cell culture began with the identification of neuroblasts in culture in a series of studies by Seecof and his collaborators.

A technique was developed for the dissociation of cells from early *Drosophila* embryos that permitted use of staged animals and relatively long-term survival.[30] Dissociations from gastrula-staged but not blastula-staged embryos yielded cells which at plating bore the ultrastructural features of undifferentiated cells yet were competent to differentiate over a period of about a week into process-extending cells with the ultrastructural characteristics of neurons. Some of the cells in these cultures were identified as neuroblasts on the basis of their large size and the appearance over time of their progeny, seen as a closely associated cluster of small cells that resembled neurons.[31] At least the first few divisions of the neuroblasts, as followed in time-lapse photographs, appeared to follow the pattern of asymmetric division that occurs *in vivo.*

Because neurogenesis continues in the postembryonic nervous system, the behavior of neuroblasts also could be studied in cultures generated from *Drosophila* larvae (third instar).[32] At this age, neuroblasts derived from different regions of the nervous system can be compared because it becomes possible to dissociate just the nervous system, in whole or in part, rather than the whole embryo. Several classes of neurons similar to those found in embryo cultures appeared, among them "Type I" cells. These cells, which were large and gave rise to clusters of smaller cells with processes typical of neurons, also stained positively with an antibody against horseradish peroxidase (HRP$^+$), unlike neuroblasts *in vivo.*[33] The reason for this discrepancy, which also appears in embryo cultures, is not understood, but adds a measure of doubt about the absolute fidelity of neuroblast behavior *in vitro.*

Furst and Mahowald[34] continued the examination of cultured neuroblasts in cultures generated from *Drosophila* embryos (4 to 6 hours postfertilization). They developed a technique based on centrifugation and differential adhesion that produced cultures enriched in neuroblasts. The blast cells and their progeny were HRP$^+$, neurites extended from the cell clusters, and most cell clusters stained intensely for AChE, thus providing further evidence that the progeny were capable of differentiating in culture. The clonal pattern of neurogenesis seen *in vivo* was also seen in culture when the cells were plated at high dilution to ensure isolation of single neuroblasts. When the antimitotic agent hydroxyurea was added to the culture medium,[35] the size of the resulting clusters was diminished. In addition, larger clusters in which the founding neuroblast would have undergone the greatest number of divisions incorporated [^3H]-thymidine for longer intervals.[34]

Taken together, these studies suggest that the behavior of neuroblasts isolated in culture generally paralleled their behavior *in vivo.* They set the stage for studies of regulation of neurogenesis using neuroblasts dissociated from mutant embryos. For example, *pecanex* is a maternal-effect mutation characterized *in vivo* by hypertrophy of the brain and nerve cord.[36] In a recent report, LaBonne and Furst[37] compared the number of neuronal cell clusters in cultures made from wild-type and *pecanex* embryos to determine whether the hypertrophy seen *in vivo* was due to failure to regulate the number of neuroblast divisions or the number of epithelial cells destined to become neuroblasts. Their finding of an increase in the number of cell clusters in cultures from mutant embryos, with no difference between wild-type and mutant in size of the clusters, supports the hypothesis that the *pecanex* mutation results in an increase in the number of neuroblasts rather than additional rounds of division.

The study of neurogenesis by analysis of cultured neuroblasts from genetically perturbed embryos is clearly in its infancy. Nevertheless, results such as these suggest that this culture system provides a means for detailed exploration of the regulation of neurogenesis that complements ongoing *in vivo* studies. Also, the ability of the progeny to differentiate at least partially provides the opportunity to examine in isolation factors that might determine or influence the phenotype of developing neurons.

B. DEVELOPMENT OF NEUROTRANSMITTER PHENOTYPE

Considerable progress has been made in examining the development of transmitter phenotype. Some of the cells in the *Drosophila* embryo (gastrula) culture system, which were apparently undifferentiated at plating, were shown to develop the ability to synthesize and accumulate acetylcholine,[38] and to manufacture enzymes active in the metabolism of acetylcholine, including AChE and ChAT, in a temporal pattern that was similar to that seen *in vivo.*[39] In addition, these cells rapidly became capable of α-bungarotoxin binding. Use of various cholinergic antagonists and agonists altered the pattern of enzyme expression, suggesting a degree of plasticity in determination of transmitter phenotype.[39]

The initial expression of a transmitter in this system, however, appears to be determined by the neuroblasts rather than by cell-to-cell interactions. In cultures highly enriched for neuroblasts, neurons identified immunocytochemically as serotonergic or dopaminergic were found in small subsets of the population of nerve cell clusters, in proportions similar to those found *in vivo.*[40] The cells were plated at low enough density in sufficient volume of medium that neither direct nor humoral interactions between cells in different clusters was likely. Addition of hydroxyurea to truncate the lineages at different times after initiation of the culture revealed that serotonergic neurons arose from early divisions of their precursor cells while dopaminergic neurons arose throughout their lineage. The lineage-dependency of transmitter phenotype expression found in these studies confirm *in vivo* data in the grasshopper.[28,29,41-42]

Huff et al.[40] showed that expression of the serotonergic phenotype did not depend upon actual cell division. By using cytochalasin-B to inhibit cytokinesis, but not nuclear division, it was possible to generate multinucleated cells that were HRP$^+$ and could extend neurites.[40,43]

Some of these cells expressed the serotonergic phenotype, suggesting that the timing of transmitter-associated gene expression was likely to depend upon the number of rounds of DNA synthesis.

C. ION CHANNEL EXPRESSION

Initial efforts to study the development of neuronal excitability focused, of course, on determining that neurons cultured from embryonic tissue or developing adult nervous systems retained the ability to become electrically excitable. These studies, and subsequent ones designed to identify in certain classes of neurons the currents underlying both ligand- and voltage-gated conductances, have been described for cockroach,[5,44-45] for locust,[46] for *Drosophila*,[6,46-54] and for *Manduca*.[55-56] In short, most of the activity thus far has been directed toward characterizing the currents found in neurons dissociated from different stages of development.

With this foundation, experiments are beginning to be designed with two long-term goals in mind: first, to understand the regulation, modification, and modulation of neuronal ion channels during development; and second, to understand how the pattern of ion channel expression affects the course of development.

Not surprisingly, sodium channels have been the foci of many studies using cultured neurons from mutant *Drosophila*. The mutants, *nap^ts^* (no action potential, temperature-sensitive) and *para^ts^* (paralytic, temperature-sensitive) mutants, which are characterized by alterations in membrane excitability, have been examined in several studies. Comparison of neurons from wild-type, *nap^ts^*, and *para^ts^* larvae showed that all were sensitive to veratridine, which blocks Na^+ channel inactivation and decreases the survival rate of the neurons. The effect of veratridine could be blocked with tetrodotoxin (TTX),[47-57] a specific blocker of Na^+ channels. *nap^ts^* and *para^ts^* neurons, however, were far less sensitive to veratridine, suggesting either a decrease in channel density or a change in Na^+ channel function. Recordings of whole-cell Na^+ currents from embryonic neurons from *nap^ts^* mutants were indistinguishable from those of wild-type neurons, a surprising finding given the strong behavioral phenotype.[50] The reason was not obvious. On the other hand, cultures from larvae carrying any of three alleles of the *para^ts^* locus revealed a significant reduction in the percentage of neurons with Na^+ currents, but not K^+ currents, when compared to wild-type neurons.[51] Changes in the current-voltage and conductance-voltage curves for neurons carrying one of the alleles were consistent with a change in the gating properties of the channel, but the specific defects involved can only be resolved with single-channel recordings.

Whole-cell Na^+ currents were also examined in embryonic neurons of two other mutants with paralytic phenotypes, *sei^ts^* (seizure, temperature-sensitive) and *tip-E*.[50] Current density was lower in neurons from both mutants than in wild-type neurons, although the percentage of neurons expressing Na^+ channels did not differ among the three groups. No differences in gating, voltage dependence, or TTX sensitivity were seen, suggesting that the *sei^ts^* and *tip-E* loci are important in regulation of channel density but not structure of the channel itself.

Several of these studies noted, almost in passing, an interesting finding with regard to a potential role for Na^+ channels in regulation of neurite outgrowth. No difference was detected between wild-type and *nap^ts^* or *para^ts^* neurons in the pattern of outgrowth, at least as determined by length of primary neurite, nor did application of TTX alter the pattern.[47,57] No obvious or gross delays were seen in *tip-E* or *sei^ts^* embryo development or in the morphological development of the mutant neurons in culture.[52] While the heterogeneity of neuron types in these cultures as well as the roughness of the measure of outgrowth or development must be taken into account in the interpretation of these data, the results suggest that Na^+ channel expression may have little effect on outgrowth of these neurons at these stages of development.

Similar studies have focused on potassium channels. Among the several types of potassium channels found in cells from larval *Drosophila* are two A-type currents.[47,48,58] As part of an effort to understand the functional aspects of voltage-dependent gating of ion channels and to understand the role of the various K^+ channels in regulating neuronal excitability, Solc and Aldrich[49] detailed the gating properties of the A_2-type K^+ channel. This channel is present in *Shaker*-deficient neurons, whereas the A_1-type channel, if normally present in neurons, would not be expressed. The similarity of gating properties of the A_1- and A_2-type channels presumably reflects similarities in the structures of the two channel types and in the genes coding for them.

Yamamoto and Suzuki[52] compared the permeability and gating properties of a noninactivating K^+ channel found in larval neurons dissociated from wild-type and *eag* mutants. *eag* has been found to cause hyperexcitability of larval presynaptic terminals and reduction of sustained outward currents in larval muscle.[47] Unexpectedly, no differences were found *in vitro* in the properties of the channel or in channel density. Several possible explanations were offered. The *eag* locus may not code for the structural protein, a different channel type that is not expressed in their culture conditions may be affected, or channel expression may be confined to the terminal processes of the neurons and therefore be difficult to detect from soma recordings.

In the long run, the voltage-clamp experiments that are relatively easy in cultured neurons but difficult or impossible *in vivo* will allow delineation of particular defects in level of channel expression or in gating properties of neurons. These results should in turn provide insight into the structure of the channels. As patterns of channel expression and the degree to which those patterns are intrinsically programmed become understood, it will be possible to examine the epigenetic events that can alter expression of particular channels and subsequently affect other developmental events.

Studies to examine the role of steroid hormones in the pattern of ion channel expression have been initiated in the *Manduca* motoneuron culture system. In *Manduca,* larval leg motoneurons undergo morphological and functional changes to accommodate the requirements of the adult.[59] These changes occur in concert with changes in the titer of 20-hydroxyecdysone (20HE), a steroid hormone known to be important in the regulation of numerous developmental events in insect nervous systems. Hayashi and Levine[56] examined the biophysical properties of identified motoneurons from larval, pupal, and adult stages that had been in culture for one day. Compared to motoneurons isolated from either larval or adult stages, those from early pupal animals showed lower current densities for both a voltage-sensitive Ca^{++} current and for a transient, inactivating outward K^+ current. The mechanism of this stage-dependent modulation is not known, but may reflect changes in the steroid hormone level. The possible relationships between the changes in current density and the behavioral and morphological changes seen *in vivo* are not yet understood.

A role for 20HE is also being explored in pupal antennal-lobe neurons from *Manduca.*[62] Neurons were dissociated from stages 2, 5, or 18 (of 18 stages of metamorphic adult development), which are before, during the rising phase of, and after the 20HE peak. After two weeks of growth in culture, the presence or level of A-like, but not delayed-rectifier potassium currents depended in some cell types on both the stage at which the neurons were dissociated and on the concentration of 20HE in which the neurons were grown, suggesting that the steroid hormone could have an effect on the expression of certain K^+ channels.

Recently, it has been shown possible to examine in culture the morphological and biophysical differentiation of a specific class of sensory neurons. Olfactory receptor neurons were dissociated from the antennae of stage-2 or -3 pupal *Manduca,*[61] that is, shortly after the birth of the receptors. Monoclonal antibodies (olfactory-specific antibody and male olfactory-specific antibody)[62,63] were used to positively identify the small, bipolar neurons with fine dendrites as receptor neurons. The ionic currents present in the receptors after 2 days and after 2 to 3 weeks *in vitro* were examined using whole-cell and single-channel patch-clamp techniques. At 2 days, a sustained voltage-dependent K^+ current was present.[64] At 2 to 3 weeks, one type of sodium channel, and at least three types of potassium channels, but no calcium channels were found. The activity of one of the K^+ currents, which resembled the delayed rectifier, could be inhibited by the application of nucleotides to the cytoplasmic side of the membrane.[65] Stimulation with female pheromone blend resulted in a change in the pattern of channel openings.[66] These baseline studies provide the framework for exploring the processes by which transduction of olfactory signals occurs and the manner in which transduction capability develops.

D. NEURITE OUTGROWTH

Two basic questions about neurite outgrowth have been approached in insect cultures: what are the mechanisms underlying the growth of neuritic processes, and what developmental interactions modify the pattern of outgrowth? A series of papers by Wu and colleagues[67-69] provide an example of an approach to the first question using *Drosophila* larval neurons.

To determine the contribution of membrane recycling pathways to neurite growth, the outgrowth patterns of cultured wild-type and *shibire*^{ts1} larval neurons were compared. *In vivo* observation[70] had previously revealed an accumulation of pit-like structures in the membrane of the presynaptic terminal, suggesting that *shibire*^{ts1} produced a defect in membrane recycling. In tissue culture[67-69] no difference in outgrowth patterns was noted at permissive temperatures. At restrictive temperatures, time-lapse video recordings revealed a reduction in the number of cells with growth cones, a reduction in the adhesion of neurites to the substrate, and a decrease in neurite initiation, elongation, and branching in mutant compared to wild-type neurons. *shibire*^{ts1} neurons exposed to the restrictive temperature on day 1 in culture showed less branching and an odd growth pattern in the primary neurite, a pattern that could reflect a defect in membrane cycling in the growth cone, in regulation of the cytoskeleton, or in expression of molecules mediating adhesion to the substrate.

Using uptake of HRP as a marker of endocytotic activity to probe the nature of the underlying defect, Masur et al.[69] found a temperature-dependent, reversible block of constitutive endocytosis. It was not possible in this study to follow HRP uptake through the entire cycling pathway and thus determine precisely the nature of the defect. Nor were growth cones specifically studied. Nevertheless, the coincidence of decreased neurite outgrowth and block of endocytosis suggests that regulation of neurite growth may depend upon regulation of membrane recycling pathways. While further studies will be needed to establish the mechanisms underlying the changes in neurite outgrowth seen in this study, the overall approach of comparing normal with genetically perturbed neurons will certainly provide insight into the processes regulating growth.

Kim and Wu,[68] for example, have recently begun to characterize the morphologies of neurons in their cultures to provide baseline information about growth patterns. Two neuronal morphologies were prominent in cultures from brain or ganglia of third instar *Drosophila* larvae, each associated with a different growth cone morphology and pattern of motility as assessed by video recordings. Growth cones of monopolar neurons had larger lamellipodia with shorter and less numerous filopodia than did multipolar neurons, and the motility of the monopolar growth cones was greater than that of multipolar neurons. Although these two neuronal morphologies certainly subsume a much larger number of *in vivo* types, the broad classification of type is probably a reasonable approach for studies probing the mechanisms of neurite outgrowth, where the assumption is made that a defect in mechanism would be manifested similarly regardless of cell type.

On the other hand, where the goal of the experiment is elucidation of the role of cellular interactions, assays using an overly broad classification of neuronal type could easily obscure a signal affecting neurite outgrowth that is developmentally important, but recognized only by a subset of neurons. This is especially likely in insects where, for example, small numbers of neurons, and in some cases individual neurons, have been shown to follow particular and precise pathways,[71] using cues to which they and not many of their neighbors respond.

Several studies that have recently been initiated ask how various cellular interactions modulate neuritic outgrowth. Two culture systems, both utilizing neurons from larval or developing adult *Manduca,* take advantage of the presence in culture of known neuronal types to ask how particular interactions influence neuronal morphology. In the first system, motoneurons were dissociated from specific ganglia where the *in vivo* morphogenetic behavior of these neurons and their muscle targets at various developmental stages is well known.[61,72] The motoneurons were identified *in vitro* by prior *in vivo* labeling with fluorescent dye, or by size. Similarly, in the second system, neurons of the developing antennal (olfactory) lobe were dissociated from precisely staged pupal animals.[55] These cultures revealed eight morphological types, each of which bore a characteristic current signature in whole-cell patch-clamp recordings.[55,73] For this system, the *in vivo* developmental history of both the sensory input[74-76] and the antennal lobe have been characterized in detail.[77-79] An important factor in the decision to begin studies in both the motoneuron and antennal-lobe systems was the availability of detailed information about the changes in hormone titer that occur throughout the larval and pupal stages.[80,81]

Initial culture studies in both systems focused on the effect of changes in hormone titer on neurite outgrowth. Insect development depends in large part on fluctuations in the titers of ecdysteroids and juvenile hormone.[82] In the case of *Manduca* where, for example, there is a large literature on the interacting roles of ecdysteroids and juvenile hormone in the development of motoneurons and the sensory neurons that innervate them,[59] the complexity of the *in vivo* environment has made it difficult to determine whether the effect of these hormones on the growth and branching patterns of motoneurons is a direct or indirect effect.

Prugh et al.[72] studied the influence of the hormone 20-hydroxyecdysone (20HE) on the outgrowth of neurites from motoneurons shown *in vivo* to respond to the presence of 20HE. Identified motoneurons were dissociated from the thoracic ganglia of early pupal animals (just prior to the onset of dendritic outgrowth *in vivo*). Compared to leg motoneurons grown in the absence of 20HE, those grown in the presence of 20HE showed significantly greater neurite outgrowth and branching (Figure 1). Juvenile hormone blocked this effect. No effect of 20HE was seen if the motoneurons were dissociated from larval ganglia at a time when their dendritic arbors normally would be regressing or if intersegmental motoneurons, which normally do not respond to increases in ecdysteroid titer *in vivo,* were plated instead of leg motoneurons.[83] The increase in neurite outgrowth was seen whether the leg motoneurons were grown in isolation or in cultures containing other neurons from the ganglia, suggesting that the effect of the ecdysteroid is direct.[72]

The cell specificity of this effect of 20HE on neuritic outgrowth was also apparent in the responses of pupal neurons dissociated from the antennal lobes.[84] Some types of antennal-lobe neurons failed to respond morphologically to an increased 20HE titer, while others showed statistically significant, but small, increases or decreases in neurite outgrowth. The direction of the effect did not depend upon whether the neuronal type belonged to the class of local interneurons or of projection (output) neurons.

Interestingly, the response of some of the olfactory neuron types to change in hormone titer *in vitro* depended upon the developmental stage from which the neurons were dissociated. *In vivo* the morphology of antennal-lobe neurons begins to change coincident with the arrival of sensory input from the antenna.[78] Most, but not all, types of antennal-lobe neurons showed greater outgrowth and/or branching when their developmental history *in vivo* included exposure to sensory axons. Furthermore, the presence of an effect depended in some types upon hormone titer of the culture medium, suggesting interesting interplay between the two developmental signals.

These studies in *Manduca* suggest that *in vitro* studies of specific subpopulations of insect neurons can be useful for examining the mechanisms underlying the effects of steroid hormones as well as the effects of sensory ingrowth on developing neurons. They also

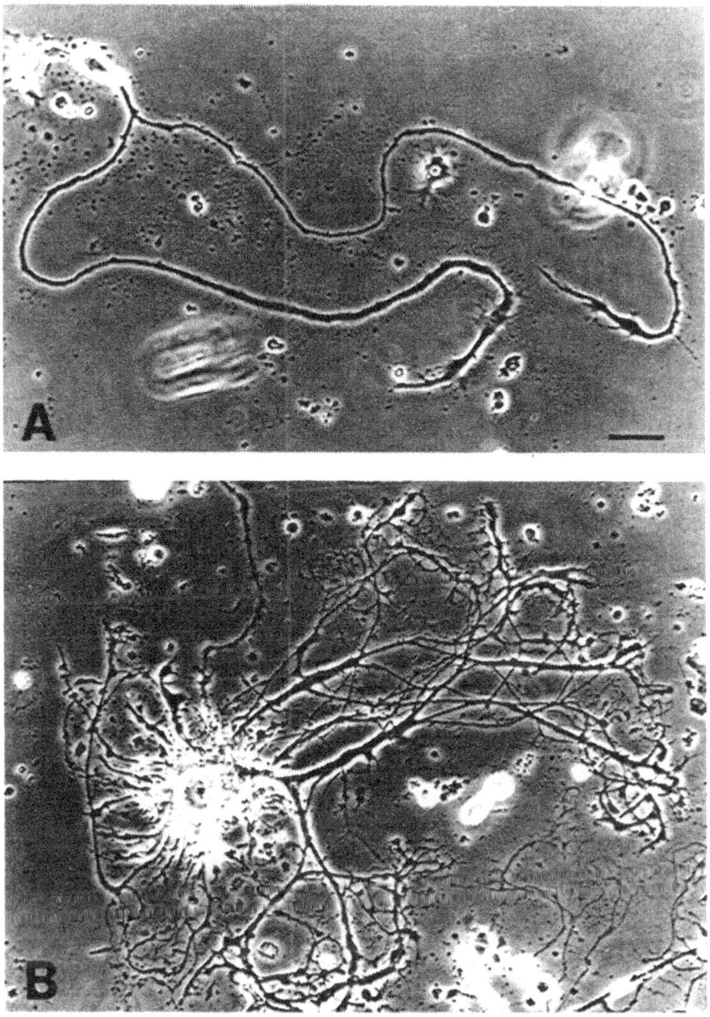

Figure 1. Neurite outgrowth in the presence and absence of the steroid hormone 20-hydroxyecdysone. Thoracic motoneurons isolated on the first day of metamorphic adult development (P0) and maintained for 6 days in medium (A) lacking hormone or (B) with hormone (1 μg/ml). Note the extensive arborization of the cell in B, typical of motoneurons grown with hormone. Micrographs courtesy of Dr. Richard B. Levine. Scale bar = 50 μm.

demonstrate that attention to staging consistency and to hormone titer of the culture media can be critical in the design of studies addressing the effects of developmental cues.

E. SYNAPTOGENESIS

The study of synapse formation in insect cultures has, so far, been an exercise in disappointment, for in general, insect neurons in primary culture have not been shown conclusively to make synapses with other neurons. While there have been reports of synaptic interaction occurring between or within cultured explants,[85,86] most workers using primary culture report only evidence of potential for synapses. From studies of *Periplaneta,*[87,91] *Musca,*[88] *Manduca,*[89] *Mellifera,*[90] *Spodoptera,*[92] and *Drosophila*[32] cultures, there are many reports of extensively overlapped neuritic processes, presumably providing ample opportunity to form synapses. With the single exception of spontaneous potentials recorded from dissociated embryonic cockroach neurons,[5] neither physiological nor ultrastructural evidence of synapses has been provided. In neurons dissociated from pupal *Spodoptera* brains,[92] embryonic cockroach ganglia,[93] and *Manduca* antennal lobes,[89] membrane thickening with associated vesicles reminiscent of presynaptic terminals *in vivo* do appear at low frequency, but without evidence of corresponding postsynaptic specializations.

Possible reasons for this general lack of synapses include the following: the medium is deficient in some necessary factor(s); the titer of a necessary hormone is inappropriate; differentiation of these neurons in culture is incomplete; or these nervous systems comprise relatively small numbers of neurons without the redundancy and parallelism common in vertebrate systems. These neurons are designed and destined to participate in interactions only with a small number of specific cells. In most of the neuronal culture systems devised to date, the neurons are drawn from the early embryos or from embryonic or larval whole brain or ganglia in which the likelihood of a neuron finding its proper partner may be remote. It may be that the synaptogenesis in insect neuronal cultures will require the use of identified neurons or subsets of neurons and their targets.

While neuron-neuron synapses appear to be rare in insect culture systems, there is evidence for the *in vitro* formation of neuromuscular junctions. An early report came from Levi-Montalcini[2] in which neuritic processes of embryonic cockroach neurons were described as overlapping with contractile myocytes. The presence of contracting myocytes that lacked contact with neurites suggested, however, that the contractions were probably myogenic rather than neurogenic. Later, in cultures from early *Drosophila* embryos, Salvaterra and his colleagues[39] found, as had others,[94-96] that myocytes developed into myotubes and were overlapped by neuronal processes. Recently, Bermudez et al.[97] demonstrated that neuromuscular junctions can develop in culture between cockroach embryonic neurons and myotubes dissociated from thoracic and abdominal muscle somites. At the ultrastructural level, these junctions were morphologically typical of *in vivo* junctions and, in addition, appeared to be functional, since spontaneous excitatory junctional potentials can be detected in intracellular recordings from the myotubes. Both culture systems, as well as the *Manduca* motoneuron-muscle cocultures currently under development by Levine and colleagues[98] hold the potential to be useful for studying the formation of neuromuscular junctions.

F. ROLE OF GLIAL CELLS

Recent evidence in both vertebrate and invertebrate nervous systems indicates that glial cells have important roles in development.[99] Ideally, one could explore developmental interactions involving glial cells in the same way one examines neuronal interactions, that is, by isolating particular interactions in the culture environment. To date this has been a difficult endeavor because conditions conducive to growth of insect glia in culture have eluded most investigators.

Dissociation of the brain or ganglia of larval, pupal, or adult insects usually produces a large number of neurons and a smaller number of cells that lack the morphology characteristic of neurons and that usually survive only briefly in culture. Based on the cellular elements present in the intact brain or ganglia, these cells could be hemocytes, connective tissue cells, tracheolar cells, or one of several classes of glia.[100,101] Identification of the types of nonneuronal cells found in dissociated embryonic nervous tissue is likely to be especially difficult, particularly in the case of glial cells for which little is known about lineage.[102-105] To date, attempts to differentiate among glial cell classes or even to distinguish glia from other nonneuronal cells by using immunocytochemical techniques have been hampered by the relative paucity of very specific antibodies and by the poor species cross-reactivity of the available antibodies. Ultrastructural criteria, such as amount of rough endoplasmic reticulum (RER) or glycogen, can be useful evidence to support the identification of an *in vitro* cell type as glial, particularly if the criteria are determined from similarly staged *in vivo* material. Where nonneuronal cells do appear in culture and the goal of the study is to examine developmental interactions, it will be critically important to determine the identity and possible developmental contribution of any nonneuronal cells present.

While most insect nervous system cultures have few or no nonneuronal cells, several groups have developed culture systems in which nonneuronal cells meet some of the criteria for glial cells. Hicks et al.[91] described cells of flattened or spindle-shaped morphology that typically were found in association with clumps of neurons in cultures of 16- to 21-day cockroach embryos. Gascuel et al.[90] reported the appearance of glia in cultures of pupal bee brains. In this case, the relative sparseness of the RER as well as the large size of the nuclei of the putative glial cells compared to glia *in vivo* and to neurons *in vitro* makes it difficult to determine conclusively which class of nonneuronal cells these are, or if glial, which class of glia.

Two other reports, both using embryonic tissue, suggest that glial cells occasionally do survive in culture. In their initial description of the ultrastructure of cultures from *Drosophila* gastrulae, Seecof et al.[30] noted that bundles of axons were often ensheathed by glial cells. Vanhems and Delbos[104] reported that glial cells born in the periphery of embryonic locust central nervous system explants were capable of migrating from the explant and differentiating into mature glia as determined by ultrastructural criteria, i.e., electron-opaque cytoplasm, presence of ribosomes and large amounts of rough endoplasmic reticulum, and neurite ensheathment.

Nearly pure cultures of different classes of glial cells can be generated by careful dissection from the developing antennal lobes of pupal *Manduca* (Figure 2). The perineurial sheath of the lobes is readily removed and the underlying subperineurial layer can be teased away. Cell-body glia are obtained by dissociating the neuronal cell-body packets. The cells that appear after dissociation of the remaining neuropil must be predominantly neuropil-associated glial cells, although there may be some tracheolar cells as well. The cell-body and neuropil-associated glial cells survive for a few days in culture if the medium is conditioned by an embryonic *Manduca* cell line (MRRL-CH1, cell line provided courtesy of Dr. Dwight Lynn). Survival can be extended to 4 to 5 days by culturing either in high-density cultures that also contain neurons or in the presence of medium conditioned by high-density neuron cultures,[106] suggesting that antennal-lobe neurons may release a glial growth factor. Use of the conditioned medium causes a change in the shape of both perineurial and subperineurial cells from flattened to spindle-shaped, although the latter retain veil-like protrusions of cytoplasm (Figure 2E, F). These cells remain mitotically active and will survive for weeks. Cultured antennal-lobe cells originally thought to be neuropil-associated glial cells[55] actually belong to the subperineurial class.

The absence of glial cells except under special conditions and the lack of an apparent requirement for them in most insect neuronal culture systems is one of the major distinctions between these cultures and those generated from vertebrate nervous systems. Nevertheless, while overall survival and differentiation of insect neurons may not be dependent upon glial cells, modulation of outgrowth patterns or other aspects of the development of mature neuronal phenotypes may depend upon the presence of glia, as suggested by *in vivo* studies.[107-109] For this reason, continued effort to find conditions supporting glial-cell growth in culture should strengthen the usefulness of insect culture systems for study of developmental interactions.

G. TROPHIC INTERACTIONS

While exploration of the role of trophic factors in vertebrate nervous systems has burgeoned and been greatly facilitated by the use of cell-culture methodology,[109-110] few cell-culture studies have been designed specifically to understand the possible roles for trophic factors in the development of insect nervous systems. Some intriguing recent data, however, suggest that trophic factors may be important in these nervous systems as well.

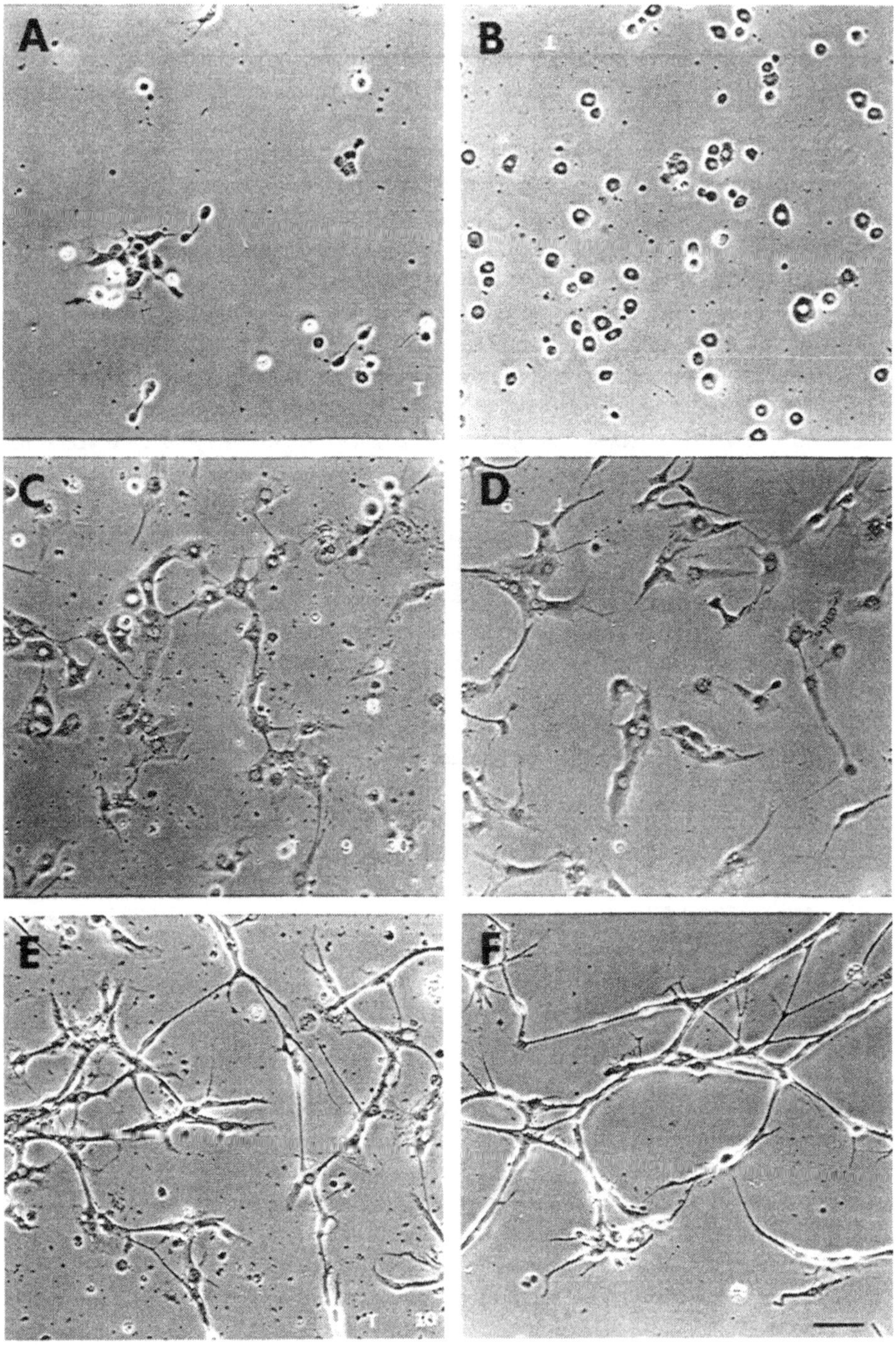

Figure 2. Perineurial and glial cells dissociated from the antennal lobes of *Manduca sexta*. The gross anatomy of the lobe permits the individual cell types to be isolated by simple dissection of the distinct layers and regions of the lobe. (A) — (D) Cells maintained for 1 day in medium conditioned by cell line generated from embryonic *Manduca*. (A) Cell-body associated glia; (B) neuropil-associated glia; (C) perineurial cells; (D) subperineurial cells. (E) — (F) 7 days *in vitro*, with conditioned medium. (E) Perineurial cells; (F) subperineurial cells. Neuropil- and cell body-associated glial cells do not survive beyond 4 to 5 days and do not appreciably change their appearance during that time. Both perineurial and subperineurial cells have similar but not identical morphologies and have changed their appearance from flattened to spindle-shaped cells. Scale bar = 50 μm.

Cultured early embryonic (gastrula) cells dissociated from *Drosophila*[111] appeared to produce and secrete a substance that elicited outgrowth from embryonic chick sensory ganglia, an effect that could be blocked by antibodies to mouse Nerve Growth Factor (NGF). Poly(A)$^+$RNA extracted from *Drosophila* contained a single band that cross-hybridized with a mouse NGF cRNA probe. Similarly, *Drosophila* neurons in culture responded to the addition of mouse NGF to the medium with an increase in cell number; an effect on neurite outgrowth was not reported. As these cultures contained a heterogeneous population of neurons as well as nonneuronal cells including myocytes, the cellular source of the factor could not be determined. Nevertheless, future biochemical and molecular characterization studies, combined with genetic analysis, should provide insight into the roles of such factors in the development of insect nervous systems.

The bee culture system recently developed by Gascuel et al.[90] provides another hint of a role for trophic (or hormonal) factors in developing insect nervous systems. Long-term survival and growth of antennal-lobe neurons from pupal bees depended on the presence of whole-brain explants in the culture dish. The nature of the factor was not examined.

In *Manduca* antennal-lobe cultures generated from animals approximately one-third of the way through metamorphic adult development, use of high-density cultures or exposure of low-density cultures to medium conditioned by high-density neuronal cultures enhanced neurite outgrowth (Figure 3). This result suggests that a neuronal factor or factors, perhaps acting in concert on the growth of glia (see above), may have a vital influence on the growth patterns that result in construction of the adult antennal lobe.

In cocultures of embryonic cockroach neurons and myotubes, both the survival of myotubes and the proportion of glutamate-sensitive neurons[112] increased in cultures containing both neurons and muscle cells. Similarly, muscle cells dissociated from developing adult legs of *Manduca* survived and differentiated into contractile myotubes when neurons dissociated from thoracic or abdominal ganglia of metamorphic adult animals were present.[98]

Attention to the possibility of trophic interactions in the regulation of development may have a benefit beyond that of greater understanding of developmentally important interactions. Identification of such factors would result in improved media that would better support the differentiation and growth of cells from insect nervous systems.

IV. CONCLUSIONS AND PROSPECTS

Previous reviews of the status of insect nervous tissue in primary culture[1,115] were colored by frustration, yet conveyed a sense of cautious optimism that neurobiologists finally would begin to make greater use of the culture techniques that have proven so powerful in studies of vertebrate neuronal tissue. While progress has been made in the intervening five years, the field as a whole has not yet blossomed. On the other hand, neurobiologists using developing insect nervous systems now have access to a number of culture systems that are mature enough to be used to probe mechanisms of development.

To sustain the usefulness of these culture systems, however, serious attention is needed in several areas:

1. The substrates appropriate for insect neurons and glia, especially given the recent vertebrate data that show that the composition of the substrate can determine the branching patterns of neurons[114,116]
2. The growth factors that are necessary for full differentiation in culture
3. The markers that permit the identification of the various types of cells present, particularly cells from embryonic tissue. This will be necessary if the promise of the culture approach, which is to isolate and examine developmental interactions among cells at the biochemical and molecular level, is to be realized.

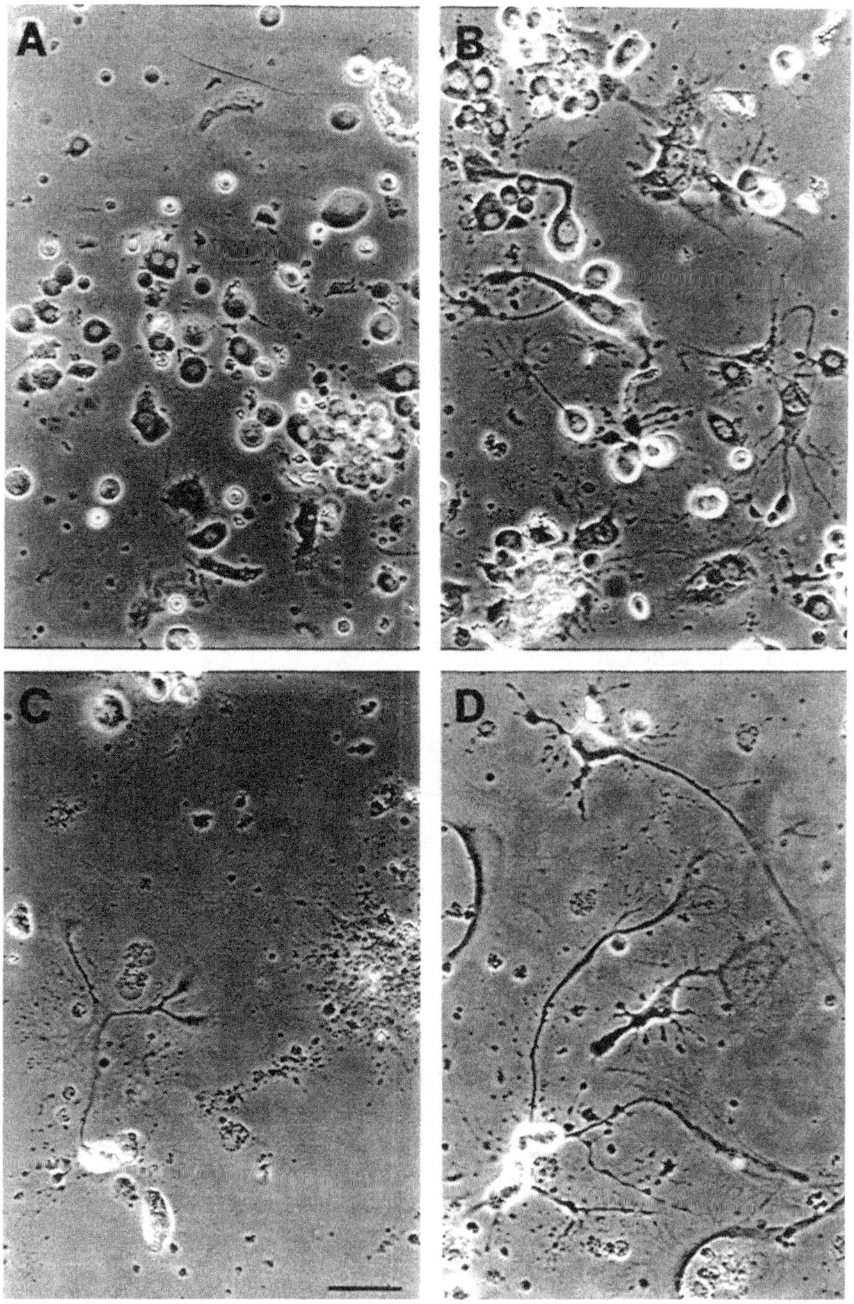

Figure 3. Comparison of antennal-lobe cultures grown in the presence or absence of medium conditioned by high-density antennal-lobe neuron-only cultures. (A) — (B) 1 day *in vitro*. All cultures were generated from stage-5 pupal antennal lobes and contained 1 antennal lobe per dish. Neurons and glial cells (A) in standard culture medium without addition of neuron-conditioned medium and (B) with 50% medium conditioned for 5 days over neuron-only cultures containing 4 antennal lobes per dish. Little process outgrowth is seen in A, while in B many of the neurons have extended thick primary neurites with large growth cones. The smaller, phase-dark cells in A are glial; the flattened cells are perineurial. Little process extension is seen in A compared to B. (C) — (D) 7 days *in vitro*. (C) standard medium; (D) with 25% medium conditioned for 5 days over neuron-only cultures containing 6 antennal lobes per dish. Glial cells have died. Fewer neurons remained in the nonconditioned cultures (C) and the overall robustness and extent of process outgrowth was greater in the conditioned cultures (D). Scale bar = 50 μm.

Krasnow et al.[115] recently developed a technique that permits purification and subsequent culturing of large numbers of different types of neurons from *Drosophila* embryos. Embryos containing a *lac*Z transgene expressed in a particular cell type are grown to the desired developmental stage, then dissociated and stained with a fluorogenic β-galactosidase substrate and purified by fluorescence-activated cell sorting. Neuroblasts and their progeny purified by this method were shown to be competent to develop *in vitro*. With the large number of *lac*Z-containing strains available or easily generated, this technique holds the potential to permit direct examination of specific developmental interactions by coupling the power of *Drosophila* genetics with the specificity of cell types that cannot be obtained by direct dissection of embryonic nervous tissue.

A number of other insect culture systems are in the developmental stage. Li and Meinertzhagen[117] have obtained cultures of *Drosophila* optic lobe and eye imaginal disks that could permit analysis of the mechanisms underlying known developmental interactions in development of the visual system,[118] such as the role of afferents on optic lobe development or the role of interactions between developing photoreceptors in the determination of cell identity. A number of mutant phenotypes are available to them. Embryonic housefly neuron cultures[88] and nymphal locust neuron cultures[119] also have recently been described.

The establishment of permanent cell lines of neural origin remains an important objective that has yet to be obtained for insects. The use of such cell lines may prove to be of significance in studies of regulatory factors of neural origin, both in terms of synthesis as well as molecular mode of action. As experience is gained with investigations of the needs of primary cultures from the insect nervous system, it is hoped that these lessons can be applied to the development of cell lines from neuroblasts or other neural sources. Highly specific molecular probes will make it possible to determine that such cell lines retain fidelity to their neural origin.

ACKNOWLEDGMENTS

The authors express their appreciation to Dr. Jon Hayashi and Dr. Cathy Krull for critical reading of the manuscript; to Dr. Richard Levine for providing Figure 1, and to Mr. Chip Hedgecock for photographic assistance. L. A. O. was supported by NIH grant NS28495. A portion of the work discussed in this chapter was conducted while H. O. was on sabbatical leave at the ARL Division of Neurobiology in the laboratory of Dr. John Hildebrand.

REFERENCES

1. **Beadle, D. J. and Hicks, D.,** Insect nerve culture, in *Comprehensive Insect Physiology, Biochemistry and Pharmacology,* Vol. 5, Kerkut, G. A. and Gilbert, L. I., Eds., Pergamon Press, Oxford, 1985, 181.
2. **Levi-Montalcini, R. and Aloe, L.,** Neuronal nets and nerve cell interactions *in vitro* in insect systems, *In Vitro,* 3, 178, 1972.
3. **Purves, D. and Lichtman, J. W.,** *Principles of Neural Development,* Sinauer Associates, Sunderland, Massachusetts, 1985, 155.
4. **Conn, P. M.,** *Methods in Neurosciences,* Vol. 2, *Cell Culture,* Academic Press, New York, 1990.
5. **Lees, G. and Beadle, D. J.,** Cockroach neuronal cultures as models for investigating the neuropharmacology of the insect central nervous system, in *Cell Culture Approaches to Invertebrate Neuroscience,* Beadle, D. J., Less, G., and Kater, S. B., Eds., Academic Press, New York, 1988, 109.
6. **Wu, C.-F.,** Neurogenetic studies of *Drosophila* central nervous system neurons in culture, in *Cell Culture Approaches to Invertebrate Neuroscience,* Beadle, D. J., Less, G., and Kater, S. B., Eds., Academic Press, New York, 1988, 149.
7. **Marks, E. P., Holman, G. M., and Borg, T. K.,** Synthesis and storage of a neurohormone in insect brains *in vitro, J. Insect Physiol.,* 19, 471, 1973.
8. **Borg, T. K. and Granger, N. A.,** Ultrastructure of the protocerebral neurosecretory cells of larval *Galleria mellonella, in situ* and after culture of the brain *in vitro, Tissue and Cell,* 9, 645, 1977.
9. **Leloup, A. M. and Marks, E. P.,** Culture organotypique du cerveau de *Manduca sexta, Ann. Endocrinol.,* 34, 766, 1973.

10. **Agui, N. and Hiruma, K.,** *In vitro* activation of neurosecretory brain cells in *Mamestra brassicae* by β-ecdysone, *Gen. Comp. Endocrinol.*, 33, 467, 1977.
11. **Oberlander, H., Berry, S. J., Krishnakumaran, A., and Schneiderman, H. A.,** RNA and DNA synthesis during activation and secretion of the prothoracic glands of saturnid moths, *J. Exp. Zool.*, 159, 15, 1965.
12. **Kambysellis, M. P. and Williams, C. M.,** *In vitro* development of insect tissues, II. The role of ecdysone in the spermatogenesis of silkworms, *Biol. Bull., Woods Hole*, 141, 541, 1971.
13. **Agui, N.,** Attempts to establish insect endocrine system *in vitro*, in *Invertebrate Tissue Culture: Research Applications*, Maramorosch, K., Ed., Academic Press, New York, 1976, 133.
14. **Marks, E. P. and Holman, G. M.,** Release from the brain and acquisition by the corpus cardiacum of a neurohormone *in vitro*, *J. Insect Physiol.*, 20, 2087, 1974.
15. **Holman, G. M. and Marks, E. P.,** Synthesis, transport, and release of a neurohormone by cultured neuroendocrine glands from the cockroach, *Leucophaea maderae*, *J. Insect Physiol.*, 20, 479, 1974.
16. **Girardie, J. and Girardie, A.,** Liberation provoquee *in vitro* du produit de neurosecretion des cellules protocerebrales medianes chez le criquet migrateur, *J. Physiol.*, 73, 707, 1977.
17. **Bollenbacher, W. E., Agui, N., Granger, N. A., and Gilbert, L. I.,** *In vitro* activation of insect prothoracic glands by the prothoracicotropic hormone, *Proc. Natl. Acad. Sci. U.S.A.*, 76, 5148, 1979.
18. **Bowen, M. F., Bollenbacher, W. E., and Gilbert, L. I.,** *In vitro* studies on the role of the brain and prothoracic glands in the pupal diapause of *Manduca sexta*, *J. Exp. Biol.*, 108, 9, 1984.
19. **Bowen, M. F., Saunders, D. S., Bollenbacher, W. E., and Gilbert, L. I.,** *In vitro* reprogramming of the photoperiodic clock in an insect brain-retrocerebral complex, *Proc. Natl. Acad. Sci. U.S.A.*, 81, 5881, 1984.
20. **Bate, C. M.,** Embryogenesis of an insect nervous system, I. A map of the thoracic and abdominal neuroblasts in *Locusta migratoria*, *J. Embryol. Exp. Morphol.*, 35, 107, 1976.
21. **Goodman, C. S.,** Embryonic development of identified neurones in the grasshopper, in *Neuronal Development*, Spitzer, N., Ed., Plenum Press, New York, 1982, 171.
22. **Taghert, P. H., Doe, C. Q., and Goodman, C. S.,** Cell determination and regulation during development of neuroblasts and neurones in grasshopper embryo, *Nature*, 307, 163, 1984.
23. **Thomas, J. B., Bastiani, M. J., Bate, M., and Goodman, C. S.,** From grasshopper to *Drosophila:* a common plan for neuronal development, *Nature*, 310, 203, 1984.
24. **Kuwada, J. Y. and Goodman, C. S.,** Neuronal determination during embryonic development of the grasshopper nervous system, *Dev. Biol.*, 110, 114, 1985.
25. **Truman, J. W. and Booker, R.,** Adult-specific neurons in the nervous system of the moth *Manduca sexta:* selective chemical ablation using hydroxyurea, *J. Neurobiol.*, 17, 613, 1986.
26. **Booker, R. and Truman, J. W.,** Postembryonic neurogenesis in the CNS of the tobacco hornworm, *Manduca sexta*, I. Neuroblast arrays and the fate of their progeny during metamorphosis, *J. Comp. Neurol.*, 255, 548, 1987.
27. **Booker, R. and Truman, J. W.,** Postembryonic neurogenesis in the CNS of the tobacco hornworm, *Manduca sexta*, II. Hormonal control of imaginal nest cell degeneration and differentiation during metamorphosis, *J. Neurosci.*, 7, 4107, 1987.
28. **Witten, J. L. and Truman, J. W.,** The regulation of transmitter expression in postembryonic lineages in the moth *Manduca sexta*, I. Transmitter identification and developmental acquisition of expression, *J. Neurosci.*, 11, 1980, 1991.
29. **Witten, J. L. and Truman, J. W.,** The regulation of transmitter expression in postembryonic lineages in the moth *Manduca sexta*, II. Role of cell lineage and birth order, *J. Neurosci.*, 11, 1990, 1991.
30. **Seecof, R. L., Alleaume, N., Teplitz, R. L., and Gerson, I.,** Differentiation of neurons and myocytes in cell cultures made from *Drosophila* gastrulae, *Exp. Cell Res.*, 69, 161, 1971.
31. **Seecof, R. L., Donady, J. J., and Teplitz, R. L.,** Differentiation of *Drosophila* neuroblasts to form ganglion-like clusters of neurons *in vitro*, *Cell Diff.*, 2, 143, 1973.
32. **Wu, C.-F., Suzuki, N., and Poo, M.-M.,** Dissociated neurons from the normal and mutant *Drosophila* larval central nervous system in cell culture, *J. Neurosci.*, 3, 1888, 1983.
33. **Jan, L. Y. and Jan, Y. W.,** Antibodies to horseradish peroxidase as specific neuronal markers in *Drosophila* and grasshopper embryos, *Proc. Natl. Acad. Sci. U.S.A.*, 79, 2700, 1982.
34. **Furst, A. and Mahowald, A. P.,** Cell division cycle of cultured neural precursor cells from *Drosophila*, *Dev. Biol.*, 112, 467, 1985.
35. **Furst, A. and Mahowald, A. P.,** Differentiation of primary embryonic neuroblasts in purified neural cell cultures from *Drosophila*, *Dev. Biol.*, 109, 184, 1985.
36. **Perrimon, N., Engstrom, L., and Mahowald, A. P.,** Developmental genetics of the 2E-F region of the *Drosophila* X-chromosome, a region rich in developmentally important genes, *Genetics*, 108, 559, 1984.
37. **LaBonne, S. G. and Furst, A.,** Differentiation *in vitro* of neural precursor cells from normal and *pecanex* mutant *Drosophila* embryos, *J. Neurogenet.*, 5, 99, 1989.
38. **Dewhurst, S. A. and Seecof, R. L.,** Development of acetylcholine metabolizing enzymes in *Drosophila* embryos and in cultures of *Drosophila* cells, *Comp. Biochem. Physiol.*, 50C, 53, 1975.

39. **Salvaterra, P. M., Bournias-Vardiabasis, N., Nair, T., Hou, G., and Lieu, C.**, *In vitro* neuronal differentiation of *Drosophila* embryo cells, *J. Neurosci.*, 7, 10, 1987.
40. **Huff, R., Furst, A., and Mahowald, A. P.**, *Drosophila* embryonic neuroblasts in culture: autonomous differentiation of specific neurotransmitters, *Dev. Biol.*, 134, 146, 1989.
41. **Taghert, P. H. and Goodman, C. S.**, Cell determination and differentiation of identified serotonin-immunoreactive neurons in the grasshopper embryo, *J. Neurosci.*, 4, 989, 1984.
42. **Doe, C. Q. and Goodman, C. S.**, Early events in insect neurogenesis, II. The role of cell interactions and cell lineage in the determination of neuronal precursor cells, *Dev. Biol.*, 111, 206, 1985.
43. **Wu, C.-F., Sakai, K., Saito, M., and Hotta, Y.**, Giant *Drosophila* neurons differentiated from cytokinesis-arrested embryonic neuroblasts, *J. Neurobiol.*, 21, 499, 1989.
44. **Christensen, B. N., Larmet, I., Shijmahara, T., Beadle, D., and Pichon, Y.**, Ionic currents in neurones cultured from embryonic cockroach *(Periplaneta americana)* brains, *J. Exp. Biol.*, 135, 193, 1988.
45. **Amar, M., Larmet, Y., Inoue, I., and Pichon, Y.**, Potassium current kinetics and single-channel currents in embryonic cultured insect neurones, *Pflugers Arch.*, 414, 145, 1989.
46. **Bermudez, I., Beadle, D. J., and Benson, J. A.**, Multiple serotonin-activated currents in isolated, neuronal somata from locust thoracic ganglia, *J. Exp. Biol.*, 165, 43, 1992.
47. **Wu, C.-F., Ganetzky, B., Haugland, F. N., and Liu, A.-X.**, Potassium currents in *Drosophila:* different components affected by mutations of two genes, *Science*, 220, 1076, 1983.
48. **Solc, C. K. and Aldrich, R. W.**, Voltage-gated potassium channels in larval CNS neurons of *Drosophila, J. Neurosci.*, 8, 2556, 1988.
49. **Solc, C. K. and Aldrich, R. W.**, Gating of single non-*Shaker* A-type potassium channels in larval *Drosophila* neurons, *J. Gen. Physiol.*, 96, 135, 1990.
50. **O'Dowd, D. K. and Aldrich, R. W.**, Voltage clamp analysis of sodium channels in wild-type and mutant *Drosophila* neurons, *J. Neurosci.*, 8, 3633, 1988.
51. **O'Dowd, D. K., Germeraad, S. E., and Aldrich, R. W.**, Alterations in the expression and gating of *Drosophila* sodium channels by mutations in the *para* gene, *Neuron*, 2, 1301, 1989.
52. **Yamamoto, D. and Suzuki, N.**, Characterization of single noninactivating potassium channels in primary neuronal cultures of *Drosophila, J. Exp. Biol.*, 145, 173, 1989.
53. **Yamamoto, D. and Suzuki, N.**, Properties of single chloride channels in primary neuronal cultures of *Drosophila, Biochim. Biophys. Acta.*, 986, 187, 1989.
54. **Byerly, L. and Leung, H.-T.**, Ionic currents in *Drosophila* neurons in embryonic cultures, *J. Neurosci.*, 8, 4379, 1988.
55. **Hayashi, J. H. and Hildebrand, J. G.**, Insect olfactory neurons *in vitro:* morphological and physiological characterization of cells from the developing antennal lobe of *Manduca sexta, J. Neurosci.*, 10, 848, 1990.
56. **Hayashi, J. H. and Levine, R. B.**, Calcium and potassium currents in leg motoneurons during postembryonic development in the hawkmoth, *Manduca sexta, J. Neurosci., J. Exp. Biol.*, 171, 15, 1992.
57. **Suzuki, N. and Wu, C.-F.**, Altered sensitivity to sodium channel-specific neurotoxins in cultured neurons from temperature-sensitive paralytic mutants of *Drosophila, J. Neurogenet.*, 1, 225, 1984.
58. **Solc, C. K., Zagotta, W. N., and Aldrich, R. W.**, Single-channel and genetic analyses reveal two distinct A-type potassium channels in *Drosophila, Science*, 236, 1094, 1987.
59. **Weeks, J. C. and Levine, R. B.**, Endocrine influences on the postembryonic fates of identified neurons during insect metamorphosis, in *Determinants of Neuronal Identity*, Shankland, M., Macagno, E., Eds., Academic Press, New York, 1992, 293.
60. **Hayashi, J. H., Oland, L. A., and Hildebrand, J. G.**, The development of potassium currents in cultured insect olfactory neurons, *Soc. Neurosci. Abstr.*, 18, 230, 1992.
61. **Stengl, M. and Hildebrand, J. G.**, Insect olfactory neurons *in vitro:* morphological and immunocytochemical characterization of male-specific antennal receptor cells from the developing antennae of male *Manduca sexta, J. Neurosci.*, 10, 837, 1990.
62. **Hishinuma, A., Hockfield, S., McKay, R., and Hildebrand, J. G.**, Monoclonal antibodies reveal cell-type-specific antigens in the sexually dimorphic olfactory system of *Manduca sexta*, I. Generation of monoclonal antibodies and partial characterization of the antigens, *J. Neurosci.*, 8, 296, 1988.
63. **Hishinuma, A., Hockfield, S., McKay, R., and Hildebrand, J. G.**, Monoclonal antibodies reveal cell-type-specific antigens in the sexually dimorphic olfactory system of *Manduca sexta*, II. Expression of antigens during postembryonic development, *J. Neurosci.*, 8, 308, 1988.
64. **Stengl, M., Zufall, H., Hatt, H., Dudel, J., and Hildebrand, J. G.**, Patch-clamp analysis of male *Manduca sexta* olfactory receptor neurons in primary cell culture, *Soc. Neurosci. Abstr.*, 15, 751, 1989.
65. **Zufall, F., Stengl, M., Franke, C., Hildebrand, J. G., and Hatt, H.** Ionic currents of cultured olfactory receptor neurons from antennae of male *Manduca sexta, J. Neurosci.*, 11, 956, 1991.
66. **Stengl, M., Zufall, H., Hatt, H., and Hildebrand, J. G.**, Olfactory receptor neurons from antennae of developing male *Manduca sexta* respond to components of the species-specific sex pheromone *in vitro, J. Neurosci.*, 12, 2523, 1992.

67. **Kim, Y.-T. and Wu, C.-F.**, Reversible blockage of neurite development and growth cone formation in neuronal cultures of a temperature-sensitive mutant of *Drosophila*, *J. Neurosci.*, 7, 3245, 1987.
68. **Kim, Y.-T. and Wu, C.-F.**, Distinctions in growth cone morphology and motility between monopolar and multipolar neurons in *Drosophila* CNS cultures, *J. Neurobiol.*, 22, 263, 1990.
69. **Masur, S. K., Kim, Y.-T., and Wu, C.-F.**, Reversible inhibition of endocytosis in cultured neurons from the *Drosophila* temperature-sensitive mutant *shibire*^{ts1}, *J. Neurogenet.*, 6, 191, 1990.
70. **Kosaka, T. and Ikeda, K.**, Reversible blockage of membrane retrieval and endocytosis in the garland cells of the temperature-sensitive mutant of *Drosophila melanogaster, shibire*^{ts1}, *J. Cell Biol.*, 97, 499, 1983.
71. **Bastiani, M. J., Doe, C. Q., Helfand, S. L., and Goodman, C. S.**, Neuronal specificity and growth cone guidance in grasshopper and *Drosophila* embryos, *Trends Neurosci.*, 8, 257, 1985.
72. **Prugh, J., Della Croce, K., and Levine, R. B.**, Effects of the steroid hormone, 20 hydroxyecdysone, on the growth of neurites by identified insect motoneurons *in vitro*, *Dev. Biol.*, 154, 331, 1992.
73. **Hayashi, J. and Oland, L. A.**, unpublished observations.
74. **Sanes, J. R. and Hildebrand, J. G.**, Structure and development of antennae in a moth, *Dev. Biol.*, 51, 282, 1976.
75. **Sanes, J. R. and Hildebrand, J. G.**, Origin and morphogenesis of sensory neurons in an insect antenna, *Dev. Biol.*, 51, 300, 1976.
76. **Sanes, J. R. and Hildebrand, J. G.**, Acetylcholine and its metabolic enzymes in the developing antennae of the moth, *Dev. Biol.*, 52, 105, 1976.
77. **Oland, L. A. and Tolbert, L. P.**, Glial patterns during early development of antennal lobes of *Manduca sexta:* a comparison between normal lobes and lobes deprived of antennal axons, *J. Comp. Neurol.*, 255, 196, 1987.
78. **Oland, L. A., Orr, G., and Tolbert, L. P.**, Construction of a protoglomerular template by olfactory axons initiates the formation of olfactory glomeruli in the insect brain, *J. Neurosci.*, 10, 2096, 1990.
79. **Hildebrand, J. G., Matsumoto, S. G., Tolbert, L. P., Schneiderman, A. M., and Camazine, S. M.**, Postembryonic development of the antennal lobes in the moth *Manduca sexta*, in *Neuronal Development: Cellular Approaches in Invertebrates*, Goodman, C. S. and Pearson, K. G., Eds., *Neurosci. Res. Prog. Bull.*, 20, 1982, 891.
80. **Warren, J. T. and Gilbert, L. I.**, Ecdysone metabolism and distribution during the pupal-adult development of *Manduca sexta*, *Insect Biochem.*, 16, 65, 1986.
81. **Vaught, S. and Tolbert, L.**, personal communication.
82. **Kerkut, G. A. and Gilbert, L. I.**, Eds., *Comprehensive Insect Physiology, Biochemistry, and Pharmacology*, Pergamon Press, Oxford, Vols. 7–8, 1985.
83. **Witten, J. L. and Levine, R. B.**, Cellular specificity of steroid influences on process outgrowth of identified motoneurons in culture, *Soc. Neurosci. Abstr.*, 17, 1320, 1991.
84. **Oland, L. A., Hayashi, J. H., and Tolbert, L. P.**, Effect of the steroid hormone 20-hydroxyecdysone on the branching patterns of cultured neurons from the developing olfactory lobe of the moth, *Soc. Neurosci. Abstr.*, 18, 230, 1992.
85. **Provine, R. R., Aloe, L., and Seshan, K. R.**, Spontaneous bioelectric activity in long term cultures of the embryonic insect central nervous system, *Brain Res.*, 56, 364, 1973.
86. **Levi-Montalcini, R. and Seshan, K. R.**, Long-term cultures of embryonic and mature insect nervous and neuroendocrine systems, in *Tissue Culture of the Nervous System*, Sato, G., Ed., Plenum Press, New York, 1973, 1.
87. **Chen, J. S. and Levi-Montalcini, R.**, Long term cultures of dissociated nerve cells from the embryonic nervous system of the cockroach *Periplaneta americana*, *Arch. Ital. Biol.*, 108, 503, 1970.
88. **Harrison, J. B., Leech, C. A., Katz, J., and Satelle, D. B.**, Embryonic and adult neurones of the housefly *(Musca domestica)* in culture, *Tiss. Cell*, 22, 337, 1990.
89. **Oland, L. A.**, unpublished observations.
90. **Gascuel, J., Masson, C., and Beadle, D. J.**, The morphology and ultrastructure of antennal lobe cells from pupal honeybees *(Apis mellifera)* growing in culture, *Tiss. Cell*, 23, 547, 1991.
91. **Hicks, D., Beadle, D. J., Giles, D. P., and Usherwood, P. N. R.**, Ultrastructure of dissociated nerve cells of *Periplaneta americana* (L.) (Dictyoptera: Blattidae) growing in culture, *Int. J. Insect Morphol. Embryol.*, 10, 225, 1981.
92. **Hicks, D., Beadle, D. J., and Giles, D. P.**, Ultrastructure of dissociated neurones from pupae of *Spodoptera littoralis* (Lepidoptera: Noctuidae) in culture, *Int. J. Insect Morphol. Embryol.*, 10, 399, 1981.
93. **Beadle, D. J., Hicks, D., and Middleton, C.**, Fine structure of *Periplaneta americana* neurons in long-term culture, *J. Neurocytol.*, 11, 611, 1982.
94. **Shields, G. and Sang, J. H.**, Characteristics of five cell types appearing during *in vitro* culture of embryonic material from *Drosophila melanogaster*, *J. Embryol. Exp. Morph.*, 23, 53, 1970.
95. **Seecof, R. L. and Unanue, R. L.**, Differentiation of embryonic *Drosophila* cells *in vitro*, *Exp. Cell Res.*, 50, 651, 1968.

96. **Shields, G., Dubendorfer, A., and Sang, J. H.**, Differentiation *in vitro* of larval types from early embryonic cells of *Drosophila melanogaster*, *J. Embryol. Exp. Morph.*, 33, 159, 1975.
97. **Bermudez, I., Lees, G., Botham, R. P., and Beadle, D. J.**, Myogenesis and neuromuscular culture junction formation in cultures of *Periplaneta americana* myoblasts and neurons, *Dev. Biol.*, 116, 467, 1986.
98. **Witten, J. L., Blair, J., Luedeman, R., and Levine, R. B.**, Nerve/muscle interactions during development of insect adult leg muscles *in vivo* and *in vitro*, *Soc. Neurosci. Abstr.*, 18, 40, 1992.
99. **Rakic, P.**, Glial cells in development, in *Glial-Neuronal Development*, Abbott, N. J., Ed., *Ann. N.Y. Acad. Sci.*, 633, 96, 1991.
100. **Lane, N. J.**, Invertebrate neuroglia-junctional structure and development, *J. Exp. Biol.*, 95, 7, 1981.
101. **Wigglesworth, V. B.**, Nervous and endocrine systems, in *The Principles of Insect Physiology*, 7th ed., Chapman and Hall, London, 1972, 181.
102. **Roots, B. I.**, A phylogenetic approach to the anatomy of glia, in *Dynamic Properties of Glial Cells*, Schoffeniels, E., Frank, B., Hertz, L., and Tower, D. B., Eds., Pergamon Press, Oxford, 1978, 45.
103. **Edwards, J. S.**, Neuronal guidance and pathfinding in the developing sensory nervous system of insects, in *Insect Biology of the Future-VBW 80*, Locke, M. and Smith, D. S., Eds., Academic Press, London, 1980, 667.
104. **Vanhems, E. and Delbos, M.**, Differentiation of glial cells and neurite outgrowth obtained from embryonic locust central nervous systems explants, *Brain Res.*, 411, 129, 1987.
105. **Winberg, M. L., Perez, S. E., and Steller, H.**, Generation and early differentiation of glial cells in the first optic ganglion of *Drosophila melanogaster*, *Development*, 115, 903, 1992.
106. **Oberlander, H. and Oland, L. A.**, unpublished observations.
107. **Tolbert, L. P. and Oland, L. A.**, Glial cells form boundaries for developing insect olfactory glomeruli, *Exp. Neurol.*, 109, 19, 1990.
108. **Jacobs, J. R. and Goodman, C. S.**, Embryonic development of axon pathways in the *Drosophila* CNS, I. A glial scaffold appears before the first growth cones, *J. Neurosci.*, 9, 2402, 1989.
109. **Walike, P.**, Novel neurotrophic factors, receptors, and oncogenes, *Annu. Rev. Neurosci.*, 12, 103, 1989.
110. **Thoenen, H.**, The changing scene of neurotrophic factors, *Trends Neurosci.*, 14, 165, 1991.
111. **Hayashi, I., Perez-Magallanes, M., and Rossi, J. M.**, Neurotrophic factor-like activity in *Drosophila*, *Biochem. Biophys. Res. Comm.*, 184, 73, 1992.
112. **Horseman, B. G., Seymour, C., Bermudez, I., and Beadle, D. J.**, The effects of L-glutamine on cultured insect neurones, *Neurosci. Lett.*, 85, 65, 1988.
113. **Thomas, W. E., Jordan, F. L., and Townsel, J. H.**, The status of the study of invertebrate neurons in tissue culture —Phylum Arthropoda, *Comp. Biochem. Physiol.*, 87A, 215, 1987.
114. **Sanes, J.**, Extracellular matrix molecules that influence neural development, *Annu. Rev. Neurosci.*, 12, 491, 1989.
115. **Reichardt, L. F. and Tomaselli, K. J.**, Extracellular matrix molecules and their receptors: functions in neuronal development, *Annu. Rev. Neurosci.*, 14, 531, 1991.
116. **Krasnow, M. A., Cumberledge, S., Manning, G., Herzenberg, L. A., and Nolan, G. P.**, Whole animal cell sorting of *Drosophila* embryos, *Science*, 251, 81, 1991.
117. **Li, C. and Meinertzhagen, I. A.**, Neurite outgrowth in primary cultures of *Drosophila* photoreceptor and optic lobe cells, *Soc. Neurosci. Abstr.*, 18, 1468, 1992.
118. **Banerjee, U. and Zipursky, S. L.**, The role of cell-cell interaction in the development of the *Drosophila* visual system, *Neuron*, 4, 177, 1990.
119. **Giles, D. P. and Usherwood, P. N. R.**, Locust nymphal neurones in culture: a new technique for studying the physiology and pharmacology of insect central neurones, *Comp. Biochem. Physiol.*, 80C, 53, 1985.

Chapter 7

PHYSIOLOGICAL AND DEVELOPMENTAL CAPACITIES OF INSECT CELL LINES

Herbert Oberlander and Stephen M. Ferkovich

TABLE OF CONTENTS

I. Introduction 129

II. Cellular Endocrinology 130
- A. Synthesis and Metabolism of Ecdysteroids 130
- B. Action of Ecdysteroids 130
  - 1. Ecdysteroid Receptors and Response Elements 130
  - 2. Stimulation of Enzymes and Glycoproteins 132
  - 3. Chitin Synthesis 133

III. Morphogenesis 134
- A. Regulation of Morphology 134
  - 1. Hemolymph and Tissue Factors 134
  - 2. Hormonal Effects 135
- B. Cell Lines as Sources of Developmental Factors 136

IV. Conclusions and Prospects 136

Acknowledgments 137

References 137

Herbert Oberlander is Director of the Insect Attractants, Behavior and Basic Biology Research Laboratory, Agricultural Research Service, U.S. Department of Agriculture, Gainesville, Florida. He received his B.A. degree in zoology in 1961 and a Ph.D. in biology at Case-Western Reserve University in 1965. NIH post-doctoral fellow, University of Zürich, 1965–1966; Assistant Professor Biology, Brandeis University 1966–1971; Sigma Xi, Phi Beta Kappa; Entomological Society of America, American Chemical Society, Tissue Culture Association. Endocrinology, chitin synthesis and tissue culture in insects.

I. INTRODUCTION

Insect tissue culture has a long history beginning in 1915 when Richard Goldschmidt described his success in maintaining moth testes in organ culture.[1] Goldschmidt was at Yale at the time and was influenced by Ross Harrison who, a few years earlier was the first to employ tissue-culture methodology. Despite this early beginning, insect tissue culture attracted only a few scientists, who met with limited success with organ culture, but none at all in establishing insect cell lines. It was indeed puzzling that during the ensuing decades vertebrate cell lines became widely used, which led not a few distinguished biologists to pronounce insects incapable of generating permanent cell lines. To consider this possibility, Paul Weiss organized a conference in 1955 on the feasibility of invertebrate tissue culture and concluded that "contrary to mammalian cells, insect cells will most likely never be grown in continuous culture..."[2] Predictably, this conclusion preceded by less than a decade the report by Tom Grace of Australia in *Nature* that he had succeeded in establishing four permanent cell lines from moths.[3]

During the 1960s and 1970s there was dramatic progress in the study of physiological and developmental problems in insect organ culture, assisted in part by the availability of Grace's culture medium, among others. At the same time tremendous strides were being made in establishing vertebrate cell lines that retained capabilities reflecting their *in vivo* origin.[4] Clearly, established cell lines that retained *in vivo* properties had distinct advantages over organ culture, especially the ready availability of large numbers of cells representing a homogenous population, not to mention an end to daily dissections. Unfortunately, virtually all of the insect cell lines that were available in the 1960s were of unknown cell type, did not respond to hormones, and were used primarily for studying the propagation of insect viruses.

In later years many cell lines were tested for hormonal responsiveness, and one embryonic cell line, the Kc line from *Drosophila melanogaster*, established by Echalier and Ohanessian in Paris, was affected both morphologically and biochemically by the insect molting hormones, ecdysone and 20-hydroxyecdysone.[5,6] However, as late as 1978 the same could not be said for any of the lepidopteran cell lines. Then, in 1979 Marks and Holman observed a morphological response to ecdysteroids in a *Manduca sexta* cell line, so that hormonal effects now could be studied in both dipteran and lepidopteran cell lines.[7]

The early work on hormonal effects on insect cell lines gave hope that more complex developmental events could be investigated in cell lines established for that purpose. It is important to consider what sorts of cellular activity can be observed *in vitro* that indicate a capacity for differentiation and morphogenesis among established cell lines. Unlike organ cultures, we cannot expect an insect cell line to give rise to completely differentiated structures, tissues, or organs. Thus, we have to look for discrete steps in a developmental pathway that may only partially be implemented in cell culture. In this chapter we will discuss cell lines that demonstrate changes in cellular morphology, regulation of multicellular organization, responsiveness to ecdysteroids, and production of specific cellular products. We believe that these are legitimate examples of physiological and developmental capacities of insect cell lines, but each system must be examined critically.

In our laboratory we reasoned that if we could develop cell lines from a developmentally committed but proliferating tissue type, then we would have a better chance of observing cellular activities *in vitro* that were characteristic of *in vivo* development. To accomplish this we chose the imaginal discs. First, imaginal discs are diploid dividing cells, present in the larval stage, that are already on a developmental pathway for adult structures, but are not yet differentiated. Second, in a sense these tissues are already in culture in the larval hemolymph. They thrive there without contributing to the activities of the caterpillar in which they reside, but grow and wait for the hormonal balance to change in favor of metamorphosis. Unfortunately, imaginal discs are neither large nor numerous. However, the imaginal discs have remarkable regulative capacities, and when wounded, proliferation is stimulated. Our laboratory took advantage of this property to establish the first cell lines from imaginal discs.[8] For the purposes of this chapter we will consider cell lines from any insect source so long as they display a recognizable developmental or physiological capacity.

II. CELLULAR ENDOCRINOLOGY

A. SYNTHESIS AND METABOLISM OF ECDYSTEROIDS

Cell lines established from imaginal discs of various Lepidoptera have shown remarkable capabilities in synthesizing ecdysteroids despite their origin from cells destined to become cuticle-secreting epithelium. For example, the imaginal disc-derived IAL-TND1 cell line, unlike other *T. ni* cell lines from ovaries or embryos, produces its own 20-hydroxyecdysone, as determined by radioimmunoassay, high performance liquid chromatography, and mass spectral analysis.[9] Moreover, Sommé-Martin et al. found that the IAL-PID2 cell line, which is also derived from imaginal discs, is capable of C20-hydroxylation of ecdysone to 20-hydroxyecdysone, as well as C26-hydroxylation, which results in 20,26-dihydroxyecdysone accumulating in the cells.[10]

Related studies on other epithelial-like cell lines have addressed the question of ecdysteroid synthetic capacity. For example, Ward et al. examined four lepidopteran and three orthopteran cell lines, and found ecdysteroid production most consistently in the UMBGE-4 cockroach chitin synthesizing cell line.[11] In addition, like the PID2 cells, both 20C- and 26C-hydroxylases were found in the cockroach cell line.[12] Similarly, Stiles et al. demonstrated that the BRL-AG-2 cell line derived from embryonic tissue of the cotton boll weevil is capable of both chitin synthesis and ecdysteroid production.[13] Interestingly, this cell line, like IAL-TND1, contained multicellular vesicles and aggregates. The relationship of epithelial-like capabilities and the ability to synthesize ecdysteroids also was found in dipteran as well as in orthopteran and lepidopteran cultures. A chitin-synthesizing cell line from *Chironomous tentans,* that also grows as multicellular vesicles, produced a material that cochromatographed with ecdysone.[14]

Taken together, these findings clearly show that cell lines that are capable of synthesizing chitin and grow as multicellular assemblages, or otherwise demonstrate an epithelial developmental determination, are capable of synthesizing ecdysteroids. These results support the overall hypothesis that there are alternate sites for ecdysteroid production *in vivo* and that these sites may "play a role in localized developmental events."[15]

B. ACTION OF ECDYSTEROIDS

1. Ecdysteroid Receptors and Response Elements

It is fair to state that without ecdysteroid responsive cell lines, progress toward unraveling the early events associated with hormonal action in insects would have been considerably delayed. Best-Belpomme, Cherbas, O'Connor, and colleagues were among the first to formally propose the use of cell lines as models for investigating the endocrine control of insect metamorphosis.[16,17,18] These authors all worked with the Kc cell line established from embryos of *Drosophila melanogaster.*[6] Recognition of the potential of these cell lines for studies of hormonal action came as the result of investigations by Courgeon, who was the first to observe effects of ecdysteroids on the morphology and rate of proliferation in the Kc cells.[6] These major responses of the Kc cell line to ecdysteroids led to the assumption that the mechanism of action of the hormone at the subcellular level in the Kc cells should be the same as that which operates *in vivo* even if the physiological meaning of the observations was not evident. Thus, while there may be some debate as to whether it is reasonable to derive insights on physiological processes from experiments performed on established cell lines, it is clearly advantageous to investigate the early effects of ecdysteroids in such systems.

A major contribution to our understanding of steroid hormone receptors in invertebrates was made as a result of the availability of high-specific-activity [^{3}H]-ponasterone A, which was used for characterizing ecdysteroid receptors in mass-isolated *D. melanogaster* imaginal discs[19] and in Kc cells.[20] The combination of a radioligand with greater affinity for the ecdysteroid receptor than 20-hydroxyecdysone, together with a plentiful supply of hormone-sensitive cells, provided the first demonstration of a single high-affinity saturable binding site for ecdysteroids. The relative K_D values are in the order of 10^{-7} M, ponasterone A, 10^{-7} to 10^{-8} M, 20-hydroxyecdysone, and 10^{-5} to 10^{-6} M, ecdysone.[21] Further improvements in this system were made by using [^{3}H] muristerone A as a radioligand in Kc cells.[22] The muristerone A receptor complex was more stable than other ecdysteroid receptor complexes and enabled the authors to characterize the receptor more fully and estimate its molecular weight at 120,000.

The view that there was only one binding site for ecdysteroids was revised as a consequence of research with another cell line from *D. melanogaster,* B II, which was used to demonstrate two high-affinity binding sites with kDas of 0.3×10^{-9} and 2×10^{-9}, based on the uptake of [^{3}H] ponasterone A by the intact cells. The presence of these high-affinity specific binding components was determined with Scatchard plots of binding data in cell-free extracts.[23] This work represents the first ecdysteroid receptor research with cell lines derived from a specific cell type, tumorous blood cells.[24] The presence of multiple binding sites for ponasterone A was later confirmed in intact adult *D. melanogaster* females.[25] Thus, *Drosophila* cell lines have been utilized to characterize ecdysteroid binding sites in concert with parallel studies *in vivo.*

Although there has been considerable emphasis on cell lines from *D. melanogaster,* other species also have been used for ecdysteroid receptor studies. For example, [^{3}H] ponasterone A was used as the radioligand in a *Chironomous tentans* cell line.[26,27] These authors demonstrated relative affinities for the binding component based on competition assays that supported the findings on *D. melanogaster* cells with regard to 20-hydroxyecdysone and ecdysone. Moreover, similar results were obtained by Wing, who used the *P. interpunctella* cell line, IAL-PID2, established by Lynn and Oberlander, to provide the first biochemical evidence for ecdysteroid receptors in lepidopteran cells.[28,29] [^{3}H] ponasterone A showed binding to nuclear extracts of the cells that was saturable and high affinity, while competitive binding with unlabeled hormone showed that 20-hydroxyecdysone and ecdysone displaced the radioligand, as reported for the dipteran systems. Thus, it is clear that a major portion of the progress that has been made in characterizing ecdysteroid receptors derives from research with established cell lines that retain their hormonal responsiveness.

Cell lines have been important tools not only in characterizing ecdysteroid receptors, but also in gaining insight into the early events that ensue in target cells after the hormone binds to its receptor. In this connection, the *D. melanogaster* cell lines have been critical biological material. Importantly, the Kc cell line displays little or no metabolism of 20-hydroxyecdysone,[20] so that the rapid degradation of this hormone *in vivo* is not a factor in the *in vitro* system. Moreover, morphological and biochemical responses of the Kc cell line to diverse ecdysteroids were similar to those reported for intact *D. melanogaster* tissues.[17]

Scientists believed that the Kc cells might provide information on the early effects of 20-hydroxyecdysone, and that looking for newly synthesized peptides would be the most useful approach. Ecdysteroid-induced peptides were characterized in the Kc cells just four hours after treatment of the cells with hormone. The ecdysteroid-induced peptides (EIPs) were designated EIP40, EIP29, and EIP28 to correspond to their molecular weights of 40,000, 29,000 and 28,000, respectively. It was possible to detect the EIPs only 30 minutes after the addition of ecdysteroid.[17,30] For an overview of ecdysteroid responsive genes in *D. melanogaster,* the reader is referred to Cherbas et al.,[31] a paper noteworthy for being the result of input from 27 coauthors.

Molecular analysis of gene expression in the EIP 28/29 region indicates that the corresponding mRNAs derive from four exons, and that differential use of alternative donor sites at the end of the second exon yields mRNAs that code for the two different proteins.[32] In a successful effort to map ecdysteroid response elements (EcREs), Cherbas et al.[33] utilized iodoponasterone A as the radioligand. They studied immobilized restriction fragments spanning a 9 kb genomic region that flanks the EIP 28/29 gene, and identified three specific fragments with autoradiography. One binding site mapped upstream of the transcription site while the other two mapped downstream. The two downstream sites were necessary for hormonal stimulation of the EIP 28/29 gene. Significantly, each of the three EcREs was able to function independently in conferring ecdysteroid responsiveness when located upstream of a gene that would otherwise not respond to hormone. Thus, over the course of two decades Courgeon's early observations of ecdysteroid responsiveness in the *Drosophila* Kc cell line led to a model system for characterizing ecdysteroid receptors and response elements.

During the same period in which Kc cells were investigated, other cell lines established from *D. melanogaster,* especially the S1, S2, and S3 lines, were important culture systems for molecular analysis.[34] These cell lines were used to identify an ecdysteroid response element in the regulatory region of a heat shock protein gene, *hsp* 27.[35] Just as with EIP 28/29 and EIP 40, the region required for ecdysteroid inducibility is located separately from the promoter region for the heat shock response. In an effort to unify these studies, Cherbas et al.[36] proposed a consensus sequence for ecdysone receptor binding, which is similar to response elements for receptors for some vertebrate hormones.[37]

Thus, a new understanding of ecdysteroid action at the molecular level has been obtained from intensive studies of *Drosophila* cell lines. At the same time, the *in vivo* significance of the EIP 28/29 gene product is not clear, although larval hemocytes and lymph gland showed ecdysteroid-induced activity of this gene *in vivo.*[38] The relationship of the small *hsp* genes to development *in vivo* is more apparent. Ecdysteroids induce the transcription of the four small heat shock genes in both the S3 cell line and in imaginal discs from pupariating larvae.[39,40] This may be the best indication that "the cultured cells represent a valid model system for studying...the developmental regulation of *hsp* gene activation."[41]

2. Stimulation of Enzymes and Glycoproteins

The same *Drosophila* cell lines that were so valuable in investigating the early effects of ecdysteroid action have also been of considerable use in characterizing hormonally induced enzymatic activities. Best-Belpomme et al.[16] reported that the activity of about 30 enzymes, including proteases, lipases, and kinases, did not change significantly when the Kc cells were exposed to 20-hydroxyecdysone, but that the activities of two enzymes, acetylcholinesterase and β-galactosidase, increased sharply.[42,43,44] In the case of the Kc line, acetylcholinesterase activity was virtually undetectable in the absence of hormone, but increased steadily after treatment with 20-hydroxyecdysone, reaching a peak on the third day. In another *Drosophila* cell line, derived from haploid embryos, untreated cells had a high level of acetylcholinesterase activity, but 20-hydroxyecdysone induced a sixfold increase in enzymatic activity even in this case.[16,45] Cherbas et al.[17] reported that a range of ecdysteroids induced acetylcholinesterase activity in Kc cells in roughly the same order of effectiveness as reported for an imaginal disc evagination assay, i.e., Ponasterone A > polypodine B > 20-hydroxyecdysone > inokosterone > ecdysone. These observations on the induction of enzymes by ecdysteroids in the Kc cells show that this cell line represents an excellent model system. The *in vivo* significance of the Kc cellular response to ecdysteroid appears to relate to hormonal effects on the lymph gland, a hematopoietic tissue, despite earlier notions that neuroblast cells may have contributed to this cell line.[18]

In terms of addressing questions of the consequences of ecdysteroid action beyond the early events, it would be helpful to measure biosynthetic activities that characterize development *in vivo*. Recent efforts along these lines have attempted to link enzymatic activities and glycoprotein synthesis *in vitro* with chitin metabolism and morphogenesis *in vivo*. Sommer and Spindler[46] approached this problem by assaying chitin-degrading enzymatic activity in Kc cells. Earlier reports had measured chitinase activity from both lepidopteran and cockroach cell cultures.[47,48] The Kc cells were maintained in a serum-free medium, because chitinase activity is present in the vertebrate serum usually added to insect tissue culture media. This fact may obscure low levels of chitin synthesis in some insect cell lines. Sommer and Spindler found that β-*N*-acetyl-D-glucosaminidase and β-*N*-acetyl-D-hexosaminidase were produced spontaneously by the Kc cells and released into the medium. The products of their activity were *N*-acetyl-D-glucosamine and *N*-acetyl-D-galactosamine, both of which could inhibit the corresponding enzymatic activity. The authors suggest that this system would be advantageous for investigating ecdysteroid effects on chitin-degrading enzymes.

Related studies were conducted by Kramerov,[49] who reported that a *Drosophila* cell line secreted glycoproteins into the culture medium. This glycoprotein (110,000 Da) consisted of 4000-Da subunits, which contained a glycopeptide moiety bound to *N*-acetyl-D-glucosamine residues. Similar results were obtained for the Indian meal moth cell line, IAL-PID2. Porcheron et al.[50] approached this problem by purifying PID2-produced glycopeptides with gel-permeation HPLC. While glycosylated molecules were observed across the range of molecular weights evaluated, the untreated PID2 cells displayed major *N*-acetyl-D-glucosamine labeling in the 1500-dalton fraction. This peak shifted to the 5000-Da region after treatment of the cell line with 20-hydroxyecdysone, a finding confirmed with dual labeling with radiolabeled leucine and *N*-acetyl-D-glucosamine. Moreover, a glycopeptide of similar molecular weight was observed in ecdysteroid-treated imaginal discs. Although the function of this glycopeptide is not known, we suspect it may play a role in cuticle formation, if only because the potent chitin synthesis inhibitor, teflubenzuron, prevented its appearance. In any event, the presence of a 5000-Da glycopeptide following ecdysteroid treatment in intact imaginal discs as well as in the cell line suggests that this molecule represents expression of differentiated function.

In addition to a role in cuticle formation, extracellular glycoproteins are reported to be important in imaginal disc morphogenesis (evagination), which depends upon cell rearrangement.[51,52] As mentioned earlier in connection with research on the small heat shock genes of the Schneider *Drosophila* cell lines, these cell lines have synthetic activities that typify imaginal discs. In this context Rickoll and Galewsky[53] demonstrated that the S3 cell line secreted glycoproteins into the culture medium, and that 20-hydroxyecdysone increased the labeling of these proteins. Moreover, three of the ecdysteroid-dependent secreted glycoproteins (116,000, 110,000, 80,000 daltons) were immunoprecipitated with an antiserum raised against cell membranes from *Drosophila* imaginal discs. Ecdysteroid-induced aggregation of the S3 cells was correlated with the appearance of some of these glycoproteins. The possibility of involvement of these glycoproteins in cuticle formation was not assessed by the authors, but there is a strong supposition that these hormone-directed syntheses during S3 cell reaggregation may reflect the cell-cell interactions that occur during imaginal disc morphogenesis.

3. Chitin Synthesis

The biochemistry of cuticle formation in insects has attracted the attention of surprisingly few scientists, perhaps because the complex structure of the cuticle obscures the molecular interactions that give rise to the insect exoskeleton. Morphological analysis points to the apical microvilli, "epidermal feet," as the site of synthesis of chitin, the major nonproteinaceous constituents of cuticle.[54] Interest in the hormonal regulation of chitin synthesis, as well as the mode of action of benzoylphenyl ureas that inhibit chitin synthesis, makes the availability of chitin-synthesizing cell lines a significant priority.

There is evidence that a chitin-like material is synthesized spontaneously by a variety of cell lines. Marks and Ward[55] checked for incorporation of radio-labeled *N*-acetyl-D-glucosamine into embryo-derived cell lines of the cockroach *Blatella germanica* (UMBGE-4), the moths *Manduca sexta* (MRRL-CH 34), and *Plodia interpunctella* (UMN-PIE), as well as cell lines derived from imaginal discs of *P. interpunctella* (IAL-PID2) and *Trichoplusia ni* (IAL-TND1). In all cases at least some of the samples produced a chitin-like material. Evidence that the UMBGE-4 product that incorporated *N*-acetyl-D-glucosamine was indeed chitin was based on its ability to withstand digestion for one hour at 100°C in 1 *N* NaOH, and that it was degraded by chitinase to chitobiose and *N*-acetyl-D-glucosamine. Material from the other cell lines meet all of these criteria for chitin.[56,57] In addition to orthopteran and lepidopteran cell lines, two dipteran cell lines, the *Drosophila* Kc line and a *C. tentans* line, showed evidence of chitin synthesis.[58]

Interestingly, cell lines that display some capability for synthesizing chitin-like materials often grow as multicellular vesicles, e.g., the UMBGE-4 cell line, the *C. tentans* epithelial cell line,[27,58] and the *T. ni* imaginal disc cell line.[8] However, this is not an absolute requirement for synthesis of chitin-like materials, since the *Drosophila* Kc cells and the IA1-PID2 cells grow as a monolayer.

The influence of ecdysteroids on chitin synthesis and related activities has been tested in several cell lines. The addition of 20-hydroxyecdysone increases the level of chitin synthesis in the UMBGE cell line,[56] although it depresses chitin synthesis in the *C. tentans* cell line.[58,59] In a related study, the stimulation by ecdysteroids of the uptake of the chitin precursor, *N*-acetyl-D-glucosamine, was investigated in the imaginal disc-derived cell line, IAL-PID2.[60] This reliable response to ecdysteroids was not inhibited by benzoylphenyl urea compounds that inhibit chitin synthesis *in vivo,*[60] although chitin synthesis in the UMBGE-4 cell line was blocked by diflubenzuron.[56] These results suggest that uptake of chitin precursors is not a site of action for benzoylphenyl ureas. Although the mode of action of such inhibitors remains obscure, chitin-producing cell lines should be important experimental models in such research.

III. MORPHOGENESIS

A. REGULATION OF MORPHOLOGY

1. Hemolymph and Tissue Factors

The most obvious morphological characteristic of the *Trichoplusia ni* cell line, IAL-TND1, was that it grew as fluid-filled multicellular spheres.[29] This was a cell line that apparently was derived from the epithelial cells of wing imaginal discs. One of the tests run on these cells compared the 2-D gel protein patterns of the cell line with imaginal discs, fat body, and hemocytes from the insect. The protein pattern for the cell line was virtually identical to intact imaginal discs and distinct from the pattern for other cell types that could have contaminated the primary cultures. This result, which was a control for confirming the source of the cell line, may be among the best evidence for expression of differentiated function by an insect cell line. Consider the definition of differentiation by Jacob and Monod, as quoted by Gilbert,[62] "Two cells are differentiated with respect from one another if, while they harbor the same genome, the pattern of proteins they synthesize is different." The maintenance of differentiative capacity was demonstrated when these spheres were transplanted back into larvae and underwent metamorphosis with the host, where they were recovered as pupal wings with tanned cuticle.[78]

Although we do not know the *in vivo* significance of the vesicle form that both the IAL-SPD1 and IAL-TND1 cultures took, we were confronted with a spontaneous change of the vesicles to amorphous aggregates. We found that larval hemolymph caused a reversal of this morphology back to the multicellular vesicle form. A factor was partially purified from larval hemolymph of the cabbage looper, *Trichoplusia ni,* that caused cell aggregates of an imaginal disc cell line, IAL-TND1, to form multicellular vesicles.[63,64] The vesicle-promoting factor (VPF) was characterized as a nonglycosylated polypeptide with a 16.9 kDa molecular weight.

Interestingly, we found that the action of VPF can be thwarted by coincubation of the cells with ecdysteroids. While the role of VPF *in vivo* remains unclear at this time, it may well represent one of a number of growth and morphogenetic factors that will be uncovered by use of tissue-culture systems.

2. Hormonal Effects

Cellular elongation is a typical response to ecdysteroids in both dipteran and lepidopteran cell lines.[6,65] For example, the *Manduca* cell line, MRRL-CH1, derived from embryos, responds to ecdysteroids by extensive elongation of the cells, and it does so in a dose-dependent manner.[66] Thus, we have three processes that may relate to expression of differentiated function. First, the cells do respond to ecdysteroids; second, there is a marked change in morphology of the cells that typifies this response, and third, as the work on subclones shows, there is maintenance of cellular diversity; i.e., some cells and their progeny respond well to 20-hydroxyecdysone, and some respond very poorly. Moreover, this response requires the synthesis of proteins for the ecdysteroid to have any effect.[67] Similar effects of ecdysteroids on cellular elongation in lepidopteran cell lines were obtained for TN-368 *(Trichoplusia ni)* and HPB-SL-26 *(Spodoptera littoralis).*[68] Thus, we have several cases in which the overt effect of the hormone on the cells is cellular elongation, and in which the other criteria for expression of differentiated function may come into play. Still, because we do not know the *in vivo* cell types represented here, nor do we know the developmental significance of this cellular elongation, there may be doubt as to the validity of this morphological response as an example of differentiated function.

Research with an imaginal disc-derived cell line derived from *Plodia interpunctella,* IAL-PID2, provided a quite different morphology, basically consisting of fibroblast-like cells. The wing discs in the primary cultures were cut repeatedly, resulting in both vesicular blebs and a monolayer outgrowth, which became the source of this cell line.[29] The PID2 cell line is a vigorous, rather typical cell line, without any striking multicellular formations. However, these cells have demonstrated some interesting properties, among these the ability to respond to ecdysteroids with decreased proliferation[69] and increased uptake of *N*-acetyl-glucosamine from the medium.[60] This ecdysteroid-dependent response is characteristic of an early effect of 20-hydroxyecdysone on intact imaginal wing discs at the start of metamorphosis. Thus, we have a clear demonstration of the expression of a developmental capacity that represents an early step in preparation for cuticle synthesis.

A detailed morphological study revealed significant changes in PID2 cells in response to ecdysteroids. Under hormone-free conditions the cells grow as a monolayer of round or spindle-shaped cells. Moreover, localization of wheatgerm agglutinin-gold particles showed that there were glycosylation sites only along the cellular extensions of untreated cells.[70] After treatment with 20-hydroxyecdysone, the cells formed epithelial-like aggregates that were interconnected and linked by desmosomes. In hormone-treated cells, WGA-binding sites were present not only along the cellular extensions, but also along the entire plasma membrane. This pronounced shift in the distribution of glycosylation sites perhaps suggests preparation for ecdysteroid-induced differentiation, including cuticle formation. Thus, the PID2 cells respond to ecdysteroids in a manner that is meaningful in terms of the development *in vivo* of the imaginal discs from which they were derived.

In some cell lines it may be obvious from the morphological response to hormones that a developmental phenomenon with *in vivo* significance has been obtained, despite not having a clear definition of their cellular origin. For example, a cell line derived from the hymenopteran egg parasitoid, *Trichogramma exiguum* (IPLB-Tex2), produces highly contractile cells in response to 20-hydroxyecdysone.[71] This observation suggests that the cell line may be composed of myoblast cells. Similarly, the spontaneous formation of contractile cells from a cell line was noted in NISES-AnPe-426 *(Antheraea pernyi)* after several months in culture.[72]

B. CELL LINES AS SOURCES OF DEVELOPMENTAL FACTORS

We will turn now to cell lines that display *in vivo* properties through the production of specific products utilized by other cells. An example of this situation comes from the cell line that was established from fat body of the gypsy moth.[73] This cell line is quite different in morphology from the imaginal disc-derived cell lines that we have been discussing. As one of a relatively small number of cell lines derived from a specific cell type, the LdFB cells may have maintained differentiated function. We speculated that the cell line might substitute for fat body in promoting the development of parasitoids grown in tissue culture. In this connection, we tested the possibility that the LdFB cells could affect the development of *Microplitis croceipes,* a beneficial insect. Embryogenesis and hatching of this parasite is clearly stimulated by fat body-conditioned culture medium.[74] Also, we demonstrated that the LdFB cells can substitute for larval host fat body in stimulating egg development of the parasite. The active substance(s) that induced germ-band development was dialyzable (<10 K), while the hatch-stimulating factor was nondialyzable (>10K). Both active materials were resistant to trypsin and to heat. Eggs held in media conditioned with the cell line responded in a dose-dependent manner. Manifestly, this is a dramatic example of a cell line retaining differentiated function by producing a specific product that has a developmental effect on another species.

In other culture systems, a combination of vertebrate growth factors and conditioned medium from insect cell lines leads to physiological functioning of cultured tissues, as in the case of the venom gland of the ant *Pseudomyrmex triplarinus.* This gland was cultured in medium formulated with epidermal growth factor, fibroblast growth factor, insulin, cAMP, cGMP, and isoproterenol,[75] but the medium had to be conditioned with established lepidopteran cell lines for long-term cell survival. The metabolic production of the venom was maintained for 12 months. Similarly, conditioned medium from a *Manduca sexta* cell line could substitute for various tissues in stimulating neuronal outgrowth and glial cell survival in cultures from the antennal lobe of *M. sexta* brains.[76]

IV. CONCLUSIONS AND PROSPECTS

We have seen examples of maintenance of differentiated function in cell lines derived from embryos, imaginal discs, and fat body. In general, the extra effort to obtain cell lines from specific tissue types rewards the investigator with cells that are more robust in their response to hormones, developmental capacity, and ability to produce specific cell products.[77] We have only seen a glimpse of what can be accomplished with cell lines that retain a rigorous fidelity to their *in vivo* origin.

We would like to close by sharing with you some of the reasons we believe there is now increasing success in obtaining cell lines from specific insect tissues that retain some measure of differentiated function. The most important ingredient is patience, not only staying with a problem but learning to ignore it artfully, employing what Tom Grace called "organized neglect" (quoted by Maramorosch[2]). Thus, one may wait a year or more in these cases for a spontaneous immortalizing event. Second, the medium must, at the least, keep the cells alive and foster proliferation — a problem solved for many systems with Grace's medium that was based in turn on Wyatt's analysis of insect hemolymph. Third, the investigator should think of the *in vitro* culture system as an organism — with a plastic or glass exoskeleton, an open circulatory system, a requirement for oxygen, and a need to be protected from desiccation, microbes, and predators (including humans); put another way, the tissue culturist must learn to think like a petri dish! Fourth, the investigator needs to understand as much as possible about the cellular ecology *in vivo,* so that conditions can be created in which the cells will respond to appropriate signals with responses that are meaningful in terms of their *in vivo* origin.

ACKNOWLEDGMENTS

This chapter was written while the senior author was on sabbatical at the Arizona Research Laboratory Division of Neurobiology, and appreciation is expressed to Prof. John Hildebrand and his colleagues at the University of Arizona. We also wish to thank Dr. Ann Fallon, University of Minnesota, Dr. Dwight Lynn, U.S. Department of Agriculture, and Dr. Patrick Porcheron, Pierre and Marie Curie University, for reviewing our manuscript.

REFERENCES

1. **Goldschmidt, R.**, Some experiments on spermatogenesis *in vitro*, *Proc. Natl. Acad. Sci. U.S.A.*, 1, 220, 1915.
2. **Maramorosch, K.**, Thomas D. C. Grace — insect tissue culture pioneer, *J. Invert. Pathol.*, 58, 151, 1992.
3. **Grace, T. D. C.**, Establishment of four strains of cells from insect tissue grown *in vitro*, *Nature*, 195, 788, 1962.
4. **Barnes, D. and Sato, G.**, Serum-free cell culture: a unifying approach, *Cell*, 22, 649, 1980.
5. **Echalier, G. and Ohanessian, A. M.**, Isolement, en cultures *in vitro*, de lignées cellulaires diploides de *Drosophila melanogaster*, *C. R. Acad. Sci.* (Paris), 268, 1771, 1969.
6. **Courgeon, A.-M.**, Action of insect hormone at the cellular level, *Exp. Cell Res.*, 74, 317, 1972.
7. **Marks, E. P. and Homan, G. M.**, Ecdysone action on insect cell lines, *In Vitro*, 15, 300, 1979.
8. **Lynn, D. E., Miller, S. G., and Oberlander, H.**, Establishment of a cell line from lepidopteran wing imaginal discs: induction of newly synthesized proteins by 20-hydroxyecdysone, *Proc. Natl. Acad. Sci. U.S.A.*, 79, 2589, 1982.
9. **Lynn, D. E., Feldlaufer, M. F., and Lusby, W. R.**, Isolation and identification of 20-hydroxyecdysone from a Lepidopteran continuous cell line, *Arch. Insect Biochem. Physiol.*, 5, 71, 1987.
10. **Sommé-Martin, G., Lafont, R., Colardeau, C., Coudouel, N., Oberlander, H., and Porcheron, P.**, Uptake and metabolism of ecdysone and 20-hydroxyecdysone in a Lepidopteran cell line derived from wing discs of *Plodia interpunctella*, *Insect Biochem. Molec. Biol.*, 22, 137, 1992.
11. **Ward, G. B., Kelly, T. J., Woods, C. W., and Marks, E. P.**, Ecdysteroid production by a continuous insect cell line, *Arch. Insect Biochem. Physiol.*, 5, 91, 1987.
12. **Ward, G. B., Beydon, P., Lafont, R., and Mayer, R. T.**, Metabolism of ecdysteroids by a chitin-synthesizing insect cell line, *Arch. Insect Biochem. Physiol.*, 15, 137, 1990.
13. **Stiles, B., McDonald, I. C., Gerst, J. W., Adams, T. S., and Newman, S. M.**, Initiation and characterization of five embryonic cell lines from the cotton boll weevil, *Anthonomus grandis* in a commercial serum-free medium, *In Vitro Cell. Dev. Biol.*, 28A, 355, 1992.
14. **Spindler, K. D. and Spindler-Barth, M.**, Ecdysteroid production and metabolism by an epithelial cell line from *Chironomus tentans*, *Naturwissenschaften*, 78, 78, 1991.
15. **Delbeque, J.-P., Weidner, K., and Hoffman, K. H.**, Alternative sites for ecdysteroid production in insects, *Invert. Reprod. Dev.*, 18, 29, 1990.
16. **Best-Belpomme, M., Courgeon, A.-M., and Echalier, G.**, Development of a model for the study of ecdysteroid action: *Drosophila melanogaster* cells established *in vitro*, in *Progress in Ecdysone Research; Developments in Endocrinology*, Hoffmann, J. A., Ed., Elsevier/North-Holland Biomed. Press, Amsterdam, 1980, 379.
17. **Cherbas, L., Cherbas, P., Savakis, C., Demetri, G., Manteuffel-Cymborowska, M., Yonge, C. O., and Williams, C. M.**, Studies of ecdysteroid action on a *Drosophila* cell line, in *Invertebrate Systems In Vitro*, Kurstak, E., Maramorosch, J., and Dubendorfer, A., Eds., Elsevier/North Holland Biomed. Press, Amsterdam, 1980, 217.
18. **Andres, A. J. and Cherbas, P.**, Tissue-specific ecdysone responses: regulation of the *Drosophila* genes EIP 28/29 and EIP 40 during larval development, 1992, in press.
19. **Yund, M. A. and Osterbur, D. L.**, Ecdysteroid receptors and binding proteins, in *Comprehensive Insect Physiology, Biochemistry and Pharmacology*, Kerkut, G. A. and Gilbert, L. I., Eds., Pergamon Press, Oxford, 1978, 473.
20. **Maroy, P., Dennis, R., Beckers, C., Sage, B. A., and O'Connor, J. D.**, Demonstration of an ecdysteroid receptor in a cultured cell line of *Drosophila melanogaster*, *Proc. Natl. Acad. Sci. U.S.A.*, 75, 6035, 1978.
21. **Bidmon, H. J. and Sliter, T. J.**, The ecdysteroid receptors, *Invert. Reprod. Dev.*, 18, 13, 1990.
22. **Landon, T., Sage, B. A., Seller, B. J., and O'Connor, J. D.**, Characterization and partial purification of the *Drosophila* Kc cell ecdysteroid receptor, *J. Biol. Chem.*, 263, 4693, 1988.
23. **Dinan, L.**, Ecdysteroid receptors in a tumorous blood cell line of *Drosophila melanogaster*, *Arch. Insect Biochem. Physiol.*, 2, 295, 1985.

24. **Gateff, E., Gissmann, L., Shrestha, R., Plus, N., Pfisten, H., Schroder, J., and zur Hausen, H.,** Characterization of two tumorous blood cell lines of *Drosophila melanogaster* and the viruses they contain, in *Invertebrate Systems In Vitro,* Kurstak, E., Maramorosch, K., and Dubendorfer, A., Eds., Elsevier/North-Holland Biomed. Press, Amsterdam, 1980, 517.
25. **Handler, A. and Maroy, P.,** Ecdysteroid receptors in *Drosophila melanogaster* adult females, *Mol. Cell Endocrinol.,* 63, 103, 1989.
26. **Turberg, A., Spindler-Barth, M., Lutz, B., Lezzi, M., and Spindler, K.-D.,** Presence of an ecdysteroid-specific binding protein ('receptor') in epithelial tissue culture cells of *Chironomus tentans, J. Insect Physiol.,* 34, 797, 1988.
27. **Turberg, A. and Spindler, K.-D.,** Properties of nuclear and cytosolic ecdysteroid receptors from an epithelial cell line from *Chironomus tentans, J. Insect Physiol.,* 38, 81, 1992.
28. **Wing, K. D. and Ramsey, J. P.,** Other hormonal agents: ecdysone agonists, in *Progress and Prospects in Insect Control,* 1989 BCPC Mono. #43, 1989, 107.
29. **Lynn, D. E. and Oberlander, H.,** The establishment of cell lines from imaginal wing discs of *Spodoptera frugiperda* and *Plodia interpunctella, J. Insect Physiol.,* 29, 591, 1983.
30. **Savakis, C., Koehler, M. M. D., and Cherbas, P.,** cDNA clones for the ecdysone-inducible polypeptide (EIP) mRNAs of *Drosophila* Kc cells, *Embo J.,* 3, 235, 1984.
31. **Cherbas, L., Benes, H., Bourouis, M., Burtis, K., Chao, A., Cherbas, P., Crosby, M., Garfinkel, M., Guild, G., Hogness, D., Jami, J., Jones, G. W., Koehler, M., Lepesant, J.-A., Martin, G., Maschat, F., Mathers, P., Meyerowitz, E., Moss, R., Pictet, R., Rebers, J., Richards, G., Roux, J., Schulz, R., Segraves, W., Thummel, G., and Vijuyragharan, K.,** Structural and functional analysis of some moulting and hormone-responsive genes from *Drosophila, Insect Biochem.,* 16, 241, 1986.
32. **Schultz, R. A., Shlomchik, W., Cherbas, L., and Cherbas, P.,** Diverse expression of overlapping genes: the *Drosophila* EIP 28/29 gene and its upstream neighbors, *Dev. Biol.,* 131, 515, 1989.
33. **Cherbas, P., Cherbas, L., Lee, S.-S., and Nakanishi, K.,** 26-[^{125}I] iodoponasterone A is a potent ecdysone and a sensitive radioligand for ecdysone receptors, *Proc. Natl. Acad. Sci. U.S.A.,* 85, 2096, 1988.
34. **Schneider, I.,** Cell lines derived from late embryonic stages of *Drosophila melanogaster, J. Embryol. Exp. Morphol.,* 27, 353, 1972.
35. **Morganelli, C. M., Berger, E. M., and Pelham, H. R. B.,** Transcript of *Drosophila* small *hsp-tk* hybrid genes is induced by heat shock and by ecdysterone in transfected *Drosophila* cells, *Proc. Natl. Acad. Sci. U.S.A.,* 82, 5865, 1985.
36. **Cherbas, L., Lee, K., and Cherbas, P.,** The induction of EIP 28/29 by ecdysone in *Drosophila* cell lines, *Invert. Reprod. Dev.,* 18, 108, 1990.
37. **Segraves, W. A. and Richards, G.,** Regulatory and developmental aspects of ecdysone-regulated gene expression, *Invert. Reprod. Dev.,* 18, 67, 1990.
38. **Cherbas, P. and Andres, A.,** The regulation of EIP 28/29 and EIP 40 by ecdysone in *Drosophila, Invert. Reprod. Dev.,* 18, 108, 1990.
39. **Ireland, R. C. and Berger, E. M.,** Synthesis of low molecular weight heat shock peptides stimulated by ecdysterone in a cultured *Drosophila* cell line, *Proc. Natl. Acad. Sci. U.S.A.,* 79, 855, 1982.
40. **Ireland, R. C., Berger, E., Sirotkin, K., Lund, M. A., Osterbur, D., and Fristrom, J.,** Ecdysterone induces the transcription of four heat-shock genes in *Drosophila* S3 cells and imaginal discs, *Dev. Biol.,* 93, 478, 1982.
41. **Mestril, R., Schiller, P., Amin, J., Klapper, H., Ananthan, J., and Voellmy, R.,** Heat shock and ecdysterone activation of the *Drosophila melanogaster hsp* 23 gene; a sequence element implied in developmental regulation, *Embo J.,* 5, 1667, 1986.
42. **Best-Belpomme, M. and Courgeon, A. M.,** Ecdysterone and acetylcholinesterase activity in cultured *Drosophila* cells, *Febs Letters,* 82, 345, 1977.
43. **Cherbas, P., Cherbas, L., and Williams, C. M.,** Induction of acetyl-cholinesterase activity by β-ecdysone in a *Drosophila* cell line, *Science,* 197, 275, 1977.
44. **Best-Belpomme, M., Courgeon, A. M., and Rambach, A.,** β-galactosidase is induced by hormone in *Drosophila melanogaster* cell cultures, *Proc. Natl. Acad. Sci. U.S.A.,* 75, 6102, 1978.
45. **Debec, A.,** Isozymic patterns and functional states of cell lines of *Drosophila melanogaster* cultured *in vitro,* II, *Wilhelm Roux Arch.,* 180, 107, 1976.
46. **Sommer, V. and Spindler, K.-D.,** Demonstration of β-*N*-acetyl-D-glucosaminidase and β-*N*-acetyl-D-hexosaminidase in *Drosophila* Kc-cells, *Arch. Insect Biochem. Physiol.,* 17, 3, 1991.
47. **Dziadik-Turner, C., Koga, D., and Kramer, K. J.,** Secretion of exo- and endo-β-*N*-acetylglucosaminidases by insect cell lines, *Insect Biochem.,* 11, 215, 1981.
48. **Bernier, I., Landureau, J. C., Grellet, P., and Jolles, P.,** Characterization of chitinase from haemolymph and cell cultures of cockroach *(Periplaneta americana), Comp. Biochem. Physiol.,* 47B, 41, 1974.
49. **Kramerov, A. A., Mukha, D. V., Metukovsky, E. V., and Gvodzev, V. A.,** Glycoproteins containing sulfated chitin-like carbohydrate moiety are synthesized in an established *Drosophila melanogaster* cell line, *Insect Biochem.,* 16, 417, 1986.

50. **Porcheron, P., Morinière, M., Coudouel, N., and Oberlander, H.,** Ecdysteroid-stimulated synthesis and secretion of an *N*-acetyl-D-glucosamine-rich glycopeptide in a lepidopteran cell line derived from imaginal discs, *Arch. Insect Biochem. Physiol.,* 16, 257, 1991.
51. **Rickoll, W. and Fristrom, J. W.,** The effects of 20-hydroxyecdysone on the metabolic labeling of membrane proteins in *Drosophila* imaginal discs, *Develop. Biol.,* 95, 275, 1983.
52. **Woods, D. F., Rickoll, W. L., Birr, C., Poodry, C. A., and Fristrom, J. W.,** Alterations in the cell surface proteins of *Drosophila* during morphogenesis, *Wilhelm Roux Arch.,* 196, 339, 1987.
53. **Rickoll, W. L. and Galewsky, S.,** 20-hydroxyecdysone increases the metabolic labeling of extracellular glycoproteins in *Drosophila* S3 cells, *Insect Biochem.,* 18, 337, 1988.
54. **Locke, M.,** A structural analysis of post-embryonic development, in *Comprehensive Insect Physiology, Biochemistry and Pharmacology, Vol. 2,* Kerkut, G. A. and Gilbert, L. I., Eds., Permagon Press, Oxford, 1985, 87.
55. **Marks, E. P. and Ward, G. B.,** Ecdysone synthesis in a cuticle-forming insect cell line, in *Endocrinological Frontiers in Physiological Insect Ecology,* Sehnal, F., Zabza, A., and Denlinger, D. L., Eds., Wroclaw Tech. Univ. Press, Wroclaw, 1988, 539.
56. **Ward, G. B., Newman, S. M., Klosterman, H. J., and Marks, E. P.,** Effect of 20-hydroxyecdysone and diflubenzuron on chitin production by a cockroach cell line, *In Vitro Cell. Dev. Biol.,* 24, 326, 1988.
57. **Marks, E. P., Balke, J., and Klosterman, H.,** Evidence for chitin synthesis in an insect cell line, *Arch. Insect Biochem. Physiol.,* 1, 225, 1984.
58. **Louderhausen, M., Kummen, V., Spindler-Barth, M., Spindler, K.-D., and Thomas, H.,** Chitin synthesis in insect cell lines, *Insect Biochem.,* 18, 631, 1988.
59. **Spindler-Barth, M., Kamman, V., and Spindler, K.-D.,** Hormonal regulation of chitin synthesis in two insect cell lines, in *Chitin and Chitosan,* Skjak-Brack, G., Anthonsen, T., and Sandford, P., Eds., Elsevier Applied Science, London, 1989, 279.
60. **Porcheron, P., Oberlander, H., and Leach, C. E.,** Ecdysteroid-regulation of aminosugar uptake in a lepidopteran cell line derived from imaginal discs, *Arch. Insect Biochem. Physiol.,* 7, 145, 1988.
61. **Oberlander, H., Silhacek, D. L., Leach, E., Ishaaya, I., and Shaaya, E.,** Benzoylphenyl ureas inhibit chitin synthesis without interfering with amino sugar uptake in imaginal wing discs of *Plodia interpunctella* (Hübner), *Arch. Insect Biochem. Physiol.,* 18, 219, 1991.
62. **Jacob, F. and Monod, J.,** as quoted in **Gilbert, S. F.,** *Developmental Biology,* 3rd ed., Sinauer Assoc., Sunderland, MA, 1991, 327.
63. **Lynn, D. E., Oberlander, H., and Ferkovich, S. M.,** Induction of vesicle formation in a cell line derived from imaginal discs, *In Vitro Cell Dev. Biol.,* 21, 277, 1985.
64. **Ferkovich, S. M., Oberlander, H., Dillard, C., and Leach, C. E.,** Purification and properties of a factor from insect haemolymph that promotes multicellular vesicle formation *in vitro, Arch. Insect Biochem. Physiol.,* 6, 73, 1987.
65. **Eide, P. E., Caldwell, J. M., and Marks, E. P.,** Establishment of two cell lines from embryonic tissue of the tobacco hornworm, *Manduca sexta* (L.), *In Vitro,* 11, 395, 1975.
66. **Lynn, D. E. and Oberlander, H.,** The effect of cytoskeletal disrupting agents on the morphological response of a cloned *Manduca sexta* cell line to 20-dihydroxy-ecdysone, *Wilhelm Roux Arch.,* 190, 150, 1984.
67. **Oberlander, H., Leach, C. E., and Lynn, D. E.,** Effects of cycloheximide on cellular elongation in a *Manduca sexta* cell line, *Wilhelm Roux Arch.,* 190, 60, 1981.
68. **Kislev, N., Segal, I., and Edelman, M.,** Ecdysteroids induce morphological changes in continuous cell lines of Lepidoptera, *Roux Arch. Dev. Biol.,* 193, 252, 1984.
69. **Silhacek, D., Oberlander, H., and Porcheron, P.,** Action of RH 5849, a non-steroidal ecdysteroid mimic, on *Plodia interpunctella* (Hübner), *Arch. Insect Biochem. Physiol.,* 15, 201, 1991.
70. **Cassier, P., Serrant, P., Garcia, R., Coudouel, M., André, M., Guillaumin, D., Porcheron, P., and Oberlander, H.,** Morphological and cytochemical studies of the effects of ecdysteroids in a lepidopteran cell line (IAL-PID2), *Cell Tissue Res.,* 265, 361, 1991.
71. **Lynn, D. E. and Hung, A. C. F.,** Development of continuous cell lines from the egg parasitoids *Trichogramma confusum* and *T. exiguum, Arch. Insect Biochem. Physiol.,* 18, 99, 1991.
72. **Inoue, H., Kobayashi, J., Kawakita, H., Miyazaki, J.-I., and Hirabayashi, T.,** Insect muscle cell line forms contractile tissue networks *in vitro, In Vitro Cell. Dev. Biol.* 27A, 837, 1991.
73. **Lynn, D. E., Dougherty, E. M., McClintock, J. T., and Lock, M.,** Development of various tissues of Lepidoptera, in *Invertebrate and Fish Tissue Culture,* Kuroda, Y., Kurstak, E., and Maramorosch, K., Eds., Springer-Verlag, Berlin, 1988, 239.
74. **Ferkovich, S. M., Dillard, C., and Oberlander, H.,** Stimulation of embryonic development in *Microplitis croceipes* (Braconidae) in cell culture media preconditioned with a fat body cell line derived from a nonpermissive host, gypsy moth, *Lymantria dispar, Arch. Insect Biochem. Physiol.,* 18, 169, 1991.
75. **Hink, W. F. and Butz, J. J.,** Primary culture of ant venom gland cells, *In Vitro Cell. Dev. Biol.,* 21, 333, 1985.

76. **Hayashi, J. and Hildebrand, J.,** Insect olfactory neurons *in vitro:* morphological and physiological characterization of cells from the developing antennal lobes of *Manduca sexta, J. Neurosci.*, 10, 848, 1991.
77. **Dinan, L., Spindler-Barth, M., and Spindler, K.-D.,** Insect cell lines as tools for studying ecdysteroid action, *Invert. Reprod. Dev.*, 18, 42, 1990.
78. **Oberlander, H.,** unpublished observations.

Chapter 8

DIFFERENCES IN TRANSFECTION EFFICIENCY BETWEEN DROSOPHILA CELL LINES

G. Echalier and F. Fourcade-Peronnet

TABLE OF CONTENTS

I. Introduction 143

II. Is the Low Percentage of Expressing Cells a Clue to Differences between Cell Lines? 146
A. Clonal Variation in Gene Transfer Efficiency 146
B. Possible Dependence of Efficient Gene Transfer on Specific Phases of the Cell Cycle 148

III. Every Step of the Transformation Process May Depend on Intrinsic Properties of the Recipient Cell Line 149
A. Variations in DNA Uptake 149
B. Variations in Intracellular Transport and Degradation of Input DNA 150
C. Variations in the Penetration of Exogenous DNA into Host Cell Nucleus 152
D. Integration into Host Cell Genome and Variations between Cell Lines 152
E. Variations in Early or Late Expression of Transfected Genes 153

IV. Conclusion(s) 155

References 155

0-8493-4597-9/94/$0.00+$.50
© 1994 by CRC Press Inc.

Guy Echalier is Full Professor of Zoology at the Faculté des Sciences de Paris (Université Paris VI). He received his Doctorat d'Etat at the Sorbonne (Faculté des Sciences de Paris), in 1957, for an experimental work demonstrating the role of the "Y organ" in the hormonal control of moulting in Crustaceans. With a post-doctoral grant from the U.S. National Academy of Sciences, he spent two years in the U.S., first in Carroll M. Williams' lab (Harvard), then in Renato Dulbecco's group (Caltech), primarily for training in liquid medium cell cultures. Back in France, he was elected Associate Professor at the Faculté des Sciences de Paris. Pioneering in the field of insect cell culture, during the early sixties, he was the first, in collaboration with Annie Ohanessian, to establish *in vitro* permanent cell lines from *Drosophila.* Current interest: *Drosophila* cultured cells as an experimental model for analyzing hormonal regulation of an eukaryotic genome; expression and mobility of retrotransposons in *Drosophila* cells.

I. INTRODUCTION

During the last decade, DNA-mediated gene transfer (DMGT) has become an irreplaceable tool for studying eukaryotic genomes and their regulations.

Because of the leading role played by *Drosophila melanogaster* in the particularly active field of developmental biology, the most frequent use of *in vitro* cultured cell lines of *Drosophila* is undoubtedly to welcome transferred genes of some interest, under transient or stable conditions, and thereby to permit a deep analysis of the control of their expression or of the functions of their products.

For transferring genes, it is essential to have at one's disposal proficient recipient cells. Then, available cell lines show considerable differences in their ability to be transfected. Moreover, transient and stable expressions of exogenous genes seem to correspond to quite different phenomena, and the variability noticed between lines can, therefore, refer to each of these two situations: a given cell line may transiently express foreign genes at a high level, whereas it is poorly susceptible to stable transformation, or vice versa.

Such observations have been amply documented in mammalian cells. For instance, it was mere chance whether mouse L cells — which, because of the occurrence of thymidine kinase-deficient sublines, were used by Wigler et al.[1] for optimizing transfection procedure with Ca phosphate — did prove especially capable of high transformation efficiency. After transfection with purified herpes simplex virus tk^+ gene, up to 10^{-5} cells (and even 10^{-4} in later works) could give rise to aminopterine-resistant colonies, in the selective HAT medium. Comparable transformation frequencies have not been obtained in most other cultured cell lines: many primate cell lines (for example, monkey kidney Vero cells), although being efficient transient expressors, remain rather difficult to stably transform. This low competence of many interesting cells is indeed one of the major limitations to a wider use of the DNA transformation method. Refinements and diversification of gene transfer techniques — with the use of other "facilitator" compounds or the recent development of the electroporation method — have made possible DMGT into cell lines so far considered as particularly restive for transformation. However, beyond those technical considerations, which might mainly affect the uptake of DNA, there still remain strong presumptions that complex intrinsic factors could contribute to the more or less efficient expression of input exogenous genes by various cell lines (as it seems well established in bacterial transformation).

As for *Drosophila* cells, a similar variability in the behavior of the different lines with regard to transfection is also well established.

For instance — to confine the discussion to the two lines used most worldwide — Schneider line 2, according to many investigators, might be a better transient expressor than our Kc line. Figure 1 shows a comparison of the expression of a CAT reporter gene, driven by one LTR from 1731 retrotransposon, 48 h after Ca P_i-mediated transfection into S2 and Kc cells, respectively. In order to eliminate possible variations due to the transfer technique, we tried several other methods. The polybrene procedure, popularized by A. Fallon,[2] proved rather successful with Kc cells, while it works poorly with S2 cells; nevertheless, it still seems true that, under optimized conditions, the level of *cat* activity in Kc cells transfected with DNA-polybrene remains lower than in Ca phosphate transfected S2 cells. Walker[3] reports similar success of polybrene-mediated transfer in KcH subline, as well as in S2, S3, and Dm1 lines, but she does not give any comparative figures.

As for stable transformations, however, both lines have been indifferently used by many groups and both were found to be efficient "producers" of foreign gene products. Unfortunately, one does not dispose of comparative figures, taking particularly into account the number of integrated copies.

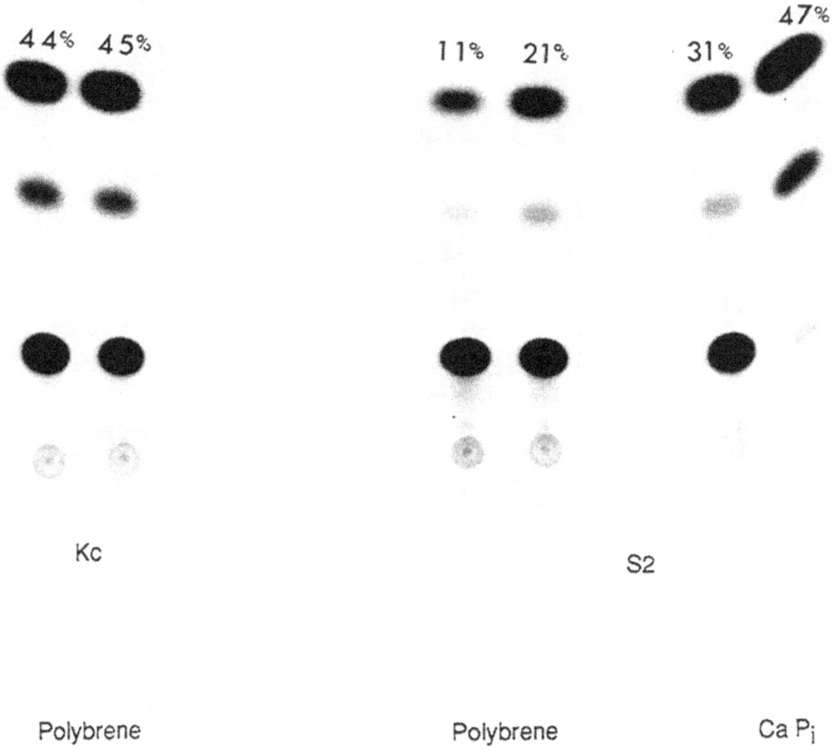

Figure 1. Differences in transient expression between Kc and S2 cells when transfected with a CAT expression plasmid (pD64-CAT) and after Ca phosphate or polybrene-mediated transfer.

As a matter of fact, most of the information about transfection efficiency of different *Drosophila* cell lines rests upon personal communications, among our small community of drosophilists, and only two published reports dealt especially with this important problem.

Burke et al.[4] made a systematic comparison between several *Drosophila* lines, by measuring their ability to incorporate tritiated guanine, a few days after transfection with the bacterial *ecogpt* gene (under the control of *copia* LTRs). As can be seen in their Table II (as reproduced in Figure 2), all of the three tested *Drosophila melanogaster* cell lines did incorporate considerably lower levels of tritiated guanine than three lines derived from *Drosophila hydei* (no more than 3 to 6% of the level observed with DH33 line and used as a reference value). The difference was less great when the *gpt* gene was driven by an *Hsp*70 promoter. This particularly high expression level of DH33 even allowed the use of autoradiography for detecting the approximately 20% of cells that, in the population, took up and expressed the transferred *gpt* gene.

More recently, Saunders et al.[5] reported that the haploid cell line 1182, grown by Debec from a mutant of *D. melanogaster,* or at least a subline from it (1182-4), could express a series of transfected genes between 20 and over 100 times more efficiently than any other

Comparison of levels of [^{3}H] guanine incorporated into cell lines transfected with CV2*gpt*

Cell type	**[^{3}H]Guanine incorporation % c.p.m. of DH33**
D1	5.7
SL2	3.3
Kc	6.2
DH14	80.0
DH15	66.0
DH33	100.0

Approximately 90 h after transfection cells were assayed for [^{3}H]guanine incorporation. The highest incorporation observed, in DH33 cells, was 12 624 c.p.m. which is equivalent to 10.0 pmol of guanine incorporated over the 20 h labeling period.

Figure 2. Table from Burke, J. F. et al., *Embo J.*, 3, 2549, 1984. With permission.

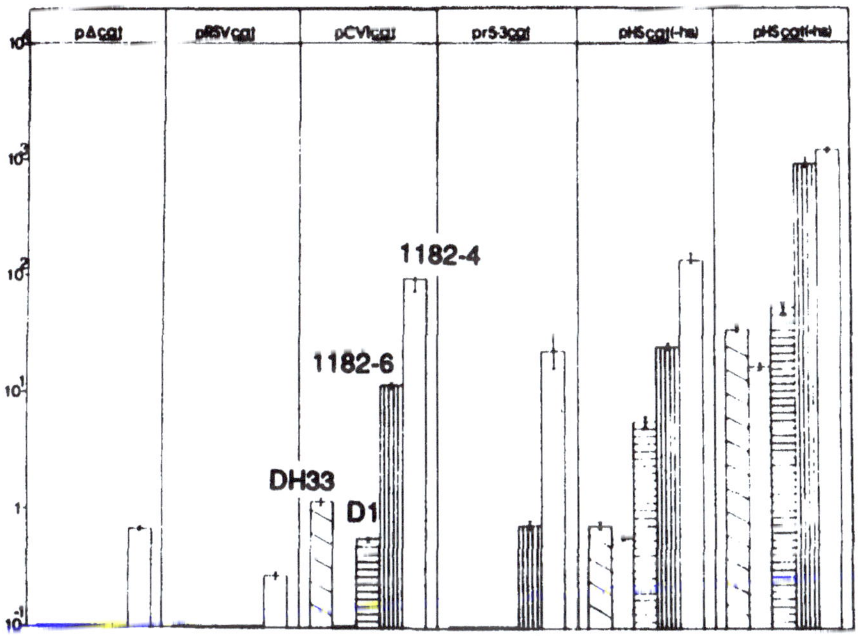

Figure 3. Activity of promoters in various cell lines. (From Saunders, S. E. et al., *NAR*, 17, 6205, 1989. With permission.)

Drosophila cell line, including *D. hydei* DH33. Moreover, those 1182 cells could be transfected by mere addition of naked DNA (that is, without any "facilitator") in serum-free media (Figure 3).

We have, of course, assayed this 1182 line and particularly its subline 4 — because it had been developed (by A. Debec) in our own laboratory — but, I regret to say, without the same success. Similarly, *D. hydei* cell line DH33 has not proved, in our hands, to be a better expressor than the Kc line. Those discrepancies might depend on our experimental system, but, more probably, they raise the difficult question of some possible "drift" in subpopulations from any given cultured cell line.

Because of important practical implications, as well as questions of theoretical interest, I would like to discuss various aspects of the existence of such marked differences in the ability of various *Drosophila* cell lines to take up, and possibly integrate and express, foreign genes. I shall use, of course, every fragmentary datum, published or not, about *Drosophila* cells, but, due to lack of information, I shall frequently refer to data borrowed from mammalian cells. The problem may be tackled in different ways.

II. IS THE LOW PERCENTAGE OF EXPRESSING CELLS A CLUE TO DIFFERENCES BETWEEN CELL LINES?

The first clue to this difficult problem lies in the fact that, among a transfected cell culture, the percentage of individual cells that transiently express the exogenous gene remains low, usually no more than a few per cent of the population, and cells that finally will integrate the gene into their genome are even much rarer (by several orders of magnitude). Those "competent" cells seem to correspond to a small specific subset of the cell population. There are, however, two possible interpretations: either every culture is a mixture of cells with genetically distinct degrees of transfection proficiency; or the intake of DNA, and its subsequent protection and integration, require that recipient cells be in a very peculiar metabolic stage of their cellular cycle. In both cases, it is easily imaginable that the proportion of competent cells, at a given time, remains low, and that it may significantly differ from one cell line to another, or even vary within the same line after several months spent under distinct conditions of growth.

As other illustrations of such possible population "drift" (see above), I would like to point out that the subline Kc107, praised by Bourouis and Jarry,[7] was especially efficient for transformation, but we did not have access to it until several years later. At that point it proved to be more suitable, at least for transient expression, than our original Kc and "home-grown" line. More recently, M. Kim and C. Coulondre,[29] in our own laboratory, found that Schneider's *Drosophila virilis* line was able to express CAT plasmids at a 60-fold higher level than S2 (unpublished data). Unfortunately, this exceptional efficiency has rapidly and unexplainedly decreased, within a few months, in the subline we have been continuously growing.

Let us consider now, with more details, the two possible explanations of those low percentages of expressing cells.

A. CLONAL VARIATION IN GENE TRANSFER EFFICIENCY

Corsaro and Pearson,[8] in a classical paper, established that (a) the transformation frequency for *tk*$^+$ varies 10- to 20-fold among randomly isolated subclones of the mouse fibroblast line LMtk$^-$CID1 and (b) recipient clones more competent for *tk*$^+$ transfer are also relatively more competent than others for the transfer of a second marker (ouabain-resistance). Moreover, the analysis of second-generation subclones indicated that clones exhibiting high competence breed true, whereas those with low competence do not (Figure 4).

We have recently repeated a similar comparison between a series of clones from Schneider's *Drosophila virilis* cell line, in order to understand the rapid decrease in expression efficiency of our subline. It can be clearly seen that *cat* enzymatic activity levels differed strikingly from one clone to another, although none of them yet exhibited as high an expression level as the subline did several months before (Figure 5).

Such results strongly suggest that genetic factors might indeed contribute to the transformation competence or expression efficiency of each line. However, correlations should be rather complex, and this does not mean — as was indicated by Wigler et al.[9] — that every daughter cell from a first "transformant" will be, ipso facto, competent in subsequent transformations, as might be expected in a too simplistic understanding of such heredity. It might have, at best, increased probability for productive transfection.

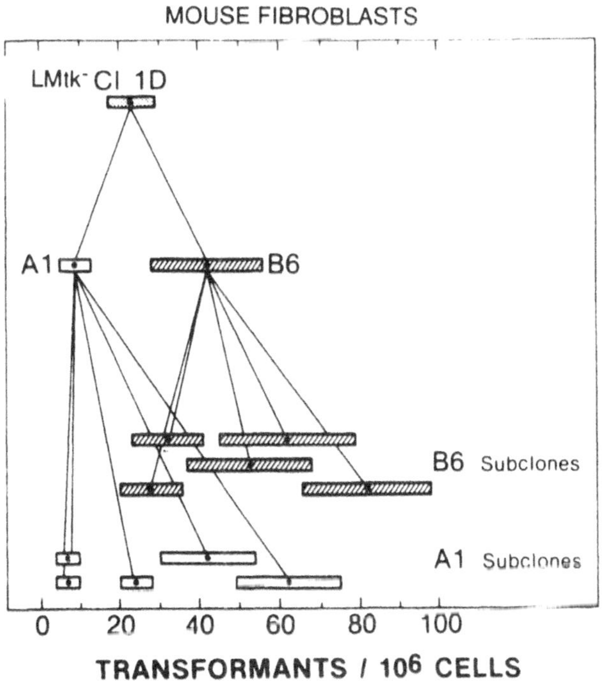

Figure 4. Mouse fibroblasts. Heritability of competence for *tk*⁺ transformation in high- and low-efficiency subclones of Cl 1D. (From Corsaro, C. M. and Pearson, M. L., *Somatic Cell Genet.*, 7, 617, 1981. With permission.)

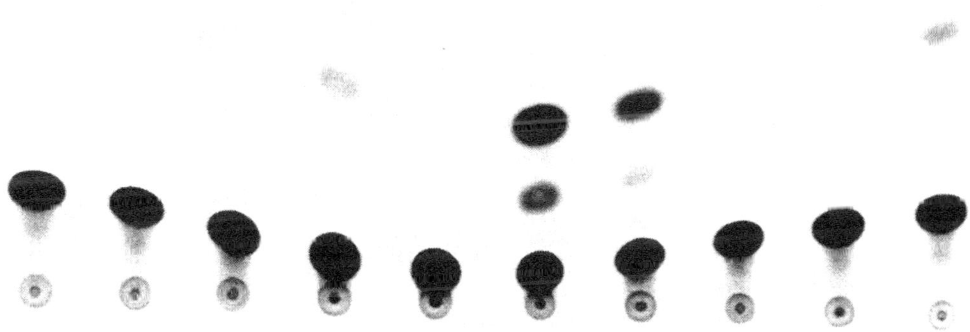

Figure 5. Differences in transient expression between a series of randomly isolated clones from Schneider's *Drosophila virilis* cell line (as measured by CAT activity, 48 h after transfection).

Late expression in cells synchronized with colcemid

Phases represented in cell population	**No. colonies in HAT per 10^6 cells**	**No. colonies in HAT per 10^6 cells/No. colonies in normal medium per 10^6 cells^a**
	(Mean ± SD)	(Mean ± SD)
G_1, G_2, S, M (asynchronous)	88.5 ± 1.5	20.0 ± 0.7
G_2, S	130.3 ± 5.8	49.7 ± 3.1
M	2.3 ± 1.3	5.9 ± 3.4
G_1	2.5 ± 0.5	7.7 ± 2.5

Figure 6. Table from Giulotto, E. and Israel, N., *BBRC*, 118, 310, 1984. With permission.

B. POSSIBLE DEPENDENCE OF EFFICIENT GENE TRANSFER ON SPECIFIC PHASES OF THE CELL CYCLE

As an alternative, the low percentage of cells that, in a transfected culture, do express early or late (that is, transiently or after more or less stable integration) some foreign gene, could be explained by the fact that efficient gene transfer and integration might require preferentially some particular metabolic state of the recipient cell. If this stage is relatively brief within the cell cycle, competent cells will be statistically rare, at any time, in nonsynchronized populations. Let us, moreover, point out that any small variation of the total length of the cell cycle or of some critical stage — which might be due to differing feeding conditions as well as to inherited factors — could be sufficient for significantly modifying the proportion of cells at the right stage during the gene transfer process.

Mammalian cell cultures can be synchronized with various methods, and several investigators could thus study the influence of the cell cycle on the efficiency of transfection. For instance, Giulotto and Israel[10] — using the classical system of tk^+ transformation of mouse L cells — concluded that the frequency of stable transformation was several fold higher when cells were exposed to DNA during S phase, whereas the efficiency of early, that is, transient, expression did not depend on the cell cycle stage at the transfer time (Figure 6).

However, in a quite different experimental system, since he delivered CAT expression plasmids into plant protoplasts and by electroporation, Nagata[11] established that tobacco protoplasts electroporated during the M phase exhibited a 3 to 5 times higher level of (transiently expressed) *cat* activity than protoplasts electroporated at other phases. It is noteworthy that the extreme rapidity of the electroporation process should permit a greater precision in this type of analysis.

Unfortunately, nobody has, so far, succeeded in efficiently synchronizing *Drosophila* cultured cells. Yet there are indirect evidences that cell cycle phases may have a real importance for transfection. It is well known, for instance, that for optimum transformation efficiency, cells must be in the exponential ("log") phase of growth, and most protocols recommend that recipient cells be seeded one day before the DNA transfer. Besides, the fact that we obtained a higher CAT plasmid expression by extending the cell exposure period up to 18 h (instead of the usual 4 h), might be explained by the recruitment of additional cells that have arrived at the proper stage.

Also probably relevant to this point might be the observations of Sinclair and Bryant:[12] an ecdysteroid treatment, for 48 h before transfection, induced a four- to fivefold increase in the expression levels of transfected genes, and this was in keeping with a higher percentage of expressing cells in the population. Because ecdysteroids are known to block *Drosophila* cells in G2 phase,[13] it might be imagined — as already suggested by Saunders and colleagues[5] —

that, after the removal of this hormonal block, many cells might enter into mitosis just around the transfection time.

III. EVERY STEP OF THE TRANSFORMATION PROCESS MAY DEPEND ON INTRINSIC PROPERTIES OF THE RECIPIENT CELL LINE

Another approach to this general problem of the variability of transfection efficiencies among *Drosophila* cell lines consists in surveying the successive steps of the transfection process — from the entry of exogenous DNA up to its possible integration into chromosomes — and in wondering whether some intrinsic cellular factor(s), characteristic of each recipient line, might not influence some critical steps and thus affect the globally evaluated cell "competence." Let us, however, emphasize from the outset that the molecular mechanisms that mediate transformation still remain poorly understood.

A. VARIATIONS IN DNA UPTAKE

One critical step is, of course, the penetration of undamaged DNA through the cell membrane. The work of reference, concerning mammalian cells and calcium phosphate method, was conducted by Loyter and colleagues.[14] They monitored the fate of the Ca phosphate-DNA complex by the use of fluorescent dyes, staining either the DNA (with DAPI) or the Ca phosphate (with chlorotetracycline). Under optimized conditions, and just after the incubation period, typical fluorescent spots were already visible in the cytoplasm of most cells of the highly competent mouse Ltk$^-$ line. They thereby confirmed the importance of a "good" calcium phosphate precipitate and showed that the number of fluorescent spots increased with DNA amount, but not beyond a limiting DNA concentration. Besides, using radiolabeled DNA, they estimated that approximately 7% of the total exogenous DNA could not be removed from intact transfected cells by a combined EDTA-nuclease treatment, and this figure should, therefore, represent the amount of DNA that was actually taken up by the cells (and not simply adsorbed on their surface).

Unfortunately, no comparable systematic work has been carried on *Drosophila* cells, but there are a few data available.

Burke et al.,[4] in the paper already quoted, compared the amounts of plasmid DNA still present in different transfected lines of *Drosophila* cells 60 h after transfection so that, strictly speaking, their figures depend not only on DNA uptake, but also on its intracellular persistence: their Southern analysis (see Figure 6) showed that *D. hydei* DH33 cells — which were too much more efficient expressors — contained three times more plasmid than *D. melanogaster* S2 line and tenfold more than D1 line. The authors concluded that plasmid uptake seems characteristic of the cell type.

In a similar semiquantitative comparison, with Southern blots, and also carried out on DNA extracted 72 h after transfection, Saunders and colleagues[5] observed, on the contrary, no significant differences between the line 1182 — which expressed transfected sequences at exceptionally high levels — and, for instance, line D1 (Figure 7).

Those conflicting results prove, at least, that DNA uptake is not the only parameter controlling productive transfection.

Nevertheless, the simple fact that transfection of different cell types may require different DNA "facilitators" strongly suggests that internalization of sufficient amounts of DNA depends indeed on physiological characteristics of the cell surface; and it is easily conceivable that endocytosis capacity might greatly vary from one cell type to another. The surprising observation of Saunders et al.[5] that 1182 cells could efficiently take up naked

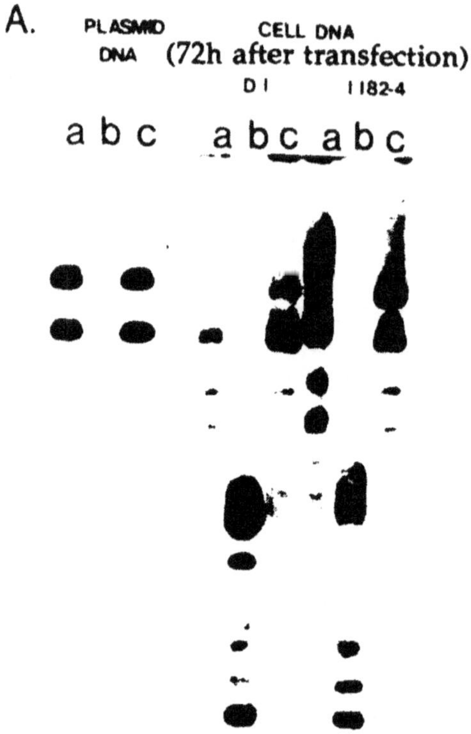

Figure 7. Southern analysis of cells transfected with pHS *cat* plasmid DNA. Genomic DNA was either (a) undigested, (b) digested with *Sau* 3A, or (c) digested with *Mbo* 1. (From Saunders, S. E. et al., *NAR*, 17, 6205, 1989. With permission.)

DNA, that is, in the absence of any facilitating or protecting agent, appears quite significant in this respect.

B. VARIATIONS IN INTRACELLULAR TRANSPORT AND DEGRADATION OF INPUT DNA

The bulk of cell-associated exogenous DNA is rapidly destroyed. Data from mammalian cells indicate that a great majority of foreign DNA remains adsorbed outside of the cell membrane[14] and will be degraded within 22 h,[15] especially in serum-complemented medium.

In addition, only a low percentage of internalized DNA will finally reach the nucleus within the following hours. Contrary to currently accepted views, quite recent results[16] suggest that DNA macromolecules do not seem to pass through the cytosol, but may be transferred directly by intermediary vesicles from the endocytosis-lysosomal compartment to the nucleus. Furthermore, they propose that transfer techniques, such as electroporation, that deliver DNA directly into the cytoplasm might thus depend on a quite different cellular pathway, and this would explain their ability to transfect cell types recalcitrant to the calcium-precipitation method.

In any case, most of DNA degradation should occur in the lysosomes, organelles rich in nuclease enzymes. Therefore, some lysomotropic compounds, such as chloroquine, might inhibit the degradation of incoming DNA and thus enhance transformation efficiency.[17] Incidentally, according to our own experience (unpublished data), the beneficial effect of chloroquine on transfected *Drosophila* cells is doubtful, and it was also found to be rather toxic by Sinclair and Bryant.[12]

To return to possible differences between various cell types, let me cite two conflicting observations that both concern mammalian cells:

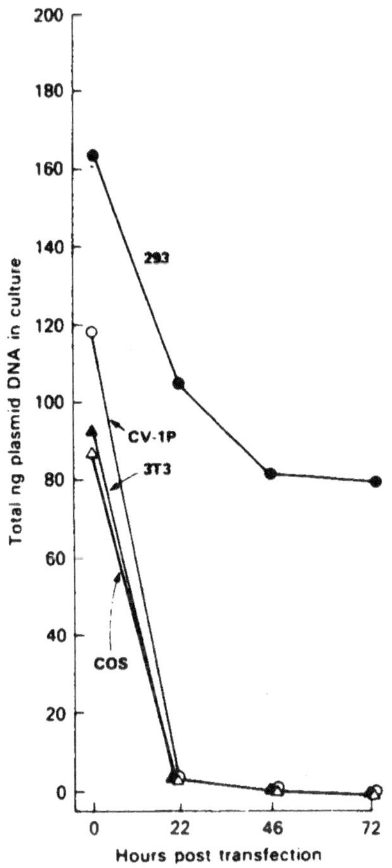

Figure 8. Stability of transfected plasmid DNA in CV-1P, COS, 293, and NIH-3T3 cells. (From Alwine, J. C., *Mol. Cell. Biol.*, 5, 1034, 1985. With permission.)

Alwine[15] reported that the particularly high enzymatic activity generated by CAT plasmids in the human line 293 seems to be due to a slightly higher initial uptake, but mainly to an enhanced stability of transfected plasmids in these cells, as compared with many other cell lines. For instance, within the first 22 h after transfection, more than 98% of the incoming plasmid DNA was lost by COS or NIH-3T3 cells, whereas, in the comparable time frame, 293 cells had lost only 30 to 40% of the transfected DNA (Figure 8).

On the other hand, Skala et al.[18] compared the stability of various plasmids in monkey kidney Vero cells — which are good transient expressors although poorly susceptible to stable transformation — and in mouse L cells which, on the contrary, are easily transformable but whose early expression is very low. Yet, in both cell types, plasmid penetration and persistence were quite comparable: about 500 plasmid molecules were found per cell, one week after transfection; and the amount of plasmids at 3 weeks was even higher in L cells than in Vero cells.

Such contrasting results suggest that differences in the stabilization of input DNA may, in a few cases, account for differences in expression among various cell lines, but that, once more, this characteristic is only one of the many variables of the whole transfection process.

As for *Drosophila* cells, there is only one report about the rapid destruction of most transfected DNA: in their pioneer work on *Drosophila* cell transfection, Morganelli and Berger[19] showed that, in Schneider's S3 cells transfected with *hsp22* gene, by the polyornithine procedure, recombinant plasmid could be detected, on Southern blots, 2 hours after

Figure 9. Southern blot analysis (21) showing the presence of intact recombinant DNA in transfected *Drosophila* S3 cells. (From Morganelli, C. M. and Berger, E. M., *Science*, 224, 1004, 1984. With permission.)

transfection, but that its amount was considerably reduced by 20 hours. Unfortunately, no special care was taken for distinguishing plasmid DNA adsorbed on the cell surface from plasmids actually internalized, so that most of the observed decrease, between 2 and 20 hours, should correspond to destruction of outside DNA (Figure 9).

C. VARIATIONS IN THE PENETRATION OF EXOGENOUS DNA INTO HOST CELL NUCLEUS

This is obviously a crucial step, since to have any chance of being transcribed and later on, possibly integrated into the host cell genome, foreign DNA sequences have to enter the nuclear compartment.

Loyter et al.,[14] who stained the calcium salt of the DNA-CaP_i complex with fluorescent chlorotetracycline, could observe that, while spots are visible in the cytoplasm of most mouse L cells at the end of the exposure time, complexes were seen in the nuclei of only a small portion (no more than 1 to 5%) of the cells.

Several authors (see, for instance, Sene and Nicolau,[20]) assumed that the dislocation of the nuclear membrane at mitosis may facilitate the entry of DNA. We thus meet again with the question of some possible preferential cell cycle stage for gene transfer. In this connection, Saunders et al.[5] supposed that the exceptionally high expression efficiency of *Drosophila* 1182 line might be correlated with some peculiarity of the division cycle in this cell line, because it was shown by Debec et al.[21] to be devoid of centriole.

It is not yet clear whether the majority of cells that succeed in transferring DNA into their nucleus are thereby able to transiently express the exogenous gene, although this was strongly suggested by the results of direct DNA injection, as performed on mammalian cells by Capecchi[22] and some other investigators: some expression of the donor DNA was observed in virtually every injected cell, and up to 20% of these latter could grow out to transformant clones. The small size of *Drosophila* cells and their loose attachment to the culture flask did not enable Echalier[30] to make similar verifications.

D. INTEGRATION INTO HOST CELL GENOME AND VARIATIONS BETWEEN CELL LINES

It seems established that, in order to be maintained beyond the first few days in the cell nucleus of a small proportion of transfected cells, the transferred DNA has to be organized into larger molecules ("transgenomes" of Ruddle): many transformant cells were shown to carry concatenated multiple copies of plasmid sequences, which could replicate and are thus reminiscent of bacterial episomes. However, because there is no mechanism of equal distribution over progeny cells, such genetic units might be progressively lost during continued culture. The still rarer switch to "stable" transformation requires some physical association with a host chromosome. Mechanistic details of the integration process remain to be elucidated, but it is obvious that the enzymatic activities for DNA metabolism of the host cells are deeply involved in the processing of foreign DNA and its association to high-molecular-weight structures. It is well known that the capacities of different cell types for DNA replication, repair, or reorganization vary considerably.*

It must be pointed out that the high frequencies of cotransformation obtained even with physically unlinked DNA sequences are significant proof that intragenomic integration of exogenous genes depends principally on intrinsic properties of the recipient cell. It is well known that virtually every cell that integrates a foreign gene will integrate also any other cotransferred gene. If one of these two markers is a selection vector, the method permits, by selecting the double transformants, successful introduction into a cell genome of any studied DNA sequence.

In *Drosophila* cells, if such simultaneous integration of two (or several) genes is well established and widely used, there is only one direct demonstration of coexpression in individual cells from cotransfected cultures: Winslow et al.,[23] during experiments involving two different expression plasmids for *Antennapedia* and *Ultra-bithorax* proteins, could detect both gene products in the same cells, by the use of fluorescein- and rhodamine-conjugated antibodies, respectively. Such a control was indeed essential for the new methodology they were initiating, namely, the analysis of intracellular interactions between two or several developmental genes, one regulator, and its possible target genes.

It is obvious that efficiency of cotransformation and, more generally speaking, the number of gene copies that can be cointroduced and maintained in a cell genome, should greatly vary with the size of DNA sequences that can be accepted by the various cell types. At least one comparative work, carried out on mammalian cells, established clearly that the length of the DNA strands that can be integrated does differ significantly as a function of the cell line: by the use of several linked markers inserted within different recombinant plasmids, Colbère-Garapin et al.[24] could show that four primate cell lines (Vero, Hela cells...) integrate, on the average, 5 kb of the plasmid, whereas mouse L cells could integrate as high as 188 kb. This observation correlates with the high cotransformation efficiency of this latter line, by contrast with poor capacities of most lines from primate origin.

E. VARIATIONS IN EARLY OR LATE EXPRESSION OF TRANSFECTED GENES

Differential expression of genes by the various tissue cells is one dogma of developmental biology. Therefore, huge differences in the expression of any given gene should naturally be expected after its transfection into different cultured cell lines, even though the tissue origin of most available *Drosophila* cell lines remains unknown.

For lack of significant information, one has to lay aside the peculiarities of the functional organization of extrachromosomal gene concatenates (in transient expression), as well as the possible influence of DNA reorganizations or of integration sites (in stable transformation), that is to say, parameters that might have been deeply influenced by the type of recipient cells.

We must admit that the regulatory sequences of transfected genes will be controlled, positively or negatively, by the specific mixture of transcriptional transactivators present in

* In this connection, it would be quite interesting to investigate the capacities for stable transformation of *Drosophila mei* and *mus* cell lines, because they derive from *Drosophila* mutants deficient in their repair systems of UV-damaged DNA.

Relative Levels of CAT Activity and *cat* Transcripts in LTR-CAT Transfected Cell Lines

(a) CAT activity

	Drosophila S-2	Murine NIH-3T3	Human Y-79
Copia	95	0	0
RSV	20	0.3	15
HIV	0.3	1.0	0.7
HIV/HIV-TAT	1.0	20	7

(b) *cat* RNA

	Drosophila S-2	Murine NIH-3T3	Human Y-79
Copia	1.8	1.6	1.4
RSV	1.6	2.4	1.7
HIV	1.0	1.0	1.0
HIV/HIV-TAT	2.3	10.8	6.3

CAT assays were conducted as described by Gorman et al. (1982). Relative CAT activities are average percent conversion values for each cell line. Percent conversion was determined by cutting out spots containing unreacted 14C-chloramphenicol and acetylated forms, and quantitating the amount of radioactivity in a liquid scintillation counter. Relative levels of *cat* RNA were determined by a slot blot procedure using nick-translated *cat* probe. Autoradiographs were scanned on a Beckman DU-8 scanner and values normalized to the level present in HIV-CAT transfected cells. Values presented are means of at least 4 replicate assays. Percentage error (SE/Mean) for all values <20%.

Figure 10. Table from McDonald, J. F. et al., in *Eukaryotic Transposable Elements,* Cold Spring Harbor Lab, Banbury Rep. 30, 219, 1988. With permission.

every cell type. One example seems quite demonstrative: a "functional dissection" of the LTRs of the retrotransposon *copia* — by the use of transient expression system with the CAT reporter gene —resulted in some discrepancies in the characterization of the most important regulatory sequences, when the analysis was carried out on *Drosophila hydei* or on *Drosophila melanogaster* cell lines.[25,26] This only means that the nature, or relative proportions, of the protein factors susceptible to interacting with the *copia* LTR control regions are significantly different in both cell lines.

Besides, it is very likely that several posttranscriptional steps — such as mRNA processing or degradation, translation, or turn-over rates of gene products — should also play an important part in the observed differences of expression. In connection with this problem, see a recent evaluation, in mammalian cells, of the most commonly used reporter genes.[28]

A quite relevant observation was reported by McDonald et al.:[27] comparing the promoter strengths of the LTRs of various retroviruses or retrotransposons, with the classical use of CAT reporter plasmids, the author established that the *cat* enzymatic activity in *Drosophila* cells transfected with a *copia*-CAT plasmid is fivefold higher than with *RSV*-CAT and 10-fold higher than with *HIV*-CAT, although their respective mRNA levels were found to be almost identical; and the reverse situation was observed in transfected mammalian cells (Figure 10). Such results should imply some differential posttranscriptional treatment of the different messengers by *Drosophila* and mammalian cell types.

IV. CONCLUSION(S)

Transfection efficiency, as estimated from frequencies of transformation and/or expression levels of transferred genes, may vary considerably, in function of the used *Drosophila* cell line, and even sometimes from one subline to another. Transfection is a complex and sequential process, every step of which may be influenced by intrinsic properties of the recipient cells.

The nature or structure of the transferred DNA sequences, and also the transfer procedure, may modify significantly all the parameters of transient expression as well as those of stable transformation, so that it is highly advisable for every experimental system to carefully (and empirically) determine the proper recipient cell line and the more suitable transfer method.

REFERENCES

1. **Wigler, M., Pellicer, A., Silverstein, S., and Axel, R.,** Biochemical transfer of single-copy eucaryotic genes using total-cellular DNA as donor, *Cell,* 14, 725–731, 1978.
2. **Fallon, A. and Willis, J. H.,** Gene transfer in insect cells, *Trends Biotech.,* 8, 217–218, 1985.
3. **Walker, V. K.,** Gene transfer in insects, *Adv. Cell Cult.,* 7, 87–124, 1989.
4. **Burke, J. F., Sinclair, J. H., Sang, J. H., and Ish-Horowicz, D.,** An assay for transient gene expression in transfected *Drosophila* cells, using (3H) guanine incorporation, *EMBO J.,* 3, 2549–2554, 1984.
5. **Saunders, S. E., Rawls, J. M., Wardle, C. J., and Burke, J. F.,** High efficiency expression of transfected genes in *Drosophila melanogaster* haploid (1182) cell line. *NAR,* 17, 6205–6216, 1989.
6. **Debec, A.,** *Drosophila melanogaster, Nature,* 274, 255–256, 1978.
7. **Bourouis, M. and Jarry, B.,** Vectors containing a prokaryotic dihydrofolate reductase gene transform *Drosophila* cells to methotrexate-resistance, *EMBO J.,* 2, 1099–1100, 1983.
8. **Corsaro, C. M. and Pearson, M. L.,** Competence for DNA transfer of ouabain resistance and thymidine kinase: clonal variation in mouse L-cell recipients, *Somatic Cell Genet.,* 7, 617–630, 1981.
9. **Wigler, M., Sweet, R., Sim, G. K., Wold, B., Pellicer, A., Lacy, E., Maniatis, T., Silverstein, S., and Axel, R.,** Transformation of mammalian cells with genes from Procaryotes and Eucaryotes, *Cell,* 16, 777–785, 1979.
10. **Giulotto, E. and Israel, N.,** DNA-mediated gene transfer is more efficient during S phase of the cell cycle, *BBRC,* 118, 310–316, 1984
11. **Nagata, T.,** Cell biological aspects of gene delivery into plant protoplasts by electroporation, *Int. Rev. Cytol.,* 116, 229, 1989.
12. **Sinclair, J. H. and Bryant, L. A.,** 20-hydroxyecdysone increases levels of transient gene expression in transferred *Drosophila* cells, *NAR,* 15, 9255–9261, 1987.
13. **Stevens, B., Alvarez, C., Bohman, R., and O'Connor, J. D.,** An ecdysteroid-induced alteration in the cell cycle of cultured *Drosophila* cells, *Cell,* 22, 675–682, 1980.
14. **Loyter, A., Scangos, G. A., and Ruddle, F.,** Mechanisms of DNA uptake by mammalian cells: fate of exogenously added DNA monitored by the use of fluorescent dyes, *PNAS,* 79, 422–426, 1982.
15. **Alwine, J. C.,** Transient gene expression control: effects of transfected DNA stability and transactivation by viral early proteins, *Mol. Cell. Biol.,* 5, 1034–1042, 1985.
16. **Orrantia, E. and Chang, P. L.,** Intracellular distribution of DNA internalized through calcium phosphate precipitation, *Exp. Cell Res.* 190, 170–174, 1990.
17. **Luthman, H. and Magnuson, G.,** High efficiency polyoma DNA transfection of chloroquine-treated cells. *NAR,* 11, 1295–1308, 1983.
18. **Skala, H., Garapin, A. C., and Colbère-Garapin, F.,** The transient expression of genes following DNA transfection: dependence on the cell line, *Ann. Inst. Pasteur/Virol.,* 137E, 13–26, 1986.
19. **Morganelli, C. M. and Berger, E. M.,** Transient expression of homologous genes in *Drosophila* cells, *Science,* 224, 1004–1006, 1984.
20. **Sene and Nicolau,** in *Liposome Drugs and Immunocompetent Cells,* Academic Press, New York, 1981.
21. **Debec, A., Szollozi, A., and Szollozi, D.,** A *Drosophila melanogaster* cell line lacking centriole, *Biol. Cell,* 44, 133–138, 1982.
22. **Capecchi, M. R.,** High-efficiency transformation by direct microinjection of DNA into cultured mammalian cells, *Cell,* 22, 479–488, 1980.
23. **Winslow, G., Hayashi, S., Krasnow, H., Hogness, D., and Scott, M.,** Transcriptional activation by the *antennapedia* and *fushi tarazu* proteins in cultured *Drosophila* cells, *Cell,* 57, 1017–1030, 1989.

24. **Colbère-Garapin, F., Ryhiner, M. L., Stephany, I., Kourilsky, P., and Garapin, A. C.,** Patterns of integration of exogenous DNA sequences transfected into mammalian cells of primate and rodent origin, *Gene*, 50, 279–288, 1986.
25. **Sinclair, J. H., Burke, J. F., Ish-Horowicz, D., and Sang, J. H.,** Functional analysis of the transcriptional control regions of the copia transposable element, *EMBO J.*, 5, 2349–2354, 1986.
26. **Sneddon, A. and Flavell, A. J.,** The transcriptional control region of the copia retrotransposon, *NAR*, 17, 4025–4035, 1989.
27. **McDonald, J. F., Strand, D. J., Brown, M. R., Paskewitz, S. M., Csink, A. M., and Vos, S. H.,** Evidence of host-mediated regulation of retroviral element expression at the posttranscriptional level, in *Eukaryotic Transposable Elements*, Cold Spring Harbor Lab, *Banbury Rep.*, 30, 219–234, 1988.
28. **Moreira, J. L., Wirth, M., Fitzek, M., and Hauser, H. J.,** Evaluation of reporter genes in mammalian cell lines, *Meth. Mol. Cell Biol.*, 3, 23–29, 1992.
29. **Kim, M. and Coulondre, C.,** unpublished data.
30. **Echalier, G.,** unpublished data.

Chapter 9

INTRINSIC STRESS RESISTANCE OF CULTURED LEPIDOPTERAN CELLS

Thomas M. Koval

TABLE OF CONTENTS

I. Introduction 159

II. X and Gamma Radiation 159
 A. Relative Resistance 159
 B. DNA Repair Processes 160
  1. Unscheduled DNA Synthesis (UDS) 160
  2. DNA Single Strand Break (SSB) Repair 161
  3. DNA Double Strand Break (DSB) Repair 162
 C. Cellular Recovery Processes 165
  1. Multiphasic Survival Curve 165
  2. Split-Dose and Liquid-Holding Experiments 166
  3. Abrogation of Enhanced Survival by RNA and Protein Synthesis Inhibition 167
  4. Radiation-Inducible Proteins 168

III. 254 nm Ultraviolet Radiation 169
 A. Relative Resistance 169
 B. DNA Repair Processes 170
  1. Photoreactivation 170
  2. Unscheduled DNA Synthesis 171
 C. Cellular Recovery Processes 171

IV. Chemical Agents 173
 A. Relative Resistance 173
 B. DNA Repair Processes 173
 C. Cellular Recovery Processes 173

V. Heat 174
 A. Relative Resistance 174
 B. Cellular Recovery Processes 176
 C. Molecular Recovery Processes 176

VI. Stress-Sensitive Mutants for Identifying Mechanisms 177
 A. Isolation 177
 B. Characterization 178

VII. Summary and Implications 179
 A. The Enigma of DNA Repair Involvement in Stress Resistance 179
 B. Inducible Mechanisms of Stress Resistance 180

0-8493-4597-9/94/$0.00+$.50
© 1994 by CRC Press Inc.

C. Prospects and Implications 180

Acknowledgments 181

References 181

Thomas M. Koval is Senior Staff Scientist at the National Council on Radiation Protection and Measurements, Bethesda, MD. He received his B.S. in Biology in 1972 at The Pennsylvania State University and M.S. and Ph.D. in Zoology (Radiation Biology) at The Ohio State University in 1974 and 1976, respectively. After serving as an NIH post-doctoral fellow in Physiology-Biophysics at the University of Illinois in 1976–1977 and a Research Associate at Allegheny General Hospital during 1977–1979, he assumed an Assistant Professorship in Radiation Therapy and Nuclear Medicine at Hahnemann Medical College in 1979 and was promoted to Associate Professor in 1982. He became Associate Research Professor of Radiology at George Washington University and Associate Staff Scientist at the National Council on Radiation Protection and Measurements in 1982. He was promoted to Staff Scientist in 1987. From 1988 to 1993 he was Professor of Oncology, Mayo Clinic and Foundation, Rochester, MN. He accepted his current position in 1993. He is a member of Sigma Xi, Radiation Research Society, Tissue Culture Association, American Society for Photobiology, American Association for Cancer Research, Environmental Mutagen Society, Health Physics Society, New York Academy of Sciences, Entomological Society of America, and American Association for the Advancement of Science. Research interests include molecular mechanisms of intrinsic cellular resistance to stress, particularly in insect cells, stress-inducible genes and proteins, and DNA repair processes.

I. INTRODUCTION

All living things must respond to a variety of stressors. The types of stress range from the background radiation levels to which all things on earth are inevitably exposed — to specific agents or conditions that affect only certain organisms or cell types. Even normal physiological processes in cells result in a large number of damaged DNA nucleotides. However, eukaryotic as well as prokaryotic cells appear to have evolved a number of different methods of reacting to the variety of stressors present in the environment. The sequence of events that follows stress exposure has been most explored in prokaryotes, especially *E. coli*.[1] The process is also moderately well understood in the lower eukaryote, yeast, and a good deal is known about certain types of stress responses, e.g., heat, even in higher eukaryotes such as insects and mammals. Across this evolutionary spectrum of cell types, the response to stress appears to involve a precisely controlled and regulated set of diverse physiological pathways. The direction and choreography of the cellular response is dependent on the nature of the stress imposed.

DNA is the ultimate source of genetic information within cells. Its fidelity is essential to maintain cellular integrity and thereby insure proper cell functioning. Unrepaired or incorrectly repaired damage to cellular DNA might reasonably be expected to influence the ability of a cell to carry out its normal activities. One may therefore deduce that cells that can most efficiently repair DNA damage would have a greater probability of survival than cells with reduced DNA repair capabilities. Most stressors, even those generally not thought of as such, e.g., the heat energy produced through normal chemical reactions involved in DNA metabolism as well as natural background levels of ionizing radiation, produce damage that is capable of being repaired by enzymatic processes.[2,3] Such processes may be constitutive or inducible. Those enzymes that are inducible might logically be expected to be especially important participants in cellular stress responses.

Various genes are activated in response to different stressors. Some are apparently specific for the inducing agent, while others appear to be more general in nature. It has been estimated that approximately 1% of the genes in yeast are DNA-damage inducible.[4] The DNA content of at least some lepidopteran cells, including those discussed extensively in this chapter, is the same as human and many other mammalian cells.[2,3,5] Assuming DNA content is proportional to the number of genes in a cell, the lepidopteran cells should have approximately the same gene number as human cells.[2,3] If, as for yeast, one percent of these genes are DNA-damage inducible, as many as 600 to 1,000 genes could be involved in such a process. Therefore, sorting out the complex regulation of the various genes induced by stress is likely to be complicated and encompass multiple regulatory pathways.

Due to the potentially critical interrelationships between inducible stress responses, the repair of DNA in maintaining genetic fidelity, and cellular survival, these topics will be explored in some detail for a variety of physical and chemical stresses. The rationale in approaching each stress agent will be to first consider the relative resistance of lepidopteran cells compared to cells of mammalian or other origin. Once this is addressed, DNA repair and/or cellular recovery mechanisms will be examined for their potential roles in stress resistance. This brief review is intended to demonstrate the remarkable intrinsic resistance of lepidopteran cells to a variety of widely differing types of stressors and to serve as a starting point in the dissection of mechanisms by which these cells are able to manifest such resistance.

II. X AND GAMMA RADIATION

A. RELATIVE RESISTANCE

A number of studies on a variety of insect species indicate that insects have considerably more inherent resistance to the lethal effects of ionizing radiation than most other organisms.[6]

This apparent resistance was believed to be a consequence of the non-dividing nature of most of the cells composing adult insects.[7] The range of lethal exposures to ionizing radiation is generally between 200 and 900 cGy for most mammals[8] and 1,000 and 100,000 cGy for most adult insects.[6,9] In general, adult insects are considered to be at least 100 times less sensitive to radiation than are vertebrates.[7] This phenomenon had been ascribed to the generalization of Bergonie and Tribondeau, which established the sensitivity of cells to radiation as being directly proportional to their reproductive activity and inversely proportional to their degree of differentiation,[10] and the absence or minute amount of cell division that occurs in the adult state of insects.[7] Supporting this idea was the general observation that dividing cells of insects, those of the gonads, were relatively more sensitive to radiation.[6,7,9] Furthermore, it has been demonstrated that doses as low as 8 cGy were sufficient to delay prophase grasshopper neuroblast cells from proceeding to metaphase.[11] These findings all circumstantially supported the hypothesis that the relative radioresistance of adult insects was due to the limited amount of cell division in these organisms.

In the mid 1970s, however, it was discovered that the TN-368 lepidopteran insect cell line exhibited a pronounced intrinsic resistance to the lethal effects of ionizing radiation.[12] These studies altered the aforementioned hypothesis since the cells were grown in culture and were actively dividing, yet retained their radiation resistance. This radioresistance was later shown to be on the order of 100-fold greater than for cultured mammalian cells[13] (Figure 1). In addition, studies involving cells from several other species of lepidoptera (moths and butterflies) have demonstrated that the marked radiation resistance is characteristic of this insect order, generally being 50 to 100 times more resistant than mammalian cells[13,14] (Figure 1). Also, the doses necessary to completely inhibit recovery of *in vitro* proliferating populations of these cells correspond to some degree with the *in vivo* doses needed to completely sterilize adult lepidoptera.[15] Degree of radiosensitivity appears to be a characteristic that is maintained by other insect orders, as well. A variety of insect cell lines representing several different orders were examined for their radiation sensitivity.[13] Cells from dipteran insects (flies and mosquitos) were much more sensitive to ionizing radiation than the lepidopteran cells, but were still 3 to 9 times more resistant than mammalian cells.[13,16] Orthopteran (roaches and grasshoppers), coleopteran (beetles and weevils), and acarine arachnid (ticks and mites) cells were intermediate in radiosensitivity between dipteran and lepidopteran cells.[13,17] These findings are consistent with those of others who have addressed *in vivo* radiosensitivity of various insect orders.[18] The above studies, therefore, unequivocally demonstrated a pronounced intrinsic radioresistance in insect cells that was not dependent on the reproductive state or potential of the cells.

B. DNA REPAIR PROCESSES

1. UNSCHEDULED DNA SYNTHESIS (UDS)

UDS is the incorporation of radioactive precursors into the DNA of cells at a time when cells are not normally involved in DNA synthesis (non-S-phase cells). This process generally involves the replacement of bases in excised regions of parental DNA strands that have been damaged by any of a number of physical or chemical mutagens. UDS is often measured autoradiographically by the incorporation of tritiated thymidine into the DNA of cells whose normal semiconservative DNA replication has been inhibited by hydroxyurea. In using this technique to determine whether X-ray induced excision repair might explain the radioresistance of the TN-368 cells, it was found that the TN-368 cells displayed approximately 12 times the amount of unscheduled DNA synthesis per nucleus in response to X-irradiation than V-79 Chinese hamster cells (Figure 2).[5] This translates to more than a 20-fold difference on a per DNA basis since V-79 cells have 1.5 to 2 times more DNA per cell.[5] The extent of UDS after irradiation can be correlated with the excision repair of DNA base damage since the repair of single- and double-strand DNA breaks is largely completed in the initial 30 to 60 minutes while UDS continues for several hours. Although this method is a sensitive and quantitative

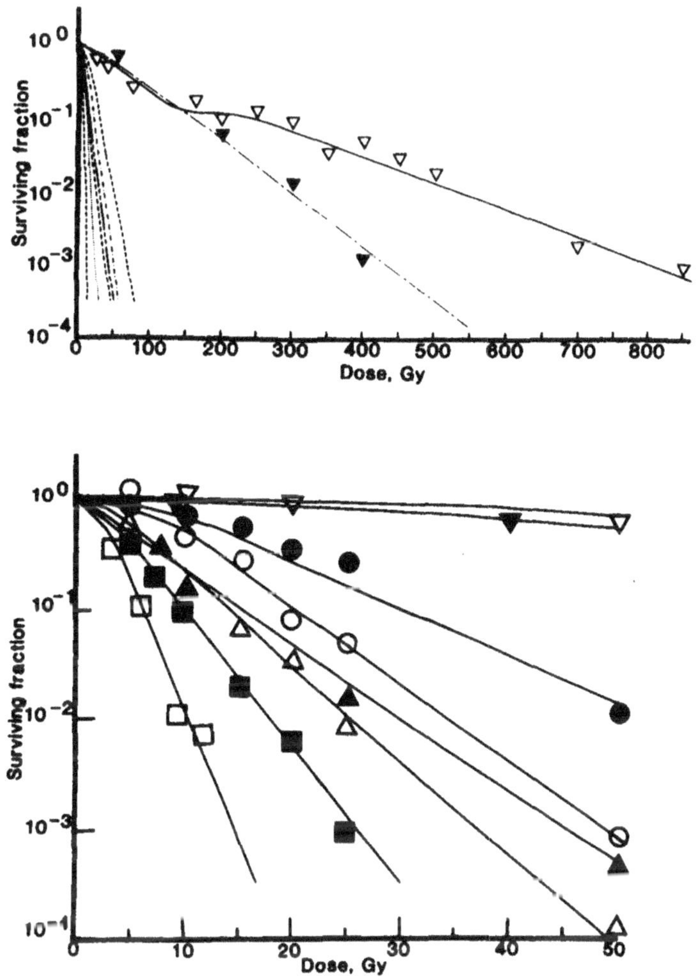

Figure 1. Lepidopteran (∇, TN-368; ▼, IPLB-SF-1254), dipteran (●, RU-ASE-2A; ○, RU-TAE-14; ▲, WR69-DM-2; △, WR69-DM-1; ■, ATC-10), and mammalian (□, V-79) cell survival comparisons. Lower curve is blowup of the low-dose portion of the upper curve. (From Koval, T. M., *Proc. Natl. Acad. Sci. U.S.A.*, 80, 4752, 1983.)

assay for estimating the amount of DNA excision repair synthesis, it is dependent on several cellular factors, such as the amount of DNA damaged, the amount of DNA per cell, the specific activity of cellular thymidine pools, the average size of repaired regions, and the number of repaired sites per unit length of DNA.[19] Both TN-368 and V-79 cells labeled as a result of normal scheduled DNA synthesis have similar grain counts, after accounting for the difference in DNA contents, which suggests that the specific activities of their cellular thymidine pools are similar. While a greater amount of DNA damage in the TN-368 cells could account for the increased UDS, investigations have demonstrated that the amount of damage for a given dose is similar in both cell types.[20] Evidence therefore suggests that the resulting high yield of UDS in TN-368 cells after X-irradiation is due to their having either a larger average size for X-ray repaired regions or a greater number of repaired regions per unit length of DNA. This issue remains unresolved.

2. DNA Single Strand Break (SSB) Repair

DNA single strand breaks are a common type of damage induced by ionizing radiation. The inability of cells to repair such damage has been associated with numerous biological

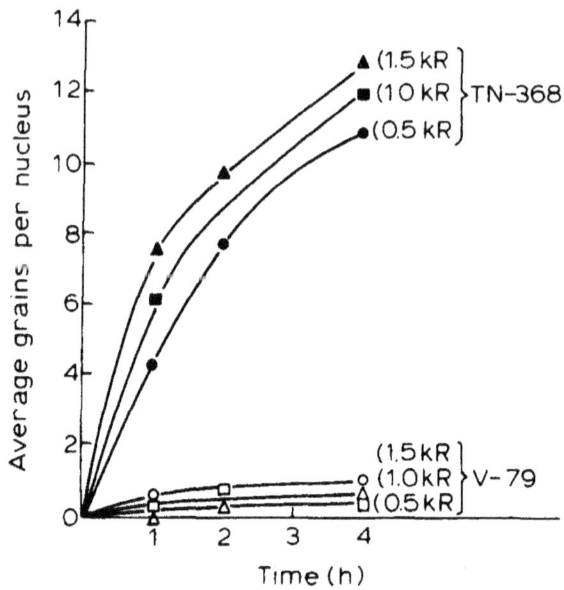

Figure 2. X-ray-induced unscheduled DNA synthesis in TN-368 and V-79 cells following doses of 5-15 Gy (1 Gy = 0.1 kR). (From Koval, T. M. et al., *Mut. Res.*, 49, 431, 1978. With permission of Elsevier Science Publishers B.V.)

dysfunctions including cell death, aging, and chromosomal aberrations. The rejoining of DNA single-strand breaks (SSB) has been demonstrated in all systems studied including bacteria,[21] cultured mammalian cells,[22] and isolated plant protoplasts.[23] The rate of DNA SSB rejoining after a 100 Gy X-irradiation, as measured by alkaline sucrose gradient sedimentation, was somewhat higher in TN-368 cells than in V-79 cells, with most breaks in the TN-368 cells being repaired within 10 min and in V-79 cells within 20 min[24] (Figure 3). However, there were slightly fewer breaks initially in the TN-368 cells than in the V-79 cells. This may have been due to the relatively large free amino acid composition of the insect cell culture medium, which could have been responsible for scavenging primary radiation-induced free radicals and also altering the character of the radiation damage to DNA. Therefore, even though repair of DNA SSB was more rapid in the insect cells, the V-79 cells also rejoined the breaks in a relatively short period of time, and considering confounding factors such as the culture medium, it is difficult to attribute the radioresistance of TN-368 cells to DNA SSB rejoining capabilities.

3. DNA Double Strand Break (DSB) Repair

As DNA double strand breaks are believed by many to be the critical molecular lesion leading to the death of cells exposed to ionizing radiation, it may also be reasonable to expect that the rejoining of such lesions would be more proficient in the lepidopteran cells. The induction and rejoining of DNA DSB were therefore compared in TN-368 and V-79 cells using the technique of non-denaturing filter elution.[20,25] Elutions were quite similar in both cell lines at doses up to approximately 90 Gy. Above this dose, DNA from TN-368 cells appeared to elute to a slightly greater extent than DNA from V-79 cells. This suggests that the induction of DSB by gamma rays was slightly greater in TN-368 than V-79 cells. Other research groups have also reported dose response differences between various mammalian cell lines.[26,27] The initial rates of repair of double strand breaks induced by 90.2 Gy were nearly identical for both cell types, with 50 percent of the breaks rejoined in approximately 12 min (Figure 4). The TN-368 cells reached a final level of about 83% of the DSB rejoined by 1 h postirradiation. The V-79 cells rejoined about 87% of the breaks by 1 h and continued rejoining to a final level of

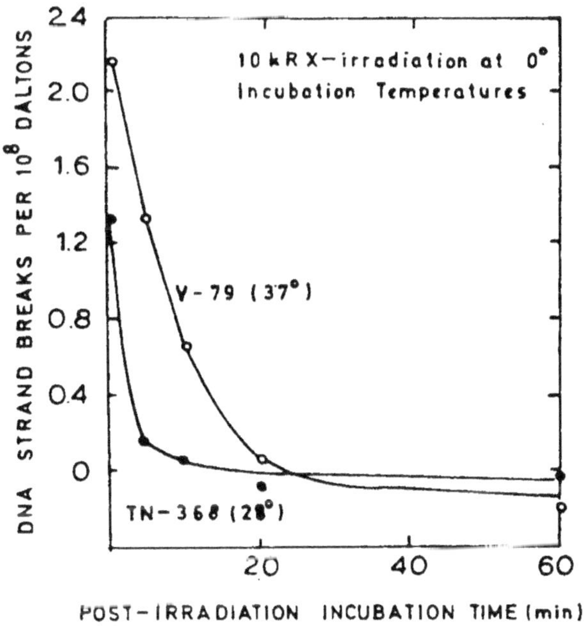

Figure 3. DNA single strand break rejoining in TN-368 and V-79 cells (1 Gy = 0.1 kR). (From Koval, T. M. et al., *Int. J. Radiat. Biol.*, 35, 183, 1979. With permission of Taylor and Francis, Ltd.)

approximately 92% by 2-h postirradiation. No further rejoining above these levels occurred in either cell line through at least 6 hours postirradiation.

Mindful of the accumulated evidence of many investigators indicating that unrepaired or misrepaired DNA DSB are the critical molecular lesions resulting in the death of cells following exposure to ionizing radiation, it was expected that the highly radioresistant TN-368 cells would repair DNA DSB more rapidly and/or to a greater extent than more sensitive cell lines. In fact, after a dose of 90.2 Gy, the TN-368 cells rejoined DSB at approximately the same rate as the much more radiosensitive V-79 cells. Similarly, it was found that DSB are rejoined at about the same rate in both normal human fibroblasts and the radiosensitive fibroblasts of individuals having ataxia telangiectasia, with 50% rejoined within 10 min postirradiation.[26] Both of these studies indicate that the rate of repair of DSB is not necessarily the major factor associated with relative radioresistance. The final extent of DSB rejoining is often thought to be a more critical parameter in determining cell lethality. However, the radioresistant TN-368 cells rejoined DSB to a slightly lesser extent than V-79 cells following a dose of 90.2 Gy. This, again, is not consistent with the hypothesis that unrejoined DNA DSB are responsible for lethality and cannot explain the radioresistance of the lepidopteran cells.

Since maximal rejoining occurred in both cell lines by 3 hours postirradiation, the amount of rejoining at 3 hours was evaluated at various dose levels (Figure 5). For V-79 cells, the percentage of DSB rejoined at 3 hours postirradiation was inversely proportional to dose from 22.6 to 226 Gy. This is consistent with the findings of other groups using V-79 and other mammalian cells.[28,29] The amount of DSB rejoining in TN-368 cells also declined with increasing gamma ray doses; however, the decline was much smaller than for V-79 cells, dropping from 86 to 78% for doses of 22.6 to 226 Gy, respectively. Over the same dose range, the amount of DSB rejoining in V-79 cells decreased from 96 to 67%. Although this indicates a greater capacity of the TN-368 cells to rejoin DSB at very high doses, it essentially eliminates DNA DSB rejoining capacity as a basis for the marked resistance at lower doses.

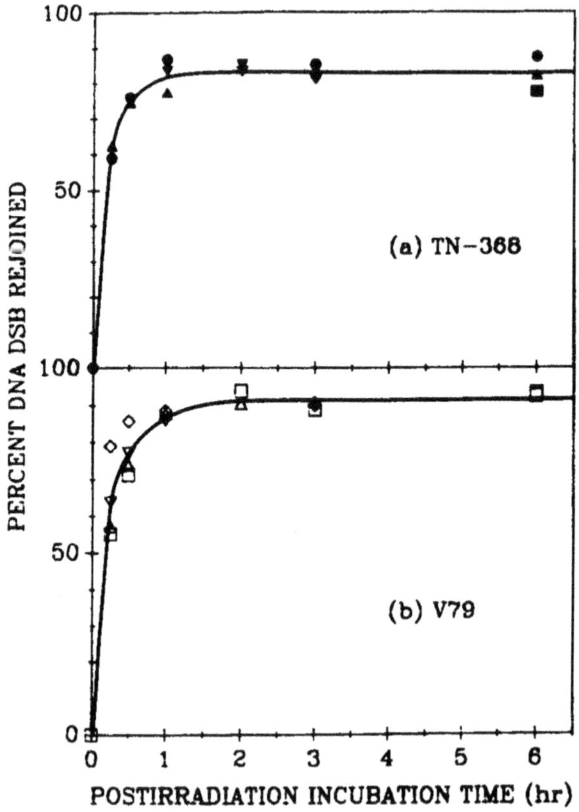

Figure 4. Kinetics of DNA DSB rejoining in TN-368 and V-79 cells following a dose of 90.2 Gy. (From Koval, T. M. and Kazmar, E. R., *Radiat. Res.*, 113, 268, 1988. With permission of Academic Press, Inc.)

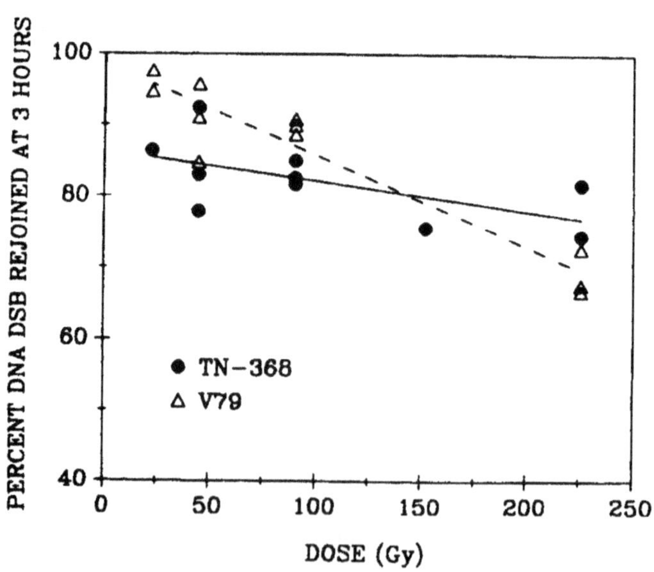

Figure 5. Percentage of DNA DSB rejoining in TN-368 and V-79 cells at 3 hours postirradiation. (From Koval, T. M. and Kazmar, E. R., *Radiat. Res.*, 113, 268, 1988. With permission of Academic Press, Inc.)

The DNA from unirradiated TN-368 cells was found to elute to a slight but consistently greater amount than the DNA from unirradiated V-79 cells. This may be due to higher steady-state (background) levels of DSB in the TN-368 cells or to the much smaller TN-368 chromosomes. In addition, if the values for the amount of DSB rejoining at 3 h are extrapolated back to the ordinate (Figure 5), approximately 1 and 14% of the DSB for the V-79 and TN-368 cells, respectively, remain at zero dose. This again may indicate that the TN-368 cells can tolerate a higher background level of DSB and function in a relatively normal manner, even with 14% residual DNA DSB. This would be consistent with the pronounced radioresistance of the TN-368 cells. While there does not appear to be a correlation between chromosome number and radiosensitivity in insects,[15] the small size and holokinetic nature of lepidopteran chromosomes, which have been discussed elsewhere,[13,14] may play a role in contributing to radioresistance. These factors would also be compatible with the speculation that the TN-368 cells are able to remain viable in the presence of a relatively large number of DSB.

C. CELLULAR RECOVERY PROCESSES

In the face of chronic environmental alterations such as elevated temperature or constant exposures to toxic chemical agents or ionizing radiation, cells have evolved multiple biochemical mechanisms to preserve their existence. Stress responses have been studied for decades in prokaryotic cells, in which at least four inducible systems have been described: the S.O.S. response, the adaptive response, the oxidative stress response, and the heat stress response.[1] Except for yeast, stress-inducible responses are much less well characterized in eukaryotic cells. However, the heat-stress response has been examined in detail in *Drosophila*[30,31] as well as other insects, mammals, and other eukaryotic organisms.[32] In addition, more than 15 years ago, Ducoff discussed the evidence for inducible DNA repair processes in insects.[33] Recently, various investigators have begun to pursue the isolation and identification of genes and proteins involved in the inducible responses of mammalian cells.[34-38] While several groups have now reported on inducible processes in eukaryotic cells, none have demonstrated a more pronounced impact on cell survival in response to ionizing radiation than that reported for TN-368 cells.[38-44] The following sections provide a number of lines of evidence for a radiation-inducible process(es) in the TN-368 cell line.

1. Multiphasic Survival Curve

Typical radiation cell survival curves (a plot representing the fractional or percent survival of cells versus the amount of stress or treatment delivered)[45] are most commonly composed of a shoulder followed by an exponential straight-line response as dose increases.[45] Although the width of the shoulder and the slope of the exponential portion may vary, the basic shape of the curve usually does not. A multiphasic survival response having an inflection point or shoulder on the resistant portion of the curve has been observed for yeast[46,47] and algal[48,49] cells, and only recently, using special techniques, for mammalian cells.[50,51] In the earlier cases the multiphasic nature has been shown to be a consequence of the cellular radioresistance of a small number of contaminating cells representing a more resistant portion of the cell cycle, i.e., dividing cells in yeast and S-phase cells in a primarily G_2-phase algal cell population. In the recent latter case, with mammalian cells, inducible recovery is believed to be responsible for the curve shape. Another response that is somewhat comparable has been observed in *Escherichia coli,*[52] protozoa,[53] and yeast,[54] in which the survival curves were interpreted as resulting from the induction of a repair system not operating in unirradiated or slightly irradiated cells.

The multiphasic nature of the TN-368 cell survival curve was examined in detail approximately a decade ago.[39,55] It was hypothesized that the multicomponent response could probably be accounted for by one of three explanations: (1) the existence of a genetically resistant subpopulation of cells within the TN-368 line; (2) the presence of a radioresistant fraction due to a particular metabolic state within a genetically uniform cell population; or (3) the occurrence of an inducible repair process that enhances cellular survival. Investigations aimed at understanding the responsible underlying factors demonstrated that TN-368 cells exhibited the multiphasic X-ray survival response under both oxic and anoxic conditions and at dose rates between 9.1 and 202 Gy per minute.[39] This indicated that the degree of oxygenation, a potential mediator of radiation effects, did not modify the multiphasic nature of the response and suggested that radiolytic oxygen depletion was not a factor. The D_0 values for each slope (D_0 is the dose required to reduce survival by e^{-1} on the exponential portion of a cell survival curve, i.e., to 37% of its previous value[45]) and the width of the inflection point region all increased proportionately for the nitrogen curve over that of air, the oxygen enhancement ratio being approximately the same for all curve components.

The surviving fraction (the relative proportion of cells surviving a stress or treatment compared to those not treated) of cells irradiated with 606 Gy in air was well into the logarithmic region of the most resistant portion of the survival curve. If the multiphasic nature of the curve were due to the presence of various distinct genetic subpopulations having different radiosensitivities, the progeny of cells recovering from this dose would have been expected to have very few, if any, cells representative of the more sensitive curve components. However, upon examining the survival of the progeny of such previously irradiated cells, the same multiphasic survival response was obtained as for previously unirradiated cells.[39]

In mammalian cell populations having a short G_1 phase, cells in late S are typically the least sensitive to ionizing radiation, while those in mitosis are the most sensitive.[56] A similar but much more subtle pattern was observed in the TN-368 cells. Cell cycle studies, using partially synchronized TN-368 cells obtained by release from a 15-hour exposure to hydroxyurea, were performed to determine whether the multiphasic survival curve might be a result of cellular variations in radiosensitivity during the cell cycle. Although greatest radioresistance occurred in late S-phase and greatest sensitivity in mitosis, there was only a relatively small amount of variation throughout the cell cycle.[39] Therefore, cell cycle variations in radiosensitivity were also ruled out as a primary contributor to the multiphasic shape of the survival curve. It was postulated at this point that an inducible or activated repair system, which is dormant in unirradiated or low-dose irradiated cells, was the most likely explanation for the TN-368 multiphasic survival response.[39]

2. Split-Dose and Liquid-Holding Experiments

As one means of providing further evidence for an inducible recovery process in the TN-368 cells, experiments involving split doses and liquid holding following irradiation were performed.[41] For liquid-holding experiments, TN-368 cells entering stationary growth phase were irradiated with gamma rays and then incubated for various periods of time before cell dilution and plating for colony formation. As observed for mammalian and other cells, the surviving fraction increased several fold over cells diluted and plated immediately after irradiation. For split-dose experiments, the cells were incubated for various periods of time between two equivalent split doses of radiation and then plated for colony formation immediately following the second dose (Figure 6). Again, in a manner similar to mammalian cells, survival was considerably greater in TN-368 cells treated in this manner than cells plated immediately following a single dose equal to the sum of the split doses. Both liquid-holding and split-dose processes exhibited similar biphasic recovery kinetics and reached maximal levels by 6 h. The survival levels attained by cells that underwent these processes were quite high compared to mammalian cells. The insect cells reached survival values of 0.99 for 61 Gy to 0.82 for 306 Gy when 6 h of liquid holding was permitted to occur, and from 0.97 for 92 Gy to 0.77 for 306 Gy when 6 h separated equivalent split doses. A sharp decline occurred at a dose of 428 Gy, resulting in survival levels of 0.09 and 0.13 after 6 h of liquid holding and 6 h between split doses, respectively. The similar recovery kinetics of both split-dose and

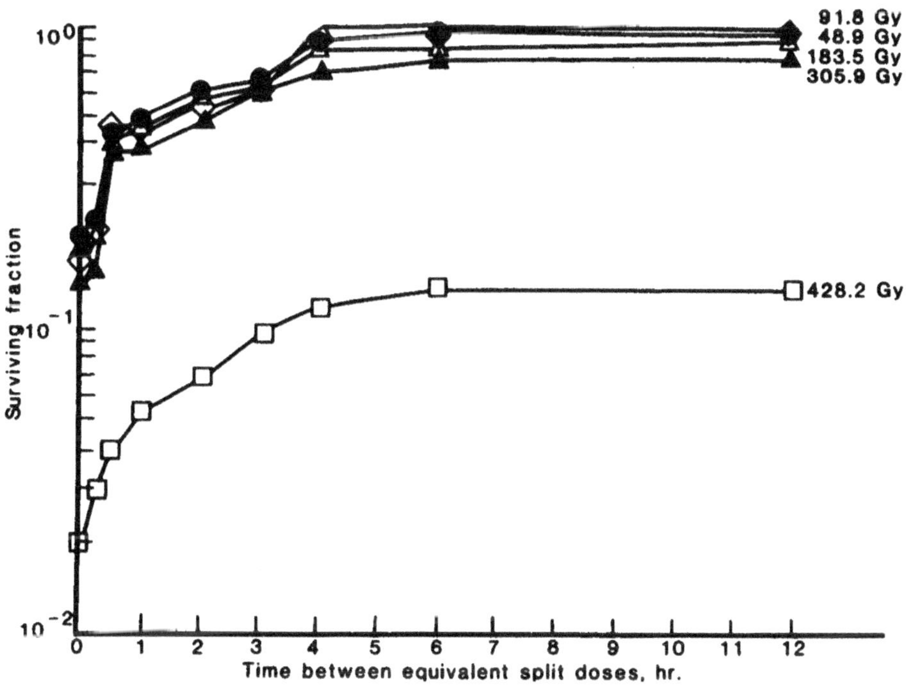

Figure 6. Split-dose recovery kinetics for TN-368 cells. (From Koval, T. M., *Radiat. Res.*, 115, 413, 1988. With permission of Academic Press, Inc.)

liquid-holding experiments suggested that both processes reflected the same underlying mechanism.

Although mammalian cells also undergo the recovery processes described above, these studies demonstrated that TN-368 cells exhibited a type of recovery that led to unexpectedly high levels of survival beyond that accounted for by conventional cellular recovery processes. The split-dose experiments especially presented further evidence of the existence of an inducible recovery process. For example, survival of cells irradiated with 150 Gy was near 0.15. Compare this with the survival of cells receiving 306 Gy in 2 splits of 153 Gy, which was approximately 0.77. If the damage produced by the first of the 153 Gy splits was completely repaired, one might optimistically expect the survival to be dependent on the second dose alone, or near 0.15. Instead, the survival was approximately 5 times greater, indicating not only that the damage induced by the first dose had been repaired to a large extent, but also that the damage induced by the second dose was repaired to a greater extent than if the cells had received only one of the split doses. These studies therefore further support the existence of an inducible process in the TN-368 cells that is responsible for the increased cellular survival in the presence of otherwise lethal ionizing radiation damage.

Also worth emphasizing is that the 6-h liquid-holding and split-dose recovery points decreased very slightly from 61 to 306 Gy, but then dropped sharply at 428 Gy. This steep drop in survival paralleled a similar response in the same dose range of the multiphasic curve obtained by irradiating and immediately plating TN-368 cells. This further suggests that the broad inflection point on the cell survival curve is due to an inducible process and that the drop in survival between 306 and 428 Gy reflects the saturation of this process (Figure 7).

3. Abrogation of Enhanced Survival by RNA and Protein Synthesis Inhibition

Convincing evidence for an inducible recovery process in the TN-368 cells emanated from experiments in which protein and RNA synthesis were inhibited during split-dose and liquid-holding recovery conditions[40] (Figure 8). When the cells were incubated in cycloheximide or

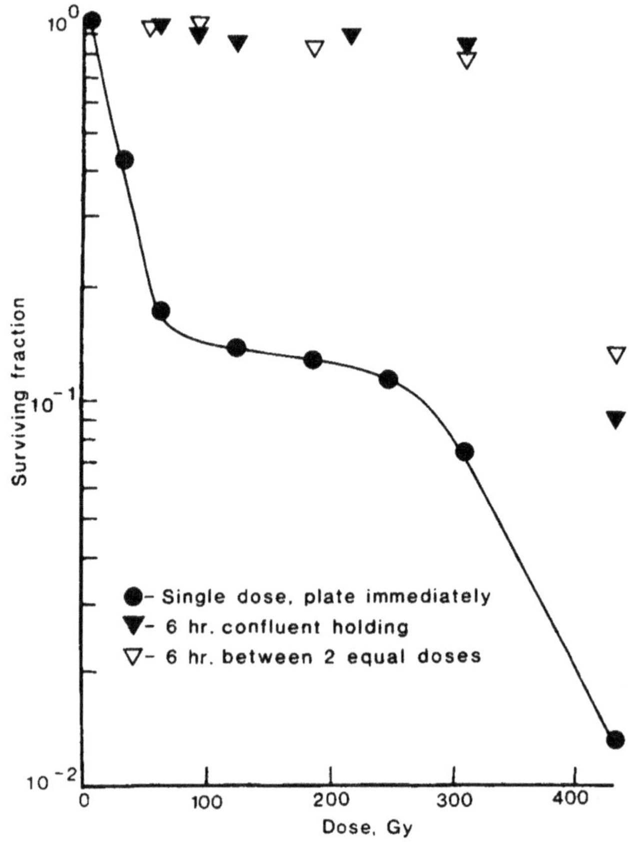

Figure 7. Survival differences between irradiating TN-368 cells with a single dose and plating immediately versus allowing 6 h of liquid (confluent) holding before plating or 6 h between equivalent split doses. (From Koval, T. M., *Radiat. Res.*, 115, 413, 1988. With permission of Academic Press, Inc.)

actinomycin-D following irradiation, but prior to cell dilution and plating (liquid-holding experiments), no increase in survival was observed. Correspondingly, when the cells were incubated in cycloheximide during the interval between split radiation doses, no increase in survival occurred. The concentrations of cycloheximide and actinomycin-D that were used in these studies inhibited protein and RNA synthesis, respectively, and caused little or no toxicity to control cells, as determined by lack of effect on colony-forming ability and absence of microscopically determined cytopathic effects (morphological alterations). In addition, when cycloheximide was added prior to the first irradiation, but removed during the interval between split doses, survival increased as for cells treated with radiation only, indicating that the lack of survival increase for the reciprocal treatment was not due to synergistic effects between the radiation and cycloheximide. The absence of an increase in survival in the presence of either cycloheximide or actinomycin-D therefore serves as evidence that protein and RNA synthesis are essential for the development of the enhanced survival observed in the liquid-holding and split-dose recovery studies.

4. Radiation-Inducible Proteins

The above studies provided compelling evidence for the inducibility of a process resulting in enhanced survival in irradiated TN-368 cells. Therefore, two-dimensional polyacrylamide gel electrophoresis analyses were performed to determine the general pattern of protein

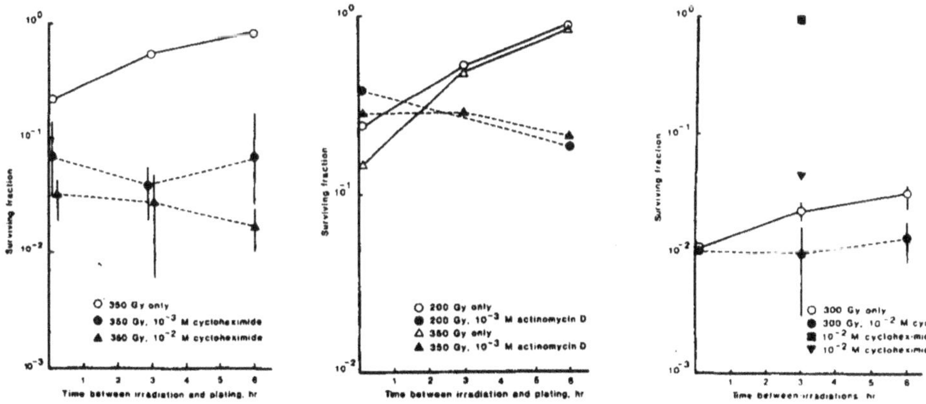

Figure 8. TN-368 cells entering plateau growth were irradiated and incubated in medium with or without (A) cycloheximide or (B) actinomycin-D for various times before plating for colony formation in drug-free medium. In (C), exponentially growing cells were irradiated and plated in medium with or without cycloheximide during a period between split radiation doses and then plated for colony formation. (From Koval, T. M., *Mutat. Res.*, 173, 291, 1986. With permission of Elsevier Science Publishers B.V.)

synthesis following radiation. Cells were analyzed by pulse-labeling with ^{35}S-methionine and sampling at 3 hours after receiving 200 Gy. Of 1176 proteins detected, the irradiated cells produced 281 proteins not present in the untreated cells, with 35 of these being of sufficient intensity to place them in the top 50% of the spot population detected on the gels and 4 in the top 10%. The irradiated cells also discontinued production of 236 proteins that were present in the untreated cells, with 43 of these being in the top 50% and 4 in the top 10% of the spot population. In addition, among proteins present in both the untreated and irradiated cells, 1 increased nearly 50-fold, 2 increased 25- to 30-fold, 11 increased 10- to 20-fold, 24 increased 5- to 10-fold, and 153 increased 2- to 5-fold in the irradiated population.[57] Of these, 49 were in the top half and 23 in the top quarter (in intensity) of the spot population. Alternately, 7 decreased 10- to 30-fold, 24 decreased 4- to 10-fold, and 54 decreased 2- to 4-fold in the irradiated cells from levels present in the untreated cells. Thus, irradiated TN-368 cells were characterized by induced proteins that were not synthesized in unirradiated cells, as well as the absence of many normally produced proteins. These alterations appear to be intimately connected to the pronounced resistance to stress displayed by these cells.

III. 254 nm ULTRAVIOLET RADIATION

A. RELATIVE RESISTANCE

TN-368 cells are also quite resistant to 254-nm ultraviolet (UV) radiation. An early report estimated that the TN-368 cells were about 7 times more resistant to the lethal effects of UV than V-79 Chinese hamster cells.[58] However, based upon the F_0 (F_0 is the fluence required to reduce survival by e^{-1} on the exponential portion of a cell survival curve, i.e., to 37% of its previous value) of the TN-368 cell survival curve obtained in more extensive later studies, it appears that the TN-368 cells are at least 12 times more resistant than the V-79 cells.[59] At 77.5 J/m^2, the TN-368 cell F_0 is also approximately 4 times greater than that for *Drosophila* cells[60] and several times higher than that of rat kangaroo cells,[61] chick-embryo cells,[62] fish cells,[63] frog cells,[64,65] and a unicellular cyanobacterium,[66] as well as several other prokaryotes.[67]

Radiation cell survival curves are most commonly composed of a shoulder followed by an exponential straight-line response as dose increases. Just as a multiphasic survival curve has been observed for the TN-368 cells in response to X-rays,[39] a biphasic response was obtained for UV (Figure 9). A detailed discussion concerning the nature of the multiphasic survival

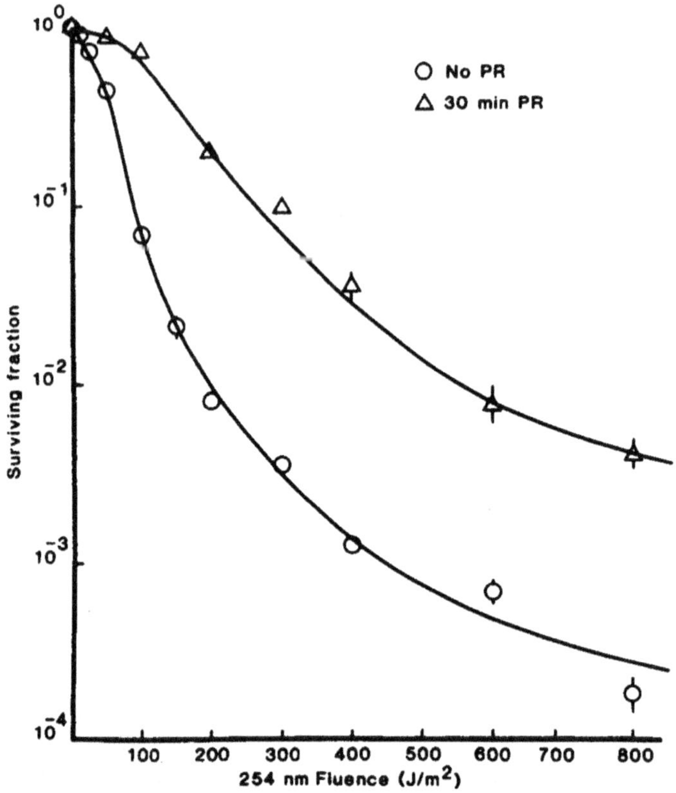

Figure 9. 254 nm UV survival curves for otherwise untreated TN-368 cells and those exposed to photoreactivating light. (From Koval, T. M., *Mutat. Res.*, 166, 149, 1986. With permission of Elsevier Science Publishers B.V.)

response for X-rays was presented in the previous section, and very strong evidence indicates that an inducible repair process is responsible.[40] Because of arguments against cell heterogeneity (discussed in Section II.C.1)[39] and the meticulous care taken to insure homogeneous distribution and dosimetry of the UV light, present speculation is that the TN-368 cells also possess a UV-inducible recovery process that is associated with the biphasic survival curve.

B. DNA REPAIR PROCESSES

1. Photoreactivation

Photoreactivation (PR) is an enzymatic process mediated by exposure to light in the approximate range of 310–480 nm[67] that is specific for the repair of cyclobutane-type pyrimidine dimers in DNA.[67,68] Such dimers represent a major component of the damage induced by 254 nm UV. Photoreactivation in insect cells was recognized over 30 years ago.[69] Other processes have since been studied, and more specifically, the destruction of pyrimidine dimers in insect cell DNA[70] and mRNA[71] by photoreactivating illumination has been reported. DNA photoreactivating enzyme has been isolated from silkworm eggs, another lepidopteran insect.[72] Figure 9 illustrates the results of the first quantitative examination of the photoreactivation of lethality in a lepidopteran cell line. The effect of photoreactivation on the shape of the TN-368 survival curve is to broaden the shoulder and decrease the slope. This is the more typical effect of photoreactivation on survival,[63,65,67] although Griggs and Bender[64] reported an increase in extrapolation number with PR of frog cells, but no change in slope. The extrapolation number of the TN-368 cells is the same in the absence and presence of photoreactivating light. The F_0, which can be used as a measure of UV resistance, is about 3 times greater in

TABLE 1
Survival Enhancement of Photoreactivated TN-368 Cells

	Fluence (J/m^2)						
	50	**100**	**200**	**300**	**400**	**600**	**800**
$SF_{(PR)}$[a] / $SF_{(no\ PR)}$	2.0	10.8	19.6	23.3	23.0	15.4	11.8

[a] Surviving fraction of cells receiving PR treatment divided by surviving fraction of cells receiving no PR following equivalent exposures to 254 nm radiation. These ratios are calculated from the data presented in Figure 9.

From Koval, T. M., *Mutat. Res.*, 166, 149, 1986. With permission.

photoreactivated over nonphotoreactivated TN-368 cells (Figure 9). In addition, depending on the 254 nm UV fluence, survival can be increased as much as 23 times by photoreactivation (Table 1). These results demonstrate that the TN-368 cells are quite proficient at repairing UV damage by PR and that nearly 70% of the damage is photoreactivable. Although *Drosophila* cells were more sensitive to UV than the TN-368 line, they were capable of photoreactivating nearly 80% of the UV damage, perhaps implying a larger role for cyclobutane-type pyrimidine dimers in the lethality of these cells.[60] Photoreactivation has also been found to reverse the majority of UV-induced blockage of DNA fork progression in lepidopteran cells.[73]

2. Unscheduled DNA Synthesis

Given the correlation between the resistance of TN-368 cells to X-rays and the associated large amount of UDS, it was obviously of interest to examine the similar comparison with UV, especially given the already described resistance of the TN-368 cells to UV light. UDS was measured in the TN-368 cells in the same manner as described for X-rays.[58] There was a greater amount of tritiated thymidine incorporation into V-79 cells than TN-368 cells at each dose of UV examined (Figure 10). However, taking into account the greater amount of DNA per V-79 cell, the amount of incorporation was about the same for a given amount of DNA in both cell types. The average number of silver grains per nucleus (due to tritium decay) increased proportionally with the dose of UV and postirradiation time for both cell types. This suggests that the repair process was not saturated in the dose range studied, and that after increasing amounts of damage, there were correspondingly greater amounts of repair. Since the survival of the TN-368 cells was much greater than that of V-79 cells following UV irradiation and the UDS was about the same in both cell types, apparently some factor other than UDS or excision repair must be responsible for the greater survival of TN-368 cells to UV. Although the TN-368 cells are quite adept at performing PR, it should be remembered that the UV resistance demonstrated, even in the absence of light (so that no PR can occur), was at least 12 times greater than the V-79 cells on the basis of F_0. This resistance is further enhanced up to 23-fold when PR is permitted to take place.

C. CELLULAR RECOVERY PROCESSES

Liquid holding is responsible for as much as a twofold increase in TN-368 cell survival.[59] This is a rather small increase when compared with a 23-fold survival enhancement due to photoreactivation. This suggests that the primary lethal lesion induced by UV light is the pyrimidine dimer and that the preferred means of repairing this lesion in the TN-368 cells is by photoreactivation. This is supported by the relatively small amount of UV-induced

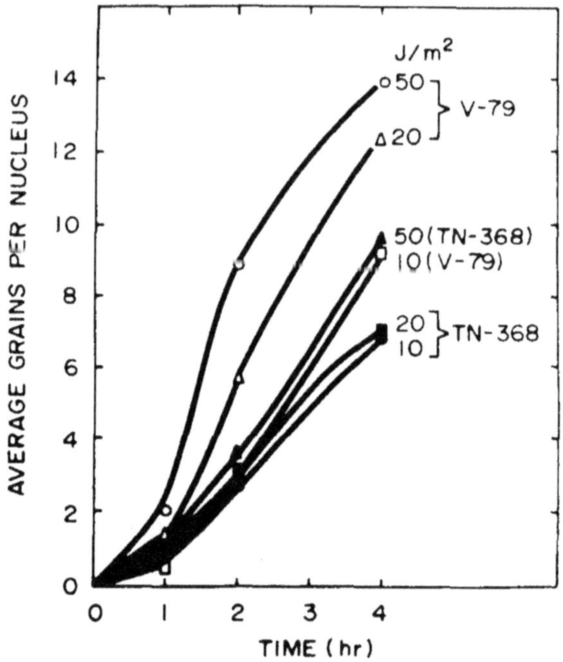

Figure 10. Unscheduled DNA synthesis in TN-368 and V-79 cells following fluences of 10 to 50 J/m². (From Koval, T. M. et al., *Genetics*, 87, 513, 1977. With permission of Genetics Society of America.)

unscheduled DNA synthesis,[58] a process which may be a reflection of the same molecular repair mechanism as liquid-holding recovery, i.e., excision repair.[74,75]

The loss of photoreactivability during liquid holding has been studied by several investigators.[61,62,76,77] The general finding of these studies was that there is a loss of PR ability with extended periods of liquid holding. With the TN-368 cells, an investigation was made not only of PR after liquid holding, but also liquid holding following PR (Table 2). As already discussed, photoreactivation enhances survival approximately 10 times more than liquid-holding treatment. When PR follows a 5-hour liquid-holding period, survival is reduced slightly, but significantly, from that obtained when PR occurs immediately following 254-nm irradiation. This result is compatible with the above-mentioned observations on chick-embryo fibroblasts,[62] rat kangaroo cells,[61] fish cells,[76] and *E. coli.*[77] To explain this finding, it has been suggested that the initial steps of excision repair take place during liquid holding and that they can preclude subsequent photoreactivation.[77] However, others have correlated the loss of photoreactivability with cell growth or cell cycle progression.[62,76]

The TN-368 cells repeatedly demonstrate a further significant enhancement of survival beyond that of immediate photoreactivation alone when liquid holding follows photoreactivation (Table 2). This observation, together with the previous ones, would be consistent with the following scenario.[59] The integrity of some dimers is altered by a dark-repair process during liquid holding. This then precludes the PR of these altered dimers. Therefore, due to the inability of photoreactivating enzyme to use the altered dimers as a substrate, survival for PR alone would be greater than PR following liquid holding. When PR occurs immediately after UV, most dimers are repaired by this method, and the increase in survival during the post-PR liquid-holding period is due either to the repair of nondimer photoproducts or to the dark repair of dimers that were previously inaccessible to the action of photoreactivating enzyme. Thus, survival for liquid holding following PR would surpass PR alone.

TABLE 2
TN-368 Fractional Survival Values after PR, Liquid Holding, and Combined Treatments

Fluence	Treatment[a]				
(J/m²)	UV	PR	LH	PH + LH	LH + PR
100	0.093 ± 0.009	0.668 ± 0.013	0.192 ± 0.016	0.790 ± 0.070	0.585 ± 0.009
200	0.028 ± 0.002	0.296 ± 0.002	0.045 ± 0.002	0.396 ± 0.012	0.285 ± 0.028
400	0.0037 ± 0.0007	0.053 ± 0.011	0.0054 ± 0.00002	0.066 ± 0.005	0.052 ± 0.006

[a] UV, 254-nm UV only and plate immediately for colony formation; PR, 30-min PR after 254-nm UV and then plate; LH, 5-h liquid holding in HBSS in the dark and then plate; PR + LH, 30-min PR followed by 5-h liquid holding and plate LH + PR, 5-h liquid holding followed by 30-min PR and plate.

From Koval, T. M., *Mutat. Res.*, 166, 149, 1986. With permission.

IV. CHEMICAL AGENTS

A. RELATIVE RESISTANCE

Radioresistant TN-368 lepidopteran insect cells were examined with respect to their sensitivity to a variety of chemical agents including methyl methanesulfonate (MMS), *N*-methyl-*N'*-nitro-*N*-nitrosoguanidine (MNNG), propane sultone (PS), mitomycin C (MMC), and 4-nitroquinoline 1-oxide (4NQO).[78] Based on survival ability, the TN-368 cells were at least 10- to 100-fold more resistant to these agents than most mammalian cells, as assessed by several investigators in a number of mammalian cell systems[79-91] (Figure 11). In all cases, the resistance was demonstrated by an extension of the shoulder region and a nearly parallel shifting of the exponential portion of the survival curve to the right. This could be due, in part, to the capability of excluding these chemical agents from the cell, a decreased accessibility of the agents to DNA, an enhanced potential to scavenge reactive chemical species as has already been suggested for these cells,[5] or a greater capacity to repair damage induced by these chemicals than the capacity of mammalian cells. Results with another potent chemical agent, doxorubicin, indicate that both influx of this agent into and efflux from TN-368 cells were similar to V-79 cells and could not explain the TN-368 resistance.[92]

B. DNA REPAIR PROCESSES

As might be expected based on their resistance to such a wide variety of chemicals and their ability to repair the various types of DNA damage discussed for X-rays and UV, the TN-368 cells were also quite capable of rejoining damaged DNA following treatment with doxorubicin and repairing cross-linking damage induced by 8-methoxypsoralen and 365 nm light.[93] However, these types of repair do not appear to be performed more rapidly or to a greater extent than in V-79 cells. It has been reported, though, that mosquito cells are more adept at repairing DNA strand breaks than mammalian cells after treating with bleomycin, an X-ray-mimetic drug.[94] Therefore, as indicated earlier, insect cells may be superior in repairing X-ray-like DNA damage, but about the same as mammalian cells in repairing various other types of DNA damage.

C. CELLULAR RECOVERY PROCESSES

Concentrations of these agents, which reduced survival to about 10%, were used to assess recovery ability assayed by colony formation in liquid-holding and split-dose experiments. Liquid-holding experiments were performed by exposing cells in the plateau phase of growth for 1 h to 8 mM MMS, 50 μM MNNG, 9 mM PS, 100 μM MMC, or 175 μM 4NQO, removing

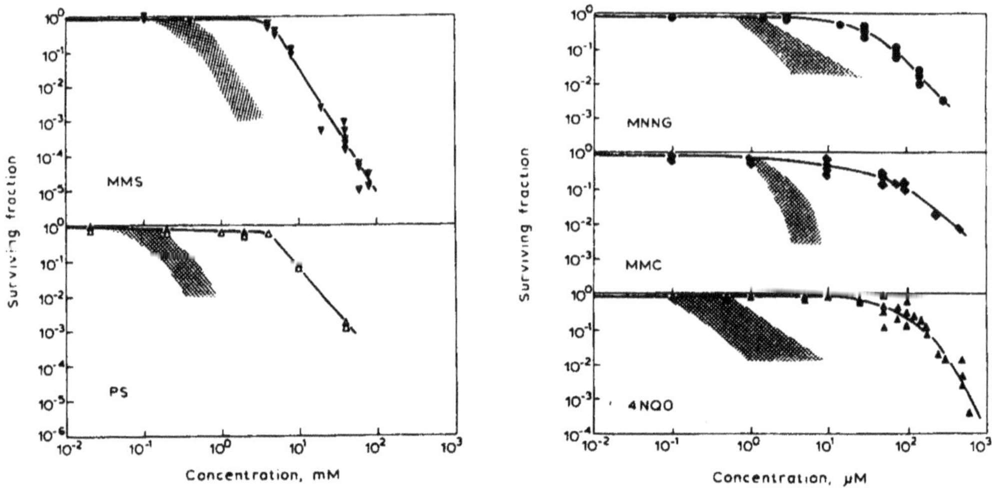

Figure 11. Survival of TN-368 cells treated in exponential growth for 1 h with the indicated chemical agents. Shaded area represents general range of survival for a variety of mammalian cell types. (From Koval, T. M., *Mutat. Res.*, 262, 219, 1991. With permission of Elsevier Science Publishers B.V.)

the drug and incubating cells in spent medium for 6 h, and plating for colony formation. Split-dose experiments were performed by exposing exponentially growing cells to the above drug concentrations for 1 h, incubating in fresh medium for 6 h, exposing the cells to the agent for an additional hour, and plating. The TN-368 cells were able to recover from MMS, MNNG, and PS in both types of experiments. Recovery from 4NQO was observed in liquid-holding experiments and not assessed in split-dose experiments. In all cases where recovery was observed, survival enhancement was approximately twofold (Table 3). Recovery from MMC (a cross-linking agent) exposure was not observed in either type of experiment. In addition, recovery from 8-methoxypsoralen plus UVA light (PUVA), another cross-linking treatment, was not observed. Therefore, even though the cells are quite resistant to killing by these cross-linking agents, they may have difficulty in recovering from such treatments.

V. HEAT

DNA is believed to be the major target for killing by each of the agents discussed thus far. DNA damage is not generally suspected, however, to be the primary mechanism for heat-induced cell killing. Therefore, on this basis, studies with heat have the potential of being quite informative from a mechanistic point of view.

A. RELATIVE RESISTANCE

In order to assess their sensitivity to heat, exponentially growing TN-368 cells having a normal growth temperature of 28°C were heated by immersion of culture flasks in precision-controlled (±0.05°C) water baths at 33, 37, 41.5, and 44°C for various periods of time before assessing survival by plating cells for colony formation[95] (Figure 12). These cells were much more resistant to heat treatment than *Drosophila*, WR69-DM-1, cells, a dipteran line.[31] Both cell lines were routinely cultured at 28°C. However, the heating time at 33°C that resulted in 50% cell survival was approximately 30 min for the *Drosophila* cells but slightly more than 48 h for TN-368 cells, nearly 100 times longer. Similar relationships existed at higher temperatures as well. At 37°C, 50% cell survival was reached after only a slightly longer exposure than at 33°C, still about 1/2 h for the *Drosophila* cells, while it took about 29 h to reach the same survival level in TN-368 cells. At 42°C, *Drosophila* cell survival fell to 50% in about 3 min, whereas at 41.5°C, TN-368 cell survival did not reach 50% until approximately

TABLE 3
Liquid-Holding and Split-Dose Recovery Factors[a] for TN-368 Cells Treated With Chemical Agents

Drug (concentration)	Liquid-holding recovery factor	Split-dose recovery factor
MMS, 8 m*M*	2.3, 2.8, 1.9, 1.9 ($\bar{x}$ = 2.2)	1.7, 3.6 ($\bar{x}$ = 2.6)
PS, 9 m*M*	3.2, 1.9, 1.7, 3.8 ($\bar{x}$ = 2.6)	2.0
MNNG, 50 μ*M*	2.6, 1.9, 1.4 ($\bar{x}$ = 2.0)	2.1
MMC, 110 μ*M*	1.0, 0.9, 0.7, 0.8 ($\bar{x}$ = 0.9)	1.1, 1.3, 0.9, 0.9 ($\bar{x}$ = 1.0)
4NQO, 175 μ*M*	1.5, 2.2 ($\bar{x}$ = 1.9	b

[a] Recovery factors represent the ratio of the surviving fraction resulting from either reincubating cells for 6 h before plating or between split-dose fractionation to the survival from a single dose plated immediately.

[b] Data not obtained.

From Koval, T. M., *Mutat. Res.*, 262, 219, 1991. With permission.

3 h. Even at 44°C, TN-368 cell survival fell to 50% only after about 15 min. The general shape of the heat survival curves also presents quite a contrast between the TN-368 and *Drosophila* cells. The TN-368 cells display a typical survival profile at each temperature studied, with an initial shoulder area followed by exponential killing as heating time increases (Figure 12). *Drosophila* cells, on the other hand, were characterized by a complex heat (33, 37, and 42°C) survival pattern with exponential killing followed by a broad inflection point or plateau in the 10^{-2} to 10^{-3} survival region.[31] At 42°C, this inflection point was followed by further exponential killing and another plateau near the 10^{-5} survival level. Although D_0 values were not calculated for the *Drosophila* cells, due to their complex survival curves, the slope of the exponential portion of survival curves was much steeper for *Drosophila* cells than TN-368 cells at all temperatures.

The TN-368 cells demonstrated sensitivity to heating at 41.5 and 44°C that is somewhat similar to mammalian cell heat sensitivity. For instance, survival was reduced to 50% of control after 3 h of heating at 41.5°C for TN-368 cells, or after about 1 to 4 hours of heating at the same temperature for mammalian cells.[96,97] The D_0 values for the TN-368 cells are also within the range observed for mammalian cells.[97-99] These similarities in thermal sensitivity become remarkable when it is considered that heating to 41.5 and 44°C results in temperature elevations of 13.5 and 16°C for the TN-368 cells (maintained at 28°C) compared to 4.5 and 7°C for mammalian cells (maintained at 37°C). Therefore, on this basis, the TN-368 cells demonstrate a greater relative resistance to damage by heat than mammalian cells. However, depending on the mechanisms involved in thermal damage, it is possible that the absolute temperature of the heat stress could be more important in considering cell survival than the magnitude of the temperature elevation.

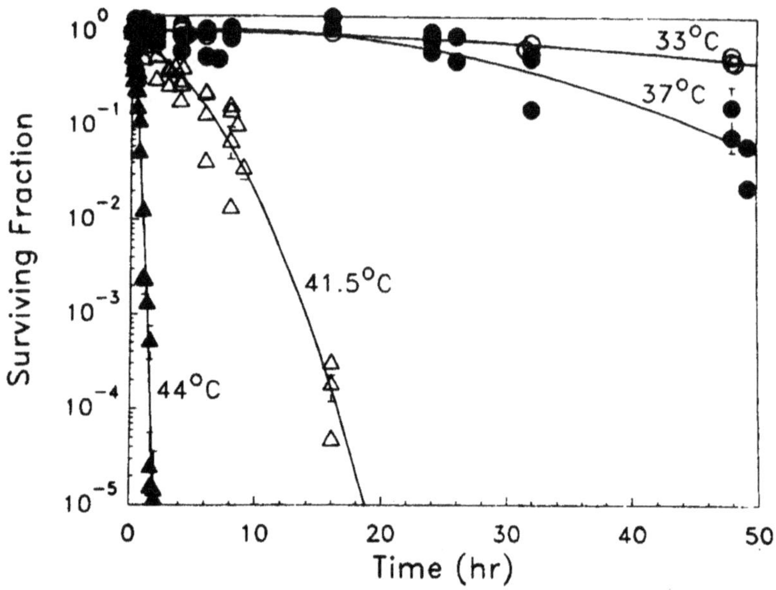

Figure 12. Survival of exponentially growing TN-368 cells heated for various times at 33, 37, 41.5 and 44°C. (From Koval, T. M. and Suppes, D., *Int. J. Radiat. Biol.*, 61, 425, 1992. With permission of Taylor & Francis, Ltd.)

B. CELLULAR RECOVERY PROCESSES

Thermotolerance, a classical induced response, was demonstrated in the TN-368 cells at 33, 37, 41.5, and 44°C. This is in contrast to the slight thermotolerance observed in *Drosophila* cells at 33°C, but not at 37 or 42°C.[31] Thermotolerance was assessed by fractionated heating experiments for both cell types; however, both the initial heating and the later follow-up heating were performed at the same temperature for the *Drosophila* cells, while the initial heating was usually followed by a later "increased-temperature" heating for the TN-368 cells. The increased-temperature heating was necessary in the case of the TN-368 cells because relatively long heating times were required to result in a decrease in survival at 33, 37, and 41.5°C. Thus, thermal resistance was more practically measured by survival at 44°C following the initial heating to induce thermotolerance at 33, 37, 41.5, or 44°C (Figure 13). It would have been otherwise impossible to detect thermotolerance at 33 and 37°C and difficult at 41.5°C. For example, a 30-min pretreatment 2 h before 120 min of heating at 44°C increased survival more than 10,000-fold over cells receiving only 120 min at 44°C. It is worth noting that the thermotolerance induction conditions in the *Drosophila* experiments (30 min at 33, 37, and 42°C followed by 1, 3, and 6 h incubation at 28°C before heating again at 33, 37, or 42°C, respectively) were more damaging to the cells than the induction conditions in the TN-368 studies (30 or 60 min at 33 and 37°C, 30 min at 41.5°C, and 15 or 30 min at 44°C followed by 2 h at 28°C before heating again at 44°C for various times). The corresponding survival levels were about 0.5 to 0.01 for the *Drosophila* cells and 1.0 to 0.2 for the TN-368 cells. Therefore, higher tolerance levels are not correlated to more heat killing by the priming heat dose, as might be expected. While it remains to be seen whether these induction conditions would make a difference in development of thermotolerance, it is clear that several distinct differences exist regarding the survival response to heat in these two cultured insect cell systems. Additionally, the TN-368 cells develop a significant amount of thermotolerance in fractionated heating experiments at 44°C in a manner quite comparable to several types of mammalian cells.[98-101]

C. MOLECULAR RECOVERY PROCESSES

The patterns of heat-induced inhibition of overall DNA and protein synthesis, measured by uptake of ^{3}H-thymidine and ^{3}H-leucine, are depicted in Figure 14 for the TN-368 cells. Results

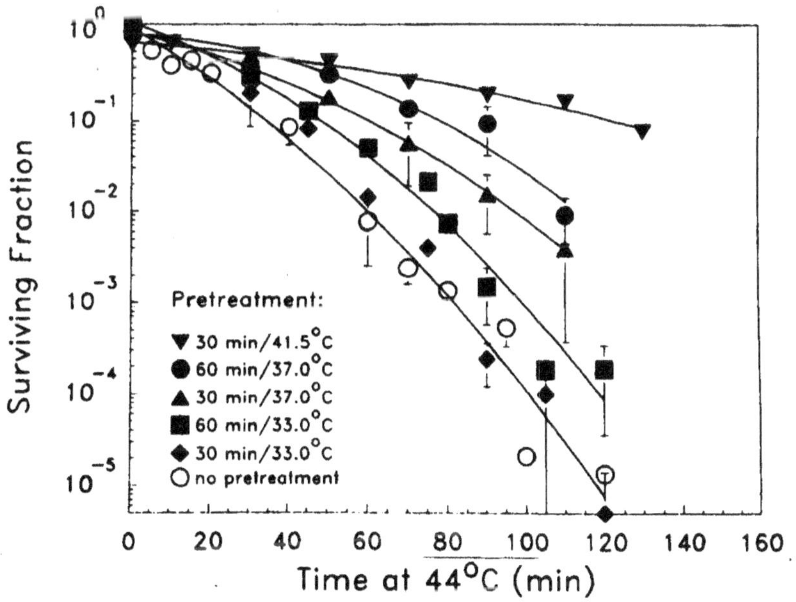

Figure 13. Survival representing the heating of exponentially growing TN-368 cells for either 30 or 60 minutes at 33, 37, or 41.5°C, incubating at 28°C for 2 h, and then heating for various times at 44°C. (From Koval, T. M. and Suppes, D., *Int. J. Radiat. Biol.*, 61, 425, 1992. With permission of Taylor & Francis, Ltd.)

are generally similar to those reported for various mammalian and human cells.[102-105] Inhibition of both types of synthesis increased with time and temperature. Recovery was somewhat faster for protein synthesis than DNA synthesis, again in line with the previous studies.[102,103] When a priming heat exposure was used to induce thermotolerance prior to a second heat treatment, DNA synthesis was reduced to a level as low as or below the level that would have been produced by the second treatment alone (Figure 15). However, after a 41.5°C priming dose, protein synthesis was less inhibited and its recovery was more rapid in tolerant cells than in cells receiving only a single 44°C treatment (Figure 15). For 37 and 44°C priming regimens, the tolerant cells displayed similar protein synthesis kinetics to the nontolerant cells. The degree of inhibition in the tolerant cells was therefore dependent on the time and temperature of the priming heat exposure. This also is relatively consistent with other reports,[105,106] although higher tolerance levels did not correlate with the amount of heat killing by the priming dose. Investigations with other eukaryotic cell systems have also demonstrated that alterations in protein synthesis are not necessary for the expression of thermotolerance.[107,108] Therefore, despite the marked heat resistance observed in the TN-368 cells, fundamental alterations in DNA and protein synthesis appear to be comparable to mammalian and human cells. The teleological significance of a rapid and prolonged depression in DNA synthesis (to allow time for cellular repair and recovery) and the perhaps more intricate balance in protein synthesis inhibition and recovery have been discussed at length elsewhere.[106,109-111]

VI. STRESS-SENSITIVE MUTANTS FOR IDENTIFYING MECHANISMS

A. ISOLATION

In an effort to better approach the underlying basis for resistance, X-ray and UV-sensitive mutants of the TN-368 cell line have been isolated.[112] Mutations were presumably induced in the parent cells by ethylmethane sulfonate (EMS). The radiation-sensitive lines were selected as those cells that survived exposure to a relatively low dose of irradiation followed by conditions intended to kill DNA-repair competent cells. Approximately 8,000 wells of 96-well plates were screened for viable cells following the mutant isolation procedure, and 447 wells

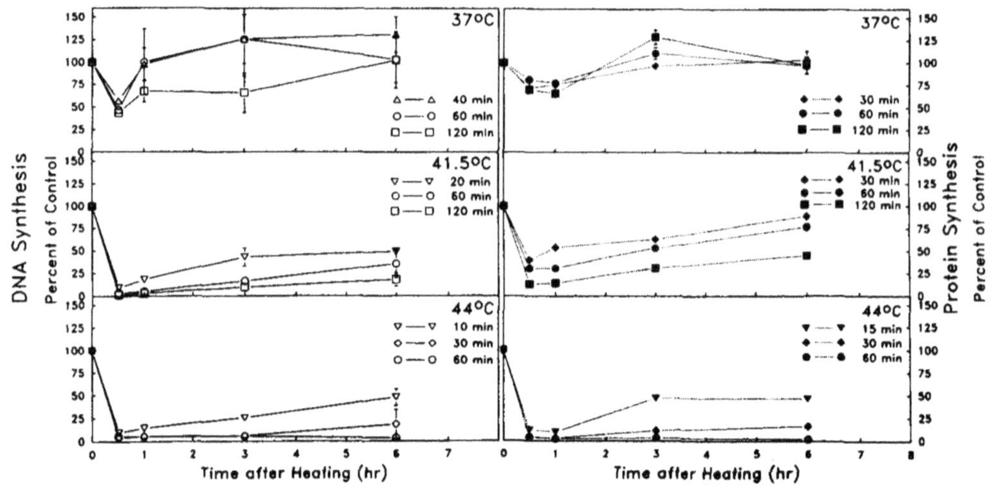

Figure 14. Relative DNA and protein synthesis in TN-368 cells following heat exposures of 37, 41.5, and 44°C for various times. (From Koval, T. M. and Suppes, D., *Int. J. Radiat. Biol.*, 61, 425, 1992. With permission of Taylor & Francis, Ltd.)

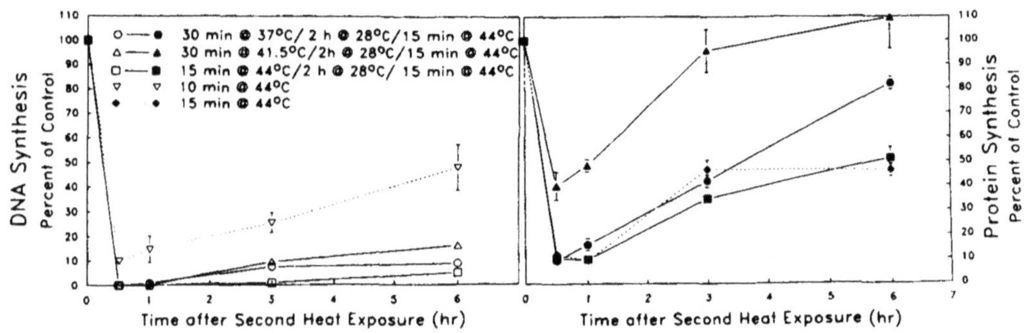

Figure 15. DNA and protein synthesis in TN-368 cells treated with a priming heat dose to induce thermotolerance, returned to incubation at 28°C for 2 h, and then evaluated after 15 minutes at 44°C. (From Koval, T. M. and Suppes, D., *Int. J. Radiat. Biol.*, 61, 425, 1992. With permission of Taylor & Francis, Ltd.)

contained potential mutants. Of these, 31 were sensitive in preliminary screening to either X-rays or UV light, depending on which type of radiation was used in the mutant selection procedure. The 31 presumed mutants were subcultured to higher densities and examined in more detail using a conventional colony formation assay to assess sensitivity, leading to the identification of two X-ray-sensitive and two UV-sensitive lines, which were determined to be stably sensitive. These mutants have been cultured for more than a year and have been subcultured over 140 times. Radiosensitivity has been maintained over this period. In addition, each sensitive line was frozen at an early passage number. The magnitude of the difference in sensitivity between the variants and the parent is not great for either of these physical agents. (The largest difference in survival that was observed occurred at the highest dose or fluence examined and was on the order of tenfold for each variant.) However, the difference has been observed in virtually all repeats of these experiments and is statistically significant.

B. CHARACTERIZATION

In general, the isolated mutant cell lines have growth and culture characteristics similar to the parent TN-368 line.[112] Growth curves are similar for the five cell lines, and population

TABLE 4
Cross-Sensitivity of TN-368 Mutants to Other Agents[a]

	X-rays	UV radiation	Heat	PUVA	MMC
Cs2	X	X		X	N.D.[b]
Cs7	X	[c]	X	X	N.D.[b]
UV10	X	X		X	X
UV19	X	X	X	X	X

[a] X represents sensitivity compared to parent TN-368 cell line.
[b] Not done.
[c] Normal at 100 J/m² but sensitive at 400 J/m².

From Koval, T. M., *Radiat. Res.*, 127, 58, 1991. With permission.

doubling times derived from the growth curves vary from 20.5 h for the parent cell line to a maximum of 26.5 h for the Cs7 variant. The absolute plating efficiency of the mutant lines is also slightly reduced compared to the parent line. Perhaps of greatest significance is that the chromosome number of each of the sensitive lines is reduced, to varying degrees (Cs7 by just over 15% to Cs2 by more than 25%), in comparison to the parent TN-368 line. This was likely enhanced to some extent by the griseofulvin used in the isolation procedure.[113] In addition to having a smaller mean number of chromosomes, each variant except for UV10 also has a lower bottom range than the parent line. A preliminary examination by flow cytometry indicated that the percentage of S-phase cells in an exponentially growing population is similar, about 5 to 8%, in each of the cell lines. However, this analysis is somewhat complicated due to the highly varying ploidy within the cells of each line as well as between the lines.

The cross-sensitivity (compared to the parent line) of the isolated strains to X-rays, 254-nm UV radiation, 44°C heating, PUVA, and MMC was tested. A summary of the data is provided in Table 4. Both X-ray-sensitive cell lines, Cs2 and Cs7, were also sensitive to PUVA. However, while Cs2 was also sensitive to 254 nm UV radiation, it was normal with respect to 44°C heating. The Cs7 cells were to some extent the opposite: sensitive to 44°C heating, but mixed in sensitivity to UV radiation. That is, although the Cs7 cells appeared to be normal in sensitivity to 100 J/m^2 of 254-nm UV radiation, they were sensitive at 400 J/m^2. Both UV-sensitive lines, UV10 and UV19, were also sensitive to X-rays, PUVA, and MMC. The UV10 cells were normal in sensitivity to 44°C heating while UV19 was sensitive. The only mutant that was sensitive to all of the agents tested was UV19. Based on the presumption that DNA damage is not the basis for cell killing by heat as mentioned earlier, it was interesting that one line of each category of mutant cells, Cs7 of the X-ray-sensitive and UV19 of the UV-sensitive, was sensitive to 44°C heating. This is perhaps a strong implication that at least a portion of the deficiency(ies) responsible for cellular sensitivity in the mutants is of a relatively general nature and not specifically a defect in DNA repair. This is consistent with and extends considerably similar conclusions reached using *Drosophila*.[114]

VII. SUMMARY AND IMPLICATIONS

A. THE ENIGMA OF DNA REPAIR INVOLVEMENT IN STRESS RESISTANCE

Our current understanding of the molecular mechanism(s) of lepidopteran cell resistance to the various physical and chemical agents discussed is clouded somewhat by the apparent inconsistencies in the data obtained. On one hand, the lepidopteran cells demonstrate an exceptional capacity to perform X-ray induced excision repair and are somewhat better at rejoining radiation-induced DNA single-strand breaks, which would be expected to greatly enhance the probability of survival. On the other hand, repair of radiation-induced DNA double strand breaks and UV-induced excision repair are not considerably different in magnitude from V-79 or other mammalian cells. While the lepidopteran cells are quite proficient at photoreactivating 254 nm light-induced 5- to 6-pyrimidine dimers, they are outperformed to some degree in this regard by dipteran cells, which are vastly greater in sensitivity to the 254 nm light. PUVA and MMC experiments indicate TN-368 cells do not follow the same posttreatment recovery for these agents as they do for nearly all other agents examined. However, the TN-368 cells are much more resistant to these agents than mammalian cells. In addition, they display about the same amount of DNA-protein cross-linking repair as mammalian cells. Although these findings are not necessarily incompatible, they are difficult to completely reconcile in terms of a direct relationship between DNA-repair capability and relative stress resistance.

B. INDUCIBLE MECHANISMS OF STRESS RESISTANCE

It has been demonstrated that cells cultured from insects are considerably less sensitive to ionizing radiation-induced killing than cultured mammalian cells. Furthermore, the degree of radioresistance corresponds to insect order, with lepidoptera being by far the most resistant. While the role of DNA repair processes remains somewhat a paradox, it appears increasingly certain that an inducible recovery mechanism in lepidopteran cells is at least partially responsible for their radiation resistance. This inducible system is dependent upon transcriptional and translational activities that are not present in unirradiated cells or cells receiving less than some minimal amount of radiation necessary for activation of the process; i.e., the inducible enhancement of survival is abrogated by inhibiting RNA and protein synthesis. Thus far, similar inducible mechanisms have been associated with resistance of the lepidopteran cells to 254 nm UV light and heat exposures, as well.

C. PROSPECTS AND IMPLICATIONS

One major advantage of the lepidopteran cell line model is that using mammalian or other eukaryotic cells results in very few survivors at the doses of various stressors required for accurate studies of molecular lesions.[115,116] This results in nonsurviving cells being the source of most molecular data when studying DNA repair or other subcellular mechanisms. The TN-368 cells exhibit such a pronounced radiation resistance that large numbers of surviving cells are present even at the relatively high doses necessary to evaluate molecular damage. Hence, surviving TN-368 cells might provide more relevant information regarding mechanisms that are valuable in a cell's ability to survive various stress exposures than nonsurviving mammalian cells. Experiments indicate that the amount of molecular damage induced by a given stress is about the same magnitude in both TN-368 and mammalian cells.[20] Therefore, lower amounts of damage cannot explain the higher survival levels in the lepidopteran cells. Another advantage of using the TN-368 cells, which is a consequence of their resistance, is the broad dose range used in survival studies with various stressing agents. The broad dose range makes the system more amenable than mammalian or other cells in detecting subtleties dealing with the cell survival response. For example, the multiphasic survival response of TN-368 cells to X-irradiation was first reported more than a decade ago.[55] Since this response was not typical of that observed previously from mammalian cells, several years were spent defending this data and eliminating mechanisms other than inducible recovery to explain the observation as indicated in Section II.C.1.[39] In 1990, an identical multiphasic response to radiation was presented using hamster cells, evident only by utilizing special equipment and carefully collecting data at a large number of points in the low-dose range.[50] In pursuing their studies, this group arrived at the same conclusion as had been reported previously for the TN-368 cells, that the shape is due to an inducible process.[50,51] Therefore, the lepidopteran cell system can be valuable in predicting and understanding mechanisms relevant to other eukaryotic cells.

By studying the response of lepidopteran cells to a variety of stresses as described in this chapter, it is hoped that information will be obtained that may be valuable in arriving at some understanding of cell and molecular mechanisms of stress resistance. The specific molecular basis for such resistance may have widespread implications from an evolutionary perspective as well as having an impact on our fundamental understanding of the complex mechanisms involved in regulating the molecular response to genotoxic stress in eukaryotic cells. In addition, the underlying mechanisms of stress resistance may have important ramifications regarding (1) the development of resistance of tumor cells to therapeutically administered regimens of ionizing radiation and chemotherapeutic agents and (2) the general philosophy regarding protection from hazardous and toxic agents in the environment; i.e., how detrimental are very low levels of such agents if they induce cellular mechanisms that convey resistance to further exposures. Furthermore, a detailed understanding of the stress resistance process could provide the basis for intervention to obstruct the process, thereby having great practical relevance to such diverse fields as insect pest control and the radiotherapy of human tumors. If, for example, a "key" gene or protein were identified that played a critical role in the intricate regulation of a resistant cell's response to stress, methods might be devised to exploit this finding. However, until we understand the role of the cell's response to stress, which pathways are common to a variety of stresses and which are specific to only certain stresses, as well as the identity of the genes or proteins involved in these response mechanisms, this potential method of manipulation of the stress response is untenable. A major goal of the investigations discussed is to focus on these areas with the intent of understanding and learning how to intercede in the response of lepidopteran cells to different stressors. It is believed that such studies will be of benefit to several areas of agricultural and medical science.

ACKNOWLEDGMENTS

Much of the work discussed was supported in part by USPHS grant R01 CA34158 awarded by the National Cancer Institute, DHHS, Bethesda, MD.

REFERENCES

1. **Walker, G. C.,** Mutagenesis and inducible responses to deoxyribonucleic acid damage in *Escherichia coli*, *Microbiol. Rev.*, 48, 60–93, 1984.
2. **Lewin, B.,** *Genes IV*, Oxford University Press, New York, and Cell Press, Cambridge, MA, 1990.
3. **Alberts, B., Bray, D., Lewis, J., Raff, M., Roberts, K., and Watson, J. D.,** *Molecular Biology of the Cell*, Garland Publishing, New York, 1989.
4. **Ruby, S. W. and Szostak, J. W.,** Specific *Saccharomyces cerevisiae* genes are expressed in response to DNA-damaging agents, *Mol. Cell. Biol.*, 5, 75–84, 1985.
5. **Koval, T. M., Myser, W. C., Hart, R. W., and Hink, W. F.,** Comparison of survival and unscheduled DNA synthesis between an insect and a mammalian cell line following X-ray treatments, *Mutat. Res.*, 49, 431–435, 1978.
6. **Ducoff, H. S.,** Causes of death in irradiated adult insects, *Biol. Rev.*, 47, 211–240, 1972.
7. **O'Brien, R. D. and Wolfe, L. S.,** *Radiation, Radioactivity and Insects*, Academic Press, New York, 1964.
8. **Bond, V. P., Fliedner, T. M., and Archambeau, J. O.,** *Mammalian Radiation Lethality*, Academic Press, New York, 1965.
9. **Casarett, A. P.** *Radiation Biology*, Prentice-Hall, Englewood Cliffs, New Jersey, 1968.
10. **Bergonie, J. and Tribondeau, L.,** De quelques resultants de la radiotherapie et essai de fixation d'une technique rationelle, *C.R. Acad. Sci.*, 143, 983–985, 1906.
11. **Carlson, J. G.,** A detailed analysis of X-ray induced prophase delay and reversion of grasshopper neuroblasts in culture, *Radiat. Res.*, 37, 1–14, 1969.
12. **Koval, T. M., Myser, W. C., and Hink, W. F.,** Effects of X-irradiation on cell division, oxygen consumption, and growth medium pH of an insect cell cultured *in vitro*, *Radiat. Res.*, 64, 524–532, 1975.
13. **Koval, T. M.,** Intrinsic resistance to the lethal effect of X-irradiation in insect and arachnid cells, *Proc. Natl. Acad. Sci. U.S.A.*, 80, 4752–4755, 1983.

14. **Koval, T. M.** Radiosensitivity of cultured insect cells: I. Lepidoptera, *Radiat. Res.*, 96, 118–126, 1983.
15. **LaChance, L. E., Schmidt, C. H., and Bushland, R. C.,** Radiation-induced sterilization, in *Pest Control: Biological, Physical, and Selected Chemical Methods,* Kilgore, W. W. and Doutt, R. L., Eds., Academic Press, New York, 1967, 147–196.
16. **Koval, T. M.** Radiosensitivity of cultured insect cells: II. Diptera, *Radiat Res.*, 96, 127–134, 1983.
17. **Koval, T. M.,** unpublished results, 1986.
18. **LaChance, L. E. and Graham, C. K.,** Insect radiosensitivity: dose curves and dose-fractionation studies of dominant lethal mutations in the mature sperm of 4 insect species, *Mutat. Res.*, 127, 49–59, 1984.
19. **Hart, R. W. and Setlow, R. B.,** Correlation between deoxyribonucleic acid excision repair and life-span in a number of mammalian species, *Proc. Natl. Acad. Sci. U.S.A.*, 71, 2169–2173, 1974.
20. **Koval, T. M. and Kazmar, E. R.,** DNA double-strand break repair in eukaryotic cell lines having radically different radiosensitivities, *Radiat. Res.*, 113, 268–277, 1988.
21. **Town, C. D., Smith, K. C., and Kaplan, H. S.,** Repair of X-ray damage to bacterial DNA, *Curr. Top. Radiat. Res. Q.*, 8, 351–399, 1973.
22. **Lett, J. T., Caldwell, I., Dean, C. J., and Alexander, P.,** Rejoining of X-ray induced breaks in the DNA of leukaemic cells, *Nature,* 214, 790–792, 1967.
23. **Howland, G. P., Hart, R. W., and Yette, M. L.,** Repair of DNA strand breaks after gamma-irradiation of protoplasts isolated from cultured wild carrot cells, *Mutat. Res.*, 27, 81–87, 1975.
24. **Koval, T. M., Hart, R. W., Myser, W. C., and Hink, W. F.,** DNA single-strand break repair in cultured insect and mammalian cells after X-irradiation, *Int. J. Radiat. Biol.*, 35, 183–188, 1979.
25. **Bradley, M. O. and Kohn, K. W.,** X-ray induced DNA double strand break production and repair in mammalian cells as measured by neutral filter elution, *Nucleic Acids Res.*, 7, 793–804, 1979.
26. **van der Schans, G. P., Center, H. B., and Lohman, P. H. M.,** DNA lesions induced by ionizing radiation, *Prog. Mutat. Res.*, 4, 285–299, 1982.
27. **Radford, I. R. and Hodgson, G. S.,** ^{125}I-induced DNA double strand breaks: use in calibration of the neutral filter elution technique and comparison with X-ray induced breaks, *Int. J. Radiat. Biol.*, 48, 555–566, 1985.
28. **Blocher, D. and Pohlit, W.,** DNA double strand breaks in Ehrlich ascites tumour cells at low doses of X-rays. II. Can cell death be attributed to double strand breaks?, *Int. J. Radiat. Biol.*, 42, 329–338, 1982.
29. **Weibezahn, K. F. and Coquerelle, R.,** Radiation induced DNA double strand breaks are rejoined by ligation and recombination processes, *Nucleic Acids Res.*, 9, 3139–3150, 1981.
30. **Ashburner, M. and Bonner, J. A.,** The induction of gene activity in *Drosophila* by heat shock, *Cell,* 17, 241–254, 1979.
31. **Tomasovic, S. P. and Koval, T. M.,** Relationship between cell survival and heat-stress protein synthesis in a *Drosophila* line, *Int. J. Radiat. Biol.*, 48, 635–650, 1985.
32. **Lingquist, S.,** The heat shock response, *Annu. Rev. Biochem.*, 55, 1151–1191, 1986.
33. **Ducoff, H. S.,** Radiation-induced increase in lifespan of insects: implication for theories of mammalian aging and radiosensitivity, in *Biological and Environmental Effects of Low-Level Radiation,* Vol. 1, International, Atomic Energy Agency, Vienna, 1, 103–109, 1976.
34. **Andrews, G. K., Harding, M. A., Calvet, J. P., and Adamson, E. D.,** The heat shock response in HeLa cells accompanied by elevated expression of the c-*fos* proto-oncogene, *Mol. Cell Biol.*, 7, 3452–3458, 1987.
35. **Fornace, A. J., Jr., Alamo, I., Jr., and Hollander, M. C.,** DNA damage-inducible transcripts in mammalian cells, *Proc. Natl. Acad. Sci. U.S.A.*, 85, 8800–8804, 1988.
36. **Papathanasiou, M. A., Kerr, N. C. K., Robbins, J. H., McBride, O. W., Alamo, I., Jr., Barrett, S. F., Hickson, I. D., and Fornace, A. J., Jr.,** Induction by ionizing radiation of the *gadd*45 gene in cultured human cells: lack of medication by protein kinase C, *Mol. Cell Biol.*, 11, 1009–1016, 1991.
37. **Hollander, M. C. and Fornace, A. J., Jr.,** Induction of *fos* RNA by DNA-damaging agents, *Cancer Res.*, 49, 1687–1692, 1989.
38. **Boothman, D. A., Bouvard, I., and Hughes, E. N.,** Identification and characterization of X-ray-induced proteins in human cells, *Cancer Res.*, 49, 2871–2878, 1989.
39. **Koval, T. M.,** Multiphasic survival response of a radioresistant lepidopteran cell line, *Radiat. Res.*, 98, 642–648, 1984.
40. **Koval, T. M.,** Inducible repair of ionizing radiation damage in higher eukaryotic cells, *Mutat. Res.*, 173, 291–293, 1986.
41. **Koval, T. M.,** Enhanced recovery from ionizing radiation damage in a lepidopteran insect cell line, *Radiat. Res.*, 115, 413–420, 1988.
42. **Olivieri, G., Bodycote, J., and Wolff, S.,** Adaptive response of human lymphocytes to low concentrations of radioactive thymidine, *Science,* 223, 594–597, 1984.
43. **Shadley, J. and Wolff, S.,** Very low doses of X-rays can cause human lymphocytes to become less susceptible to ionizing radiation, *Mutagenesis,* 2, 95–96, 1987.
44. **Holbrook, N. J. and Fornace, A. J., Jr.,** Response to adversity: molecular control of gene activation following genotoxic stress, *New Biol.*, 3, 825–833, 1991.

45. **Elkind, M. M. and Whitmore, G. D.**, *The Radiobiology of Cultured Mammalian Cells*, Gordon and Breach, Science Publishers, New York, 1967.
46. **Beam, C. A., Mortimer, R. K., Wolfe, R. G., and Tobias, C. A.**, The relation of radioresistance to budding in *Saccharomyces cerevisiae*, *Arch. Biochem. Biophys.*, 49, 110–122, 1954.
47. **Elkind, M. M. and Sutton, H.**, The relationship between division and X-ray sensitivity, ultraviolet sensitivity, and photoreactivation in yeast, *Radiat. Res.*, 10, 283–295, 1959.
48. **Horsley, R. J., Fucikovsky, L. A., and Banerjee, S. N.**, Studies on radiosensitivity during the cell cycle in *Oedogonium cardiacum*, *Radiat. Bot.*, 7, 241–246, 1967.
49. **Horsley, R. G. and Pujara, C. M.**, Study of the inflexion in X-radiation survival curves for synchronized cell populations of the green alga *(Oedogonium cardiacum)*, *Radiat. Res.*, 40, 440–449, 1969.
50. **Marples, B. and Joiner, J. C.**, Cell survival at very low radiation doses, *Abstracts 38th Annual Meeting of the Radiation Research Society*, 1990, 139.
51. **Marples, B. and Joiner, M. C.**, Induced repair: an explanation of intrinsic radiosensitivity. *Abstracts 40th Annual Meeting of the Radiation Research Society*, 1992, 36.
52. **Peters, J. and Jagger, J.**, Inducible repair of near-UV radiation lethal damage in *E. coli*, *Nature (London)*, 289, 194–195, 1981.
53. **Calkins, J.**, An unusual form of response in X-irradiated protozoa and a hypothesis as to its origin, *Int. J. Radiat. Biol.*, 12, 297–301, 1967.
54. **Calkins, J. and Todd, W.**, Evidence for a triggered or activated radiation repair system in *Saccharomyces cerevisiae*, *Int. J. Radiat. Biol.*, 14, 487–491, 1968.
55. **Koval, T. M.**, An insect cell line (TN-368) characterized by a biphasic radiation response under both oxic and anoxic conditions, *Radiat. Res.*, 87, 500, 1981.
56. **Sinclair, W. K. and Morton, R. A.**, X-ray sensitivity during the cell generation cycle of cultured Chinese hamster cells, *Radiat. Res.*, 29, 450–474, 1966.
57. **Koval, T. M. and Fornace, A. J.**, unpublished results, 1992.
58. **Koval, T. M., Hart, R. W., Myser, W. C., and Hink, W. F.**, A comparison of survival and repair of UV-induced DNA damage in cultured insect versus mammalian cells, *Genetics*, 87, 513–518, 1977.
59. **Koval, T. M.**, Enhanced survival by photoreactivation and liquid-holding following UV damage of TN-368 insect cells, *Mutat. Res.*, 166, 149–156, 1986.
60. **Koval, T. M.**, Photoreactivation of UV damage in cultured *Drosophila* cells, *Experientia*, 43, 445–446, 1987.
61. **Wade, M. H. and Trosko, J. E.**, Enhanced survival and decreased mutation frequency after photoreactivation of UV damage in rat kangaroo cells, *Mutat. Res.*, 112, 231–243, 1983.
62. **Bronk, B. V., Humphries, E. H., and Rupert, C. S.**, The effect of post-UV dark incubation on survival of chick-embryo fibroblasts after photoreactivation, *Mutat. Res.*, 132, 33–39, 1984.
63. **Mano, Y., Mitani, H., Etoh, H., and Egami, N.**, Survival and photoreactivability of ultraviolet-irradiated cultured fish cells (CAF-MM1), *Radiat. Res.*, 84, 514–522, 1980.
64. **Griggs, H. G. and Bender, M. A.**, Ultraviolet and gamma-ray induced reproductive death and photoreactivation in a *Xenopus* tissue culture cell line, *Photochem. Photobiol.*, 15, 517–526, 1978.
65. **Rosenstein, B. S. and Setlow, R. B.**, Photoreactivation of ICR 2A frog cells after exposure to monochromatic ultraviolet radiation in the 252–313 nm range, *Photochem. Photobiol.*, 32, 361–366, 1980.
66. **O'Brien, P. A. and Houghton, J. A.**, Photoreactivation and excision repair of UV induced pyrimidine dimers in the unicellular cyanobacterium, *Gleocapsa alpicola* (Synechocystis PCC 6308), *Photochem. Photobiol.*, 35, 359–365, 1982.
67. **Harm, W.**, *Biological Effects of Ultraviolet Radiation*, Cambridge University Press, Cambridge, 1980.
68. **Setlow, R. B. and Setlow, J. K.**, Evidence that ultraviolet-induced thymidine dimers in DNA cause biological damage, *Proc. Natl. Acad. Sci. U.S.A.*, 48, 1250–1257, 1962.
69. **Perlitsch, M. and Kelner, A.**, The reduction by reactivating light of the frequency of phenocopies induced by ultraviolet light in *Drosophila melanogaster*, *Science*, 118, 165–166, 1953.
70. **Trosko, J. E. and Wilder, K.**, Repair of UV-induced pyrimidine dimers in *Drosophila melanogaster* cells *in vitro*, *Genetics*, 73, 297–302, 1973.
71. **Jackle, H. and Kalthoff, K.**, Photoreversible UV-inactivation of messenger RNA in an insect embryo (*Smittia* spec. chironomidae, diptera), *Photochem. Photobiol.*, 32, 749–761, 1980.
72. **Muraoka, N., Okuda, A., and Ikenaga, M.**, DNA photoreactivating enzyme from silkworm, *Photochem. Photobiol.*, 32, 193–197, 1980.
73. **Styer, S. C. and Griffiths, T. D.**, Effect of UVC light on growth, incorporation of thymidine, and DNA chain elongation in cells derived from the Indian meal moth and the cabbage looper, *Radiat. Res.*, 130, 72–78, 1992.
74. **Weichselbaum, R. R., Nove, J., and Little, J. B.**, Deficient recovery from potentially lethal radiation damage in ataxia telangiectasia and xeroderma pigmentosum, *Nature (London)*, 271, 261–262, 1978.
75. **Konze-Thomas, B., Levinson, J. W., Maher, V. M., and McCormick, J. J.**, Correlation among the rates of dimer excision, DNA repair replication, and recovery of human cells from potentially lethal damage induced by ultraviolet radiation, *Biophys. J.*, 28, 315–326, 1979.

76. **Mano, Y., Kator, K., and Egami, N.,** Loss of photoreactivation in UV-irradiated cultured fish cells under different conditions, *Photochem. Photobiol.*, 35, 753–755, 1982.
77. **Roczen, D. and Mennigmann, H.-D.,** Loss of photoreactivability during liquid-holding in *Escherichia coli* WP2, *Photochem. Photobiol.*, 32, 189–192, 1980.
78. **Koval, T. M.,** Recovery from exposure to DNA-damaging chemicals in radiation-resistant insect cells, *Mutat. Res.*, 262, 219–225, 1991.
79. **Pearson, C. M. and Styles, J. A.,** Resistance of mouse lymphoma L5178YAII cells to alkylation with methylmethanesulphonate resides in a late step of excision repair, *J. Cell. Sci.*, 68, 35–48, 1984.
80. **Crathorn, A. R. and Shackleton, J.,** Repair processes and the response of dividing and non-dividing cells to methyl methanesulphonate and dimethyl sulphate, *Chem.-Biol. Interact.*, 15, 117–130, 1976.
81. **Osmack, M.,** Repeated doses of γ-rays induce resistance to *N*-methyl-*N*′-nitro-*N*-nitrosoguanidine in Chinese hamster cells, *Radiat. Res.*, 115, 609–616, 1988.
82. **Peterson, A. R., Peterson, H., and Heidelberger, C.,** Oncogenesis, mutagenesis, DNA damage, and cytotoxicity in cultured mammalian cells treated with alkylating agents, *Cancer Res.*, 39, 131–138, 1979.
83. **Nishi, Y., Hasegawa, M. M., Taketomi, M., Ohkawa, Y., and Inui, N.,** Comparison of 6-thioguanine-resistance mutation and sister chromatid exchanges in Chinese hamster V-79 cells with forty chemical and physical agents, *Cancer Res.*, 44, 3270–3279, 1984.
84. **Nakano, S., Koga, T., Nagafuchi, S., Nakamura, M., Kounoue, E., Niho, Y., and Takaki, R.,** The relationship between molecular and cellular repair from potentially lethal damage induced by *N*-methyl-*N*′-nitro-*N*-nitrosoguanidine in Chinese hamster V-79 cells, *Mutat. Res.*, 146, 271–276, 1985.
85. **Silinskas, K. C., Kateley, S. A., Tower, J. E., Maher, V. M., and McCormick, J. J.,** Induction of anchorage-independent growth in human fibroblasts by propane sultone, *Cancer Res.*, 41, 1620–1627, 1981.
86. **Samson, L. and Schwartz, J. L.,** Evidence for an adaptive DNA repair pathway in CHO and human skin fibroblast cell lines, *Nature (London)*, 287, 861–863, 1980.
87. **Laval, F. and Laval, J.,** Adaptive response in mammalian cells: crossreactivity of different pretreatments on cytotoxicity as contrasted to mutagenicity, *Proc. Natl. Acad. Sci. U.S.A.*, 81, 1062–1066, 1984.
88. **Kaina, B.,** Enhanced survival and reduced mutation and aberration frequencies induced in V-79 Chinese hamster cells preexposed to low levels of methylating agents, *Mutat. Res.*, 93, 195–211, 1982.
89. **Smith, P. J. and Paterson, M. C.,** Defective DNA repair and increased lethality in ataxia telangiectasia cells exposed to 4-nitroquinoline-1-oxide, *Nature (London)*, 287, 747–749, 1980.
90. **Scudiero, D. A.,** Decreased DNA repair synthesis and defective colony-forming ability of ataxia telangiectasia fibroblast cell strains treated with *N*-methyl-*N*′-nitro-*N*-nitrosoguanidine, *Cancer Res.*, 40, 984–990, 1980.
91. **Hoar, D. I. and Sargent, P.,** Chemical mutagen hypersensitivity in ataxia telangiectasia, *Nature (London)*, 261, 590–592, 1976.
92. **Bonner, J. A., Christianson, T. H., and Koval, T. M.,** Correlation of doxorubicin sensitivity with the stabilization of DNA topoisomerase II complexes in an extremely doxorubicin-resistant Lepidopteran insect cell line, *Proc. Am. Assoc. Cancer Res.*, 32, 338, 1991.
93. **Bonner, J. A. and Koval, T. M.,** in preparation, 1992.
94. **Bianchi, N. O. and Lopez-Larraza, D. M.,** DNA damage and repair induced by bleomycin in mammalian and insect cells, *Environ. Mol. Mutag.*, 17, 63–68, 1991.
95. **Koval, T. M. and Suppes, D.,** Heat resistance and thermotolerance in a radiation-resistant cell line, *Int. J. Radiat. Biol.*, 61, 425–431, 1992.
96. **Dewey, W. C., Hopwood, L. E., Sapareto, S. A., and Gerweck, L. E.,** Cellular responses to combinations of hyperthermia and radiation, *Radiology*, 123, 463–474, 1977.
97. **Nielsen, O. S., Henle, K. J., and Overgaard, J.,** Arrhenius analysis of survival curves from thermotolerance and step-down heated L1A2 cells *in vitro*, *Radiat. Res.*, 91, 468–482, 1982.
98. **Gerner, E. W., Boone, R., Conner, W. G., Hicks, J. A., and Bonne, M. L. M.,** A transient thermotolerance survival response produced by single thermal doses in HeLa cells, *Cancer Res.*, 36, 1035–1040, 1976.
99. **Nielsen, O. S. and Overgaard, J.,** Influence of time and temperature on the kinetics of thermotolerance in L_1A_2 cells *in vitro*, *Cancer Res.*, 42, 4190–4196, 1982.
100. **Henle, K. J. and Dethlefsen, L. A.,** Heat fractionation and thermotolerance: a review, *Cancer Res.*, 38, 1843–1851, 1978.
101. **Gerweck, L. E. and DeLaney, T. F.,** Persistence of thermotolerance in slowly proliferating plateau-phase cells, *Radiat. Res.*, 97, 365–372, 1984.
102. **Henle, K. J. and Leeper, D. B.,** Effects of hyperthermia (45°) on macromolecular synthesis in Chinese hamster ovary cells, *Cancer Res.*, 39, 2665–2674, 1979.
103. **Warters, R. L. and Stone, O. L.,** Macromolecule synthesis in HeLa cells after thermal shock, *Radiat. Res.*, 96, 646–652, 1983.
104. **Kern, D. H., Krag, D. N., Kauffman, G. L., Morton, D. L., and Storm, F. K.,** Thermal resistance of human malignant melanoma modulated by prostaglandin E_2, *J. Surg. Oncology*, 37, 60–64, 1988.

105. **Oesterreich, S., Benndorf, R., and Bielka, H.,** The expression of the growth-related 25 kDa protein (p25) of Ehrlich ascites tumor cells is increased by hypothermic treatment (heat shock), *Biomed. Biochim. Acta,* 49, 219–226, 1990.
106. **Lee, Y. J., Perlaky, L., Dewey, W. C., Armour, E. P., and Corry, P. M.,** Differences in thermotolerance induced by heat or sodium arsenite: cell killing and inhibition of protein synthesis, *Radiat. Res.,* 121, 295–303, 1990.
107. **Tomasovic, S. P., Steck, P. A., and Heitzman, D.,** Heat-stress proteins and thermal resistance in rat mammary tumor cells, *Radiat. Res.,* 95, 399–413, 1983.
108. **Watson, K., Dunlop, G., and Cavicchioli, R. R.,** Mitochondrial and cytoplasmic protein syntheses are not required for heat shock acquisition of ethanol and thermotolerance in yeast, *FEBS Lett.,* 172, 299–302, 1984.
109. **Wong, R. S. L. and Dewey, W. C.,** Molecular studies on the hyperthermic inhibition of DNA synthesis in Chinese hamster ovary cells, *Radiat. Res.,* 92, 370–395, 1982.
110. **Warters, R. L.,** Hyperthermia blocks DNA processing at the nuclear matrix, *Radiat. Res.,* 115, 258–272, 1988.
111. **Lepock, J. R., Frey, H. E., Heynen, M. P., Nishio, J., Waters, B., Ritchie, K. P., and Kruuv, J.,** Increased thermostability of thermotolerant CHL V-79 cells as determined by differential scanning calorimetry, *J. Cell. Physiol.,* 142, 628–634, 1990.
112. **Koval, T. M.,** Gamma-ray- and UV-sensitive strains of a radioresistant cell line: isolation and cross-sensitivity to other agents, *Radiat. Res.,* 127, 58–63, 1991.
113. **DeCarli, L. and Larizza, L.,** Griseofulvin, *Mutat. Res.,* 195, 91–126, 1988.
114. **Nothel, H. and Strese, B.,** Investigations of radiosensitive and radioresistant populations of *Drosophila melanogaster, Mutat. Res.,* 103, 87–90, 1982.
115. **Elkind, M. M.,** Cells, targets, and molecules in radiation biology, in *Radiation Biology in Cancer Research,* Meyn, R. E. and Withers, H. R., Eds., Raven Press, New York, 1980, 71–93.
116. **Painter, R. B.,** DNA synthesis inhibition in *HeLa* cells as a simple test for agents that damage human DNA, *J. Environ. Pathol. Toxicol.,* 2, 65–78, 1978.

INDEX

A

AcMNPV, see *Autographa californica*
Aedes aegypti, 80
Aedes albopictus, 80–81
Aeration, 47–48
AfMNPV, see *Anagrapha falcifera*
AgMNPV, see *Anticarsia gemmetalis*
Anagrapha falcifera, 59, 61
Androctonus australis, 20
Anopheles gambiae, 81–82
Anthonomus grandis, 59
Anticarsa gemmitalis, 92 2
Anticarsia gemmetalis, 19, 62
Autographa californica, 59, 61

B

Bacillus thuringiensis, 20
  with dipteran specificity, 80–83, 84–86
  with lepidopteran specificity, 73–80, 83–84
  principles and history, 73
Baculoviruses
  as biological controls, 64–65
  classification, 35
  conclusions, 21–22
  expression vector systems, 4–5
    host-dependent expression, 10–11
    isolation/identification strategies, 8–10
    for polyhedron deficient viruses, 7–8
  foreign gene expression
    in larval insects, 18–19
    for pest control, 19–21
  foreign gene products
    C-terminus alpha-amidation, 14–15
    fatty acid acylation, 15
    glycosylation, 12–14
    oligomerization, 16
    phosphorylation, 14
    proteolytic processing, 11–12
    solubility, 15–16
    viral particle assembly, 1617
  genome organization and expression, 3–4
  as pesticides, 35–50, see also *In vitro* production; Pesticides
  production
    and cell parameters, 43–44
    harvesting and storage, 44
    infection types, 41–42
    and multiplicity of infection, 42–43
    and passage effect, 43
    viral strains, 40–41
  promoters for foreign gene expression, 5–7
  replication cycle, 35–36
  specificity of
    general properties, 59–60
    granulosis viruses, 64
    nonoccluded baculoviruses, 64
    nuclear polyhedrosis viruses, 60–64
    recombinants as biological controls, 64–65
Baculovirus-yeast shuttle vectors, 9–10
Biological controls, see Pesticides
Bioreactor production, 48, 96–99
Bluetongue virus, 17
BmSNPV, see *Bombyx mori*
Bombys mori, 63, 92

C

Cabbage looper, see *Trichoplusia ni*
Cabbage worm, see *Mamestra brassiceae*
Cell lines
  from different species, 40
  from different tissues, 40
  from same tissue, 36 2
  studies comparing, 37–39
CfMNPV, see *Choristoneura fumiferana*
Chimeric promoters, 6
Choristoneura fumiferans, 62, 74–76
Cotton bollworm, see *Heliothis zea*
CpGv, see *Cydia pomonella*
C-terminus alpha-amidation, 14–15
Culex quinquefasciatus, 82–83
Culex tarlis, 83
Cydia pomonella, 63

D

Dipterans, 80 83, 84–86, see also Individual species
DNA repair, 160–165, see also Stress resistance
Drosophila, 83
  endocrinologic studies, 130–134
  glial cells in, 116–117
  ion channel expression, 111–113
  neural system studies, 109–121
  neurite outgrowth, 111–114
  neurogenesis and differentiation, 109 110
  neurotransmitter phenotype development, 110–111
  synaptogenesis, 114–116
  transfection efficiency of cell lines, 143–155
    cell cycle dependency of, 148–149
    clonal variation in, 146–148
    DNA uptake variations and, 149 150
    early vs. late expression and, 153–154
    genome integration and, 152–153
    intracellular transport and, 150–152 2
    nuclear penetration and, 152
    review of literature, 143–146
  trophic interactions in, 117–119

E

Early gene promoters, 6
Ecdysteroids, 130–134

action, 130–134
synthesis and metabolism, 130
Estigmene acrea, 92, 94
Eubaculovirinae, 35
Expression vector systems, 4–5, see also Baculoviruses
host-dependent expression, 10–11
isolation/identification strategies, 8–10
for large-scale production, 91–96
for polyhedron deficient viruses, 7–8
promoters for, 5–7

F

Fatty acid acylation, 15

G

β-Galactosidose marker, 9
Galleria mellonella, 21, 62
Gamma radiation resistance, 159–169
Glial cells, 116–117
Glycosylation, 12–14
GmMNPV, see *Galleria mellonella*
Granulosis viruses, specificity of, 64
Gypsy moth, see *Lymantra dispar*

H

HaGV, see *Helicoverpa armigera*
HaSNPV, see *Helicoverpa armigera*
Heat resistance, 173–177
Helicoverpa armigera, 62–63
Heliothis virescens, 21, 76, 92, 94
Heliothis zea, 76, 92
Hepatitis B virus, 17
Host-dependent foreign gene expression, 10–11
Host insect promoters, 6–7
Host range, see Specificity
Hz-1, see *Helicoverpa zea*
HzSNPV, see *Helicoverpa zea*

I

Insect cell production, 89–101, see also Production techniques
In vitro production
biological factors in
cell line comparisons, 36–40
production procedures, 42–44
viral strain comparisons, 40–42
chemical/physical factors in
media, 44–47
production vessels, 47–48
temperature, 44
recombinant technology and, 48–50
In vitro studies, 61–63
host range of selected baculoviruses, 61–63 2
nervous systems, see also Nervous system studies
conclusions and prospects, 119–121
organ culture, 107–109
primary cell culture, 109–119
In vivo studies, host range of selected baculoviruses, 61–63
Ion channel expression, 111–113
Isolation techniques, 8–10

J

Juvenile hormone esterase, 20

L

Larval insect systems, 18–19
LdMNPV, see *Lymantria dispar*
Lepidopterans, 73–80, 83–84, see also individual species
stress resistance
to chemical agents, 173–177
commentary, 179–181
mutant identification, 177–179
to 254 nm ultraviolet radiation, 169–172
to X and gamma radiation, 159–169
Linearized parental viral DNA, 8–9
Lymantria dispar, 2, 61, 76, 94

M

M. sexta, 19–20
Malacosoma disstra, 76
Mamestra brassicae, 59, 61, 76–77, 92, 94
Manduca sexta, 77–78, 111–115, 134–136
MbMNPV, see *Mamestra brassicae*
Media
for *In vitro* production
commercial preparations, 46–47
crude biologicals, 45–46
general requirements, 44–45
purified biochemicals, 46
for large-scale production, 99–100
MNPVs, see Nuclear polyhedrosis viruses: multiple
Modified promoters, 6
Morphogenesis, regulation of, 134–136

N

Nervous system studies
conclusions and prospects, 119–121
organ culture: brain and neurosecretion, 107–109
primary cell culture
glial cells, 116–117
ion channel expression, 111–113
neurite outgrowth, 113–115
neurogenesis and differentiation, 109–110
neurotransmitter phenotype development, 110–111
synaptogenesis, 115–116
trophic interactions, 117–119

Neurite outgrowth, 113–115
Neurogenesis, 109–110
Neurotransmitter phenotypes, 110–111 2
Nonoccluded baculoviruses, 64
Norwalk virus, 17
Nucibaculovirinae, 35
Nuclear polyhedrosis virus, 3, see also Baculoviruses
  host range, 61–63
  multiple, 59, 60–63
  single, 59, 61–63, 64
  specificity of, 60–64

O

Oligomerization, 16
OlSNPV, see *Orgyia leucostigma*
OpMNPV, see *Orgyia leucostigma*
Orgyia leucostigma, 62, 63
Oryctes rhinoceros, 19, 63
Orygia pseudotsugata, 4

P

Pesticides
  bacullovirus techniques, 19–21
  in vitro production
    and baculovirus properties, 35–36
    chemical and physical factors and, 44–48
    genetic engineering and, 48–50
    production procedures compared, 42–44
    systems compared, 36–43
  specificity of, 64–65
  viral, 35
Phosphorylation, 14
Pieris brassicae, 78 2
Plodia interpunctella, 78–79
Plutella xylostella, 92, 93
Polyhedrin fusion transfer vectors, 7–8
Polyhedrin promoters, 5
Polyoma virus, 17
p10 promoters, 5–6, 20–21
Production media, 44 47
Production techniques
  in bioreactors, 96–99
  cell line and expression vector selection, 91–96
  in vitro, 35–50
  preprequisites for large-scale, 91–96
  and serum free media, 99–100
Production vessels, 47–48
Promoters
  chimeric and modified, 6
  early gene, 6
  host insect and nonbaculovirus, 6–7
  influence on production, 489–50
  p10, 5–6
  polyhedrin, 5
Proteolytic processing, 11–12
Pyemotes tritici, 21

R

Radiation resistance, 159–169
Recombinant techniques of pesticide production, 45–50, 108

S

S. frugiperda larvae, 19
Saltmarsh caterpillar, see *Estigmene acrea*
SeMNPV, see *Spodoptera exigua*
SfMNPV, see *Spodoptera frugipera*
Silkworm, see *Bombys mori*
SiMNPV, see *Spodoptera littoralis*
SNPVs, see Nuclear polyhedrosis virus: single
Soybean caterpillar, see *Anticarsa gemmitalis*
Specificity
  baculoviruses
    general properties, 59–60
    granulosis viruses, 64
    nonoccluded baculoviruses, 64
    nuclear polyhedrosis viruses, 60–64
    recombinants as biological controls, 64–65
  for dipterans, 80–83, 84–86
  for lepidopterans, 73–80, 83–84
Spodoptera exigua, 62, 79
Spodoptera frugipera, 61–62, 79, 91–92, 95
Spodoptera litteralis, 62, 79
Spodoptera littura, 62
Stress resistance
  to chemical agents, 173–177
  commentary, 179–181
  mutant identification, 177–179
  to 254 nm ultraviolet radiation, 169–172
  to X and gamma radiation, 159–169
SuMNPV, see *Spodoptera littura*
Suspension cultures, 47
Synaptogenesis, 115–116 2

T

Temperature
  and *in vitro* production, 44
  resistance to heat, 173–177
Thermoresistance, 173–177
TnGV, see *Trichoplusia*
TnMNPV, see *Trichoplusia ni*
TnSNPV, see *Trichoplasia ni*
Tobacco budworm, see *Heliothis virescens*
Transfection efficiency in *Drosophila*
    cell cycle dependency of, 148–149
    clonal variation in, 146–148
    DNA uptake variations and, 149–150
    early vs. late expression and, 153–154
    genome integration and, 152–153
    intracellular transport and, 150–152
    nuclear penetration and, 152
    review of literature, 143–146

Transfer vectors, 7–8, see also Expression vector systems
Trichoplusia ni, 19, 21, 79–80
 host range, 62, 63
 morphogenesis, 134–136
Trophic factors, 117–119

U

UV radiation resistance, 169–172

V

Vessels for production, 47–48
Viral particle assembly, 16–17
Virulence enhancement, 49

X

X radiation resistance, 159–169

For Product Safety Concerns and Information please contact our EU representative GPSR@taylorandfrancis.com Taylor & Francis Verlag GmbH, Kaufingerstraße 24, 80331 München, Germany

Batch number: 10731074

Printed by Printforce, the Netherlands